Intelligent Control and Learning Systems

Volume 22

Series Editor

Dong Shen, School of Mathematics, Renmin University of China, Beijing, China

The Springer book series Intelligent Control and Learning Systems addresses the emerging advances in intelligent control and learning systems from both mathematical theory and engineering application perspectives. It is a series of monographs and contributed volumes focusing on the in-depth exploration of learning theory in control such as iterative learning, machine learning, deep learning, and others sharing the learning concept, and their corresponding intelligent system frameworks in engineering applications. This series is featured by the comprehensive understanding and practical application of learning mechanisms. This book series involves applications in industrial engineering, control engineering, and material engineering, etc.

The Intelligent Control and Learning System book series promotes the exchange of emerging theory and technology of intelligent control and learning systems between academia and industry. It aims to provide a timely reflection of the advances in intelligent control and learning systems. This book series is distinguished by the combination of the system theory and emerging topics such as machine learning, artificial intelligence, and big data. As a collection, this book series provides valuable resources to a wide audience in academia, the engineering research community, industry and anyone else looking to expand their knowledge in intelligent control and learning systems.

Tao Liu · Shoulin Hao · Youqing Wang · Dewei Li

Robust Iterative Learning Control of Industrial Batch Systems

Springer

Tao Liu
School of Control Science and Engineering
Dalian University of Technology
Dalian, Liaoning, China

Shoulin Hao
School of Control Science and Engineering
Dalian University of Technology
Dalian, Liaoning, China

Youqing Wang
College of Information Science
and Technology
Beijing University of Chemical Technology
Beijing, China

Dewei Li
Department of Automation
Shanghai Jiao Tong University
Shanghai, China

ISSN 2662-5458 ISSN 2662-5466 (electronic)
Intelligent Control and Learning Systems
ISBN 978-981-96-9777-9 ISBN 978-981-96-9778-6 (eBook)
https://doi.org/10.1007/978-981-96-9778-6

This work was supported by National Science Funding of China.

This Springer imprint is published by the registered company Springer Nature Singapore Pte Ltd.
The registered company address is: 152 Beach Road, #21-01/04 Gateway East, Singapore 189721, Singapore

Preface

Batch manufacturing systems have been widely established in modern industries, such as injection molding machines, robotic manipulators, oil drillers, and pharmaceutical crystallizers. Owing to the use of historical batch run data, iterative learning control (ILC) can progressively improve the system performance from batch to batch, and therefore, has been increasingly explored and utilized in various production fields and repetitive operation devices. Since the early research in the 1980s, e.g., Arimoto et al. (1984), a lot of ILC methods for linear or nonlinear batch processes and repetitive systems operated over a finite duration have been reported in the literature. A well-recognized merit is that perfect tracking can be realized if there is no change of system dynamics for batch operation along with no external disturbance. However, time- or batch-varying process uncertainties and load disturbances are often encountered in real system operations in engineering practice. These issues could deteriorate the system performance if an ILC scheme is designed only based on the ideal system description, or even jeopardize the system stability. Besides, time delay and the input/output constraints are typically involved with various batch process operations in industry, which also bring difficulties for applying most of the previously developed ILC methods to obtain desirable control performance, or even cause failure.

To address the above issues, a series of robust ILC methods have been developed by the authors in the past two decades. Advanced ILC designs based on the widely used proportional-integral-derivative (PID) control loop are presented for the convenience of engineering application, which could not only realize perfect tracking of the desired output trajectory under repetitive process uncertainties and disturbance but also maintain robust tracking against time-varying uncertainties and disturbance. Moreover, optimization-based ILC designs are provided to deal with the input and/or output constraints of batch process operation, based on the mode predictive control (MPC) principle for process optimization. Furthermore, predictor based ILC designs are given to deal with time delay in the process input, state or output as often encountered in practice, which could obtain evidently improved control performance compared to the previous ILC methods mainly devoted to delay-free batch processes. In addition, data-driven ILC methods are also presented for application to batch operation systems with unknown dynamics and time-varying uncertainties.

Benchmark examples from the existing literature are used to demonstrate the advantages of the proposed ILC methods, along with real applications to industrial injection molding machines, 6-degree-of-freedom robotic manipulator, and refrigerated/heating circulators of pharmaceutical crystallizers.

To facilitate understanding the main contents, the technical development of the proposed ILC methods is arranged in a self-contained manner. Only knowledge of basic linear algebra and complex analysis is assumed for the readers, and some preliminary knowledge on the basic principle of ILC and robust control theory are provided before presenting each of the proposed designs and theoretical analysis. Illustrative examples and experimental applications are given for all the proposed ILC methods in an easy-to-follow style. Hence, it is believed that the monograph should be of interest to control engineers and researchers in industrial system control theory and engineering fields, and also can be used for undergraduate- and graduate-level control courses in process system engineering, industrial automation engineering, chemical engineering, mechanical engineering, electrical engineering, biomedical engineering, etc.

Dalian, China Tao Liu
Dalian, China Shoulin Hao
Beijing, China Youqing Wang
Shanghai, China Dewei Li
March 2025

Acknowledgements We sincerely want to thank the Springer editor, Dong Shen, for supporting us to prepare this monograph into the book series of Intelligent Control and Learning Systems. The authors are grateful to the financial support from the National Natural Science Foundation of China under Grants 62173058, 62361136585, 62261160385, 62327807, 62473077, the Hong Kong Research Grant Council (Grant no. N HKUST628/22), and the NSF for Distinguished Young Scholars under Grant 62225303, for visiting each other and meeting together to consult the book details.

Contents

Abbreviations and Symbols

Abbreviations

1D	One-dimensional
2D	Two-dimensional
3D	Three-dimensional
AS	Asymptotically stable
ATE	Averaged tracking error
CCL	Cone complementarity linearization
CEF	Composite energy function
D	Derivative
DDA	Double-dynamic analysis
DDC	Data-driven control
DDILC	Data-driven iterative learning control
DDSPLC	Data-driven set-point learning control
DLDM	Dynamic linearization data model
DOF	Dynamic output feedback
ESO	Extended state observer
FM	Fornasini-Marchesini
FMLSS	Fornasini-Marchesini local state space
GESO	Generalized extended state observer
GPC	Generalized predictive control
HAS	Horizontal asymptotically stable
HCI	Horizontal convergence index
I/O	Input/output
IESO	Iterative extended state observer
ILC	Iterative learning control
IMC	Internal model control
KCI	K-direction convergence index
KYP	Kalman-Yakubovich-Popov
LMI	Linear matrix inequality
LPV	Linear parameter varying

MPC	Model predictive control
MSE	Mean-square-error
NAS	Necessary and sufficient
NLMI	Nonlinear matrix inequality
NOILC	Norm-optimization iterative learning control
P	Proportional
PD	Proportional-derivative
PI	Proportional-integral
PID	Proportional-integral-derivative
PLC	Programmable logic controller
PLS	Partial least-squares
PSO	Predictive state observer
PWM	Pulse-width modulator
Q-ILC	Quadratic-criterion-based iterative learning control
QP	Quadratic programming
R2R	Run-to-run
RMSE	Root-mean-squared-error
RSSE	Root-sum-squared-error
SF	State feedback
SISO	Single-input-single-output
SOF	Static output feedback
SP	Smith predictor
SPR	Set-point related
TCI	T-direction convergence index
VAS	Vertical asymptotically stable
VCI	Vertical convergence index

General Symbols

$\Re$	Field of real numbers
$\Re^n$	Euclidean space of dimension n
$\Re^{m\times n}$	$m \times n$ real number space
$\mathbb{Z}_N$	Set of integer numbers from 0 to N
$\mathbb{Z}_+$	Field of positive integer numbers
∞	Infinity
$l_2[0, +\infty)$	Space of square summable sequences on $[0, +\infty)$
$y(t)$	Process output response with regard to t in time domain
$y(t, k)$	Process output response with regard to t in time domain and k in iteration domain
$y_s(t, k)$	Set-point command with regard to t in time domain and k in iteration domain
$y_{\mathrm{d}}(t)$	Desired output trajectory/set-point value with regard to t in time domain

$x^h(t,k)$	Horizontal system state
$x^v(t,k)$	Vertical system state
$x(t)$	System state with regard to t in time domain
$\dot{x}(t)$	Time derivative of system state
$x(t,k)$	System state with regard to t in time domain and k in iteration domain
$u(t)$	Process input with regard to t in time domain
$u(t,k)$	Process input with regard to t in time domain and k in iteration domain
$w(t)$	Load disturbance with regard to t in time domain
$w(t,k)$	Load disturbance with regard to t in time domain and k in iteration domain
$e(t)$	Output error with regard to t in time domain
$e(t,k)$	Output error with regard to t in time domain and k in iteration domain
$e_s(t,k)$	Set-point tracking error with regard to t in time domain and k in iteration domain
$\vec{a}(k)$	Vector composed by all vectors $a(t,k)$ at batch k where t is the time step
$Y(z)$	Output response in z-domain
$U(z)$	Process input in z-domain
$W(z)$	Load disturbance in z-domain
s	Laplace operator
z	Discrete-time operator
T_s	Sampling period
$\boldsymbol{I}$	Identity matrix with appropriate dimensions
$\boldsymbol{I}_n$	Identity matrix of dimension n
$\mathbf{0}$	Zero matrix (or vector) with appropriate dimensions
$\mathbf{0}_{n\times m}$	Zero matrix of dimension $n \times m$
diag $X_1,\ldots,X_m$	Block diagonal matrix with blocks $\{X_1, \ldots, X_m\}$
$A > (<)0$	The matrix A is positive (negative) definite
$A \geq (\leq)0$	The matrix A is positive (negative) semi-definite
$\square$	End of proof
$\diamond$	End of remark

Operators and Functions

$\triangleq$	Defined as
$\neq$	Not equal to
$\cup$	Union operation
Π	Continued multiplication
$\forall$	To any (all)

$\in$	Belong to
$\subset$	Subset
$\rightarrow$	Tend to
$\leftarrow$	Set as
lim	Limit
sup	Supremum
min	Minimize
max	Maximize
sign	Sign function
$\mathrm{Co}\{\cdot\}$	Convex hull
$\mathrm{CH}(\cdot)$	Conic hull
floor	Round toward negative infinity
ceil	Round toward positive infinity
round	Round to the nearest integer
$\|a\|$	Absolute value (magnitude) of $a \in \mathbb{C}$ or Euclidean norm of $a \in \mathfrak{R}^n$
$\delta f(t,k)$	Difference of $f(t, k)$ in iteration domain
$\delta_t f(t,k)$	Difference of $f(t, k)$ in time domain
$\frac{\partial f}{\partial x}$	Partial derivative of a function f with respect to x
A^{T}	Transpose of the matrix A
A^*	Complex conjugate transpose of the matrix A
A^{-1}	Inverse of the matrix A
$A^{\perp}$	Null space of the matrix A
$\mathrm{tr}(A)$	Trace of the matrix A
$\mathrm{eig}(A)$	Eigenvalue of the matrix A
$\mathrm{sym}\{A\}$	$A + A^{\mathrm{T}}$
$\rho(A)$	Spectral radius of the matrix A
$\bar{\sigma}(A)$	The largest singular value of the matrix A
$\\|A\\|$	Consistent norm of the matrix A
$\\|A\\|_2$	2-norm of a matrix $A \in L_2$ or 2-norm of a vector
$\\|A\\|_\infty$	Infinity-norm of $A \in L_\infty$
$\\|x\\|_R$	$\sqrt{x^{\mathrm{T}} R x}$
$\\|\xi(\cdot, j)\\|_{\Omega,t}$	$\\|\xi(\cdot, j)\\|_{\Omega,t} = \sqrt{\sum_{i=1}^{T} \xi^{\mathrm{T}}(i,j)\Omega\xi(i,j)}$ for a positive definite matrix Ω and a positive integer T

Chapter 1
Introduction

1.1 Control Objectives of Batch Processes

Batch processes, which are also referred to repetitive operation systems in some engineering fields like mechanical manufacturing and electrical driving, have been extensively established in various industries to yield products with high quality and efficiency, or conduct repetitive operations to accomplish specific manufacturing tasks. To enumerate a few, chemical and biological reactors, pharmaceutical crystallizers, polymer injection molding machines, robotic manipulators, semiconductor circuit board production line, materials lifting and transporting equipment, commonly belong to industrial batch processes or repetitive operation systems. Generally, an industrial batch process is described as "a process that leads to the production of finite quantities of material by subjecting quantities of input materials to an ordered set of processing activities over a finite period of time using one or more pieces of equipment." (International Society of Automation 2010). For batch runs of industrial processes, there are three basic demands in practice:

(i) Implement a sequence of processing activities to yield a desired number of products or expected output performance;
(ii) Accomplish a batch run in a finite time period, which is specifically called cycle;
(iii) Reset the initial process conditions after one batch run to zero or prescribed values for operating the next batch.

To meet the above requirements, the control objectives for batch run mainly include:

(i) Track a desired set-point value (or named trajectory) in each batch for output implementation;
(ii) Overcome the effect of repetitive load disturbance on each batch;

T. Liu et al., *Robust Iterative Learning Control of Industrial Batch Systems*, Intelligent Control and Learning Systems 22, https://doi.org/10.1007/978-981-96-9778-6_1

(iii) Maintain batch run stability along either the time direction in each batch or the batchwise direction;
(iv) Satisfy the input and output constraints of a batch process for safe operation.

Concerning batch processes with time-invariant dynamics in each cycle, the first control objective may be refined to perfectly track the desired system output trajectory for a batch run. In engineering practice, there often exist time- or batch-varying process uncertainties, along with non-repetitive type disturbance occurring during batch run. Moreover, time delay is pervasively associated with industrial batch process operation, such as the delay in heat or energy transfer, signal transmission, and output measurement (Liu and Gao 2012; Seborg et al. 2013). These issues bring difficulties for the control system design and batch performance optimization in engineering practice.

A lot of control methods were explored in the last three decades (Bonvin et al. 2006; Liu et al. 2018). Among different methods, the iterative learning control (ILC) approach has been intensively developed in the past years for various batch processes and repetitive operation systems (Ahn et al. 2007; Wang et al. 2009; Shen and Li 2019), since the pioneering work in the early 1980s (Arimoto et al. 1984). The key idea behind the ILC approach is that repetitive operation data collected from historical batches is adopted to gradually enhance the output tracking performance from batch to batch. As surveyed in the above references, a large number of the existing ILC methods had been contributed to linear or nonlinear batch processes with time-invariant dynamics in either the continuous- or discrete-time domain, aiming at perfectly tracking the set-point value or specified output profile. However, the challenges had been remained for guaranteeing robust convergence and stability under time- or batch-varying plant uncertainties, non-repetitive type disturbance, and time delay for implementation.

To tackle time- or batch-varying process uncertainties, there are only a small number of robust ILC designs explored in the references (e.g., Lee et al. 2000; Nagy and Braatz 2003; Harte et al. 2005; Tayebi and Chien 2007; Wijdeven et al. 2009; Shi et al. 2005; Liu and Wang 2012; Paszke et al. 2016; Hao et al. 2021), by incorporating the real-time feedback control with the learning type feedforward control together, i.e., the control output to the process input was updated directly by the learning controller, which was termed as direct-type ILC, based on an equivalent two-dimensional (2D) Fornasini-Marchesini (FM) or Roesser system description (Kaczorek 1985; Rogers et al. 2007). By comparison, the so-called indirect-type ILC methods were also developed for practical applications, based on the inherent feedback control structure of a batch process, such as the conventional proportional-integral-derivative (PID) loop (Wang et al. 2012; Liu et al. 2014; Hao et al. 2019, 2025), the internal model control (IMC) structure (Liu et al. 2010; Shokri-Ghaleh et al. 2021), and the model predictive control scheme (Shi et al. 2014; Li et al. 2020). A notable merit is that the feedback control design is independent of the feedforward control design in such an indirect-type ILC scheme, that is, the closed-loop controller takes full responsibility for tracking the specified output trajectory along the time direction and maintaining the closed-loop control system stability under process uncertainties, while the set-point learning controller is purely adopted to enhance the output tracking performance and convergence rate along the batch direction. It should be noted that only a few references discussed the stability conditions of a closed-loop ILC design along with robustness constraints, by using the so-called double-dynamic analysis (DDA) (Meng and Moore 2017) and norm-optimal technique (Son et al. 2016).

Concerning time delay involved with the input(s), output(s), or the internal state(s) of a batch process, there are an even smaller number of references devoted to the corresponding ILC methods. The earlier reference (Xu et al. 2001) adopted the well-known Smith predictor (SP) for predicting the delay response of a single-input-single-output (SISO) batch process of open-loop stable type, so as to combine with a classical ILC scheme for improving the system output tracking performance. A modified SP was proposed to design an ILC scheme (Tian and Song 2020) for application to open-loop unstable or integral batch processes with delay response. Another ILC design (Li et al. 2005) was presented to accommodate for constant process state and input delays, by taking a 2D linear Roesser model to describe batch operation. A phase lag compensator was suggested to design ILC for eliminating the adverse effect of the process input delay (Tan et al. 2009). In contrast, a robust design of 2D ILC based on a state predictor was explored in the time domain for linear type batch processes under the input delay and time-varying uncertainties (Shokri-Ghaleh et al. 2021). With the combination of a state feedback control law and a feedforward learning control strategy, alternative ILC designs in terms of the robust $\mathcal{H}_\infty$, use the math sign theory were presented in the references (Liu and Gao 2010; Wang et al. 2013; Tao et al. 2019) for dealing with time-varying state delays. By designing a state predictor for delay-free output prediction, a direct-type ILC design (Hao et al. 2016) and another indirect-type ILC scheme (Liu et al. 2024) were explored to overcome the adverse effect from the process input or output delay, respectively.

Based on the above brief review on the development of robust ILC methods for industrial batch processes and repetitive systems subject to time- or batch-varying plant uncertainties, delay response, and non-repetitive type disturbance from cycle to cycle, it should be noted that a series of direct- and indirect-type ILC methods were successfully developed by the authors in the past decade. It is therefore motivated to summarize these results into a systematic framework for the convenience of a study from scratch. Specifically, the book starts from an PI based ILC design in Chap. 2, followed by PID based ILC design in Chap. 3, closed-loop ILC schemes with state or output feedback in Chaps. 4 and 5, extended state observer (ESO) based ILC design for enhancing disturbance rejection performance in Chap. 6, MPC-based ILC design under the process input and/or output constraints in Chap. 7, 2D state predictor based ILC design for dealing with the process input delay in Chap. 8, predictive state observer (PSO) based ILC design for dealing with the output delay in Chap. 9, robust 2D ILC design for tackling the process state delay in Chap. 10. Finally, two alternative data-driven set-point learning control schemes are presented in Chap. 11 for application to batch processes or repetitive systems with unknown dynamics. In this way, the main contents are organized in an easy-to-follow style to facilitate a beginner to learn the robust ILC theory. Moreover, scholars and technicians in the process engineering field could easily find out the robust ILC designs suitable for their specific problems of batch process control, or even proceed with further exploration of these results in a more efficient manner.

1.2 Fundamentals of Two-Dimensional (2D) System Control

There are two commonly used 2D system descriptions, named the Roesser type model and the Fornasini-Marchesini (FM) type model, respectively.

In general, a 2D Roesser-type model is described by

$$\begin{cases} \begin{bmatrix} x^h(t+1,k) \\ x^v(t,k+1) \end{bmatrix} = \begin{bmatrix} A_{11} & A_{12} \\ A_{21} & A_{22} \end{bmatrix} \begin{bmatrix} x^h(t,k) \\ x^v(t,k) \end{bmatrix} + \begin{bmatrix} B_1 \\ B_2 \end{bmatrix} u(t,k) + \omega(t,k), \\ y(t,k) = \begin{bmatrix} C_1 & C_2 \end{bmatrix} \begin{bmatrix} x^h(t,k) \\ x^v(t,k) \end{bmatrix}, \quad t,k = 0,1,2,\ldots \end{cases} \tag{1.1}$$

where t and k denote the horizontal and vertical coordinates, respectively; $x^h(t,k) \in \Re^{n_1}$ and $x^v(t,k) \in \Re^{n_2}$ represent the horizontal and vertical state variables, respectively; $u(t,k) \in \Re^m$, $y(t,k) \in \Re^l$ and $\omega(t,k) \in \Re^{n_1+n_2}$ are the control input, the controlled output, and the external disturbance, respectively; $A_{11}, A_{12}, A_{21}, A_{22}, B_1, B_2, C_1$, and C_2 denote the nominal system matrices with compatible dimensions; The boundary conditions of a 2D Roesser-type model in (1.1) are denoted by

$$x^h(0,k),\ x^v(t,0),\ t,k = 0,1,2,\ldots$$

where $x^h(0,k) \in \Re^{n_1}$ and $x^h(t,0) \in \Re^{n_2}$ are the horizontal and vertical boundary conditions, respectively. For notational brevity, the system state is denoted by $x(t,k) \triangleq \left[[x^h(t,k)]^{\mathrm{T}}, [x^v(t,k)]^{\mathrm{T}} \right]^{\mathrm{T}}$.

A 2D Fornasini-Marchesini local state-space (FFLSS) model is generally described by

$$\begin{cases} x(t+1,k+1) = A_1 x(t,k+1) + A_2 x(t+1,k) + Bu(t,k+1) + \omega(t,k+1), \\ y(t,k+1) = Cx(t,k+1), \quad t,k = 0,1,2,\ldots \end{cases} \tag{1.2}$$

where t and k denote the horizontal and vertical coordinates, respectively; $x(t,k) \in \Re^n$ is the state vector; $u(t,k) \in \Re^m$ is the control input; $y(t,k) \in \Re^p$ is the output; $\omega(t,k) \in \Re^n$ is the external disturbance; A_1, A_2, B and C denote constant matrices with compatible dimensions. The boundary conditions are defined by

$$x(0,k),\ x(t,0),\ t,k = 0,1,2,\ldots$$

Correspondingly, the definitions on asymptotical stability of 2D systems, along with some related results, are briefly introduced below, which are fundamental to present the main contents in the book.

Definition 1.1 (Kaczorek 1985) Given any boundary conditions $x(0,k)$ and $x(t,0)$, if the unforced state response of a 2D Roesser system described in (1.1) under $u(t,k)=0$ and $\omega(t,k)=0$ satisfies

$$\lim_{t,k\to\infty} x(t,k)=0, \tag{1.3}$$

the 2D Roesser system is guaranteed to be asymptotically stable (AS).

Definition 1.2 (Shi et al. 2005) Suppose that there is a vertical boundary condition $x^v(t,0)=0$. Given any horizontal condition $x^h(t,0)$, if the unforced horizontal state response of a 2D Roesser system in (1.1) satisfies

$$\lim_{t\to\infty} x^h(t,k)=0,\ \forall k=0,1,2,\ldots, \tag{1.4}$$

the horizontal state of the 2D Roesser system is called horizontal asymptotically stable (HAS).

Definition 1.3 (Shi et al. 2005) Suppose that there is a horizontal boundary condition $x^h(0,k)=0$. Given any horizontal boundary condition $x^v(t,0)$, if the unforced vertical state response of 2D system in (1.1) satisfies

$$\lim_{k\to\infty} x^v(t,k)=0,\ \forall t=0,1,2,\ldots, \tag{1.5}$$

the vertical state of 2D Roesser system in (1.1) is guaranteed to be vertically asymptotically stable (VAS).

Lemma 1.1 (Kaczorek 1985) The unforced 2D Roesser system in (1.1) is asymptotically stable if there is a block diagonal matrix $P=\text{diag}\{P_h,P_v\}$, where $P_h\in\Re^{n_1\times n_1}$ and $P_v\in\Re^{n_2\times n_2}$, satisfying

$$A^{\mathrm{T}}PA-P<0. \tag{1.6}$$

Lemma 1.2 (Shi et al. 2005) The unforced 2D Roesser system in (1.1) is asymptotically stable if there are a function $V(\cdot)$ and a scalar $0<\rho<1$ satisfying

(i) $V(x)\geq 0$ for $x\in\Re^{n_1+n_2}$, and $V(x)=0$ if and only if $x=0$;
(ii) $V(x)\to\infty$ as $\|x\|\to\infty$;
(iii) for any boundary conditions

$$\sum_{\substack{t+k=T_0+K_0+i+1\\ T_0\le t\le T_0+i\\ K_0\le k\le K_0+i}} V(x(t,k)) < \rho \sum_{\substack{t+k=T_0+K_0+i\\ T_0\le t\le T_0+i\\ K_0\le k\le K_0+i}} V(x(t,k)),\ \forall T_0>0,\ K_0>0,\ i>0. \tag{1.7}$$

Meanwhile, the minimum of ρ is termed as the convergence index of the 2D Roesser system.

Lemma 1.3 (Shi et al. 2005) The unforced 2D Roesser system in (1.1) is HAS if there are a function $V_h(\cdot)$ and a scalar $\alpha \in (0, 1)$ satisfying

(i) $V_h(x) \geq 0$ for $x \in \Re^{n_1}$, and $V_h(x) = 0$ if and only if $x = 0$;
(ii) $V_h(x) \to \infty$ as $\|x\| \to \infty$;
(iii) for any horizontal boundary conditions and zero vertical boundary condition,

$$\sum_{k=0}^{N} V_h(x^h(t+1,k)) < \alpha \sum_{k=0}^{N} V_h(x^h(t,k)),\ \forall t>0,\ N>0. \tag{1.8}$$

Meanwhile, the minimum of α is termed as the horizontal convergence index (HCI) of the 2D Roesser system.

Lemma 1.4 (Shi et al. 2005) The unforced 2D Roesser system in (1.1) is VAS if there are a function $V_v(\cdot)$ and a scalar $\beta \in (0, 1)$ satisfying

(i) $V_v(x) \geq 0$ for $x \in \Re^{n_2}$, and $V_v(x) = 0$ if and only if $x = 0$;
(ii) $V_v(x) \to \infty$ as $\|x\| \to \infty$;
(iii) for any vertical boundary conditions and zero horizontal boundary condition,

$$\sum_{t=0}^{N} V_v(x^v(t,k+1)) < \beta \sum_{t=0}^{N} V_v(x^v(t,k)),\ \forall k>0,\ N>0. \tag{1.9}$$

Meanwhile, the minimum of β is termed as the vertical convergence index (VCI) of the 2D Roesser system.

In contrast, the stability of a 2D FMLSS system in (1.2) is evaluated by three definitions and a lemma below.

Definition 1.5 (Kaczorek 1985) Under any bounded boundary conditions $x(t, 0)$ and $x(0, k)$, $t, k = 0, 1, 2, \ldots$, the unforced 2D FMLSS system in (1.2) becomes asymptotically stable if

$$\lim_{t,k\to\infty} x(t,k) = 0. \tag{1.10}$$

Definition 1.6 (Liu and Wang 2012) The unforced 2D FMLSS system in (1.2) under $u(t, k+1) = 0$ and $\omega(t, k+1) = 0$ is regarded to be asymptotically stable in the time direction (briefly, T-stable), if the following condition holds for any integer $N > 0$, any finite T-boundary $x(0, k)$, $(k = 0, 1, 2, \ldots)$ and a finite K-boundary $x(t, 0)$,

$$\lim_{t\to\infty} \sum_{k=0}^{N} \|x(t, k)\| = 0. \tag{1.11}$$

Definition 1.7 (Liu and Wang 2012) The unforced 2D FMLSS system in (1.2) under $u(t, k+1) = 0$ and $\omega(t, k+1) = 0$ is regarded to be asymptotically stable in the batch direction (briefly, K-stable), if the following condition holds for any integer $N > 0$, any finite K-boundary $x(t, 0)$, $(t = 0, 1, 2, \ldots)$ and a finite K-boundary $x(0, k)$,

$$\lim_{k\to\infty} \sum_{t=0}^{N} \|x(t, k)\| = 0. \tag{1.12}$$

Lemma 1.6 (Shi et al. 2005) An unforced 2D FMLSS system described in (1.2) under $u(t, k+1) = 0$ and $\omega(t, k+1) = 0$ becomes 2D stable if there exist scalars $\rho \in (0, 1)$, $\rho_t \in (0, 1)$, $\rho_k \in (0, 1)$ and a function $V(\cdot)$ such that

(i) $V(x) \geq 0$ for $\forall x \in \Re^n$, and $V(x) = 0$ if and only if $x = 0$;
(ii) $V(x) \to \infty$ as $\|x\| \to \infty$;
(iii) for any boundary conditions, there stands

$$\sum_{\substack{t+k=T_0+K_0+i+1 \\ T_0\leq t\leq T_0+i \\ K_0\leq k\leq K_0+i}} V(x(t, k)) < \rho \sum_{\substack{t+k=T_0+K_0+i \\ T_0\leq t\leq T_0+i \\ K_0\leq k\leq K_0+i}} V(x(t, k)), \ \forall T_0 > 0, \ K_0 > 0, \ i > 0, \tag{1.13}$$

and accordingly, the minimum of ρ is termed as the 2D convergence index of the 2D FMLSS system;

(iv) for any integer $N > 0$, any T-boundary condition of $x(0, k)$, and a zero K-boundary condition $x(t, 0) = 0$, there stands

$$\sum_{k=0}^{N} V(x(t+1, k)) \leq \rho_t \sum_{k=0}^{N} V(x(t, k)), \ \forall t > 0, \tag{1.14}$$

and accordingly, the minimum of ρ_t is termed as the T-direction convergence index (TCI) of the 2D FMLSS system;

(v) for any integer $N > 0$, any K-boundary condition of $x(t, 0)$, and a zero T-boundary condition $x(0, k) = 0$, there stands

$$\sum_{t=0}^{N} V(x(t,k+1)) \leq \rho_k \sum_{t=0}^{N} V(x(t,k)), \ \forall k > 0, \tag{1.15}$$

and accordingly, the minimum of ρ_k is termed as the K-direction convergence index (KCI) of the 2D FMLSS system.

1.3 Performance Criteria

For assessing the control performance of a batch operation system, a fundamental criterion is referred to as no steady-state tracking error along the batch direction, i.e.,

$$\lim_{k\to\infty} e(t,k) = \lim_{k\to\infty} \left[y_{\mathrm{d}}(t) - y(t,k)\right] = 0, \ \forall t \in \mathbb{Z}_N, \tag{1.16}$$

where $y(t,k)$ and $y_{\mathrm{d}}(t)$ denote the system output at time t in the batch k and the desired reference trajectory, respectively; N is the time length in each batch.

Moreover, to further evaluate the batch tracking performance, the following error criteria are typically adopted in engineering practice:

1. The averaged tracking error (ATE)

$$\mathrm{ATE}(k) = \frac{1}{N}\sum_{t=1}^{N} |e(t,k)|. \tag{1.17}$$

2. The root-mean-square-error (RMSE)

$$\mathrm{RMSE}(k) = \sqrt{\frac{1}{N}\sum_{t=1}^{N} e^2(t,k)}. \tag{1.18}$$

3. The root-sum-squared-error (RSSE)

$$\mathrm{RSSE} = \sqrt{\sum_{t=1}^{N} e^2(t)}. \tag{1.19}$$

4. The mean-square-error (MSE)

$$\mathrm{MSE}(k) = \frac{1}{N}\sum_{t=1}^{N} e^2(t,k). \tag{1.20}$$

To counteract the adverse effect from the external disturbance to the controlled output along both the time and batch directions, a commonly used 2D control objective is in the form of

$$J_{\mathrm{BP}} = \sum_{t=0}^{N_1=N} \sum_{k=0}^{N_2\to\infty} \left(\gamma^{-1} \|y(t,k)\|_2^2 - \gamma \|\omega(t,k)\|_2^2\right) \tag{1.21}$$

for the 2D Roesser system in (1.1), or

$$J_{\mathrm{BP}} = \sum_{t=0}^{N_1=N} \sum_{k=0}^{N_2\to\infty} \left(\gamma^{-1} \|y(t,k+1)\|_2^2 - \gamma \|\omega(t,k+1)\|_2^2\right) \tag{1.22}$$

for the 2D FMLSS system in (1.2), where $\gamma > 0$ is a 2D performance index to be minimized.

References

Ahn H-S, Chen YQ, Moore KL (2007) Iterative learning control: brief survey and categorization. IEEE Trans Syst Man Cybern-Part C: Appl Rev 37:1099–1121

Arimoto S, Kawamura S, Miyazaki F (1984) Bettering operation of robots by learning. J Robot Syst 1(2):123–140

Bonvin D, Srinivasan B, Hunkeler D (2006) Control and optimization of batch processes. IEEE Trans Control Syst Mag 26(1):34–45

Hao S, Liu T, Paszke W, Gałkowski K (2016) Robust iterative learning control for batch processes with input delay subject to time-varying uncertainties. IET Control Theory Appl 10(15):1904–1915

Hao S, Liu T, Gao F (2019) PI based indirect-type iterative learning control for batch processes with time-varying uncertainties: a 2D FM model based approach. J Process Control 78:57–67

Hao S, Liu T, Paszke W, Gałkowski K, Wang Q-G (2021) Robust static output feedback based ILC design with a finite-frequency-range 2-D $\mathcal{H}_\infty$ specification for batch processes subject to nonrepetitive disturbances. Int J Robust Nonlinear Control 31(12):5745–5761

Hao S, Liu T, Rogers E, Wang Y, Paszke W (2025) PI-based set-point learning control for batch processes with unknown dynamics and nonrepetitive uncertainties. IEEE Trans Autom Control 70(6):3649–3664

Harte TJ, Hätönen J, Owens DH (2005) Discrete-time inverse model-based iterative learning control stability, monotonicity and robustness. Int J Control 78(8):577–586

International Society of Automation (2010) Batch Control. Part 1, Models and Terminology, ANSI/ISA–88.00.01–2010. ISA, Research Triangle Park, N.C

Kaczorek T (1985) Two-dimensional linear system. Springer, Berlin

Lee JH, Lee KS, Kim WC (2000) Model-based iterative learning control with a quadratic criterion for time-varying linear systems. Automatica 36(5):641–657

Li X-D, Chow TWS, Ho JKL (2005) 2-D system theory based iterative learning control for linear continuous systems with time delays. IEEE Trans Circ Syst-I: Regular Pap 52(7):1421–1430

Li D, He S, Xi Y, Liu T, Gao F, Wang Y, Lu J (2020) Synthesis of ILC-MPC controller with data-driven approach for constrained batch processes. IEEE Trans Ind Electron 67(4):3116–3125

Liu T, Gao F (2010) Robust two-dimensional iterative learning control for batch processes with state delay and time-varying uncertainties. Chem Eng Sci 65(23):6134–6144

Liu T, Gao F, Wang YQ (2010) IMC-based iterative learning control for batch processes with uncertain time delay. J Process Control 20(2):173–180
Liu T, Gao F (2012) Industrial process identification and control design: step-test and relay-experiment-based methods. Springer, London
Liu T, Wang Y (2012) A synthetic approach for robust constrained iterative learning control of piecewise affine batch processes. Automatica 48(11):2762–2775
Liu T, Wang X, Chen J (2014) Robust PID based indirect-type iterative learning control for batch processes with time-varying uncertainties. J Process Control 24(12):95–106
Liu K, Chen YQ, Zhang T, Tian S, Zhang X (2018) A survey of run-to-run control for batch processes. ISA Trans 83(1):107–125
Liu T, Hao S, Wang Y, Na J (2024) Predictive state observer based set-point learning control for batch manufacturing processes with delay response. IEEE Trans Ind Electron 71(1):788–797
Meng D, Moore KL (2017) Convergence of iterative learning control for SISO nonrepetitive systems subject to iteration-dependent uncertainties. Automatica 79(5):167–177
Nagy ZK, Braatz RD (2003) Robust nonlinear model predictive control of batch processes. AIChE J 49(7):1776–1786
Paszke W, Rogers E, Gałkowski K (2016) Experimentally verified generalized KYP based iterative learning control design. Control Eng Pract 53:57–67
Rogers E, Galkowski K, Owens DH (2007) Control systems theory and applications for linear repetitive processes. Springer, Berlin
Seborg DE, Edgar TF, Mellichamp DA, Doyle FJ III (2013) Process dynamics and control, 4th edn. Wiley, Hoboken
Shen D, Li X (2019) A survey on iterative learning control with randomly varying trial lengths: model, synthesis, and convergence analysis. Annu Rev Control 48(1):89–102
Shi J, Gao F, Wu T-J (2005) Robust design of integrated feedback and iterative learning control of a batch process based on a 2D Roesser system. J Process Control 15(8):907–924
Shi J, Zhou H, Cao Z, Jiang Q (2014) A design method for indirect iterative learning control based on two-dimensional generalized predictive control algorithm. J Process Control 24(10):1527–1537
Shokri-Ghaleh H, Ganjefar S, Shahri A (2021) Robust iterative learning control for uncertain continuous-time system with input delay and random iteration-varying uncertainties. IET Control Theory Appl 15(13):1749–1761
Son T, Pipeleers G, Swevers J (2016) Robust monotonic convergent iterative learning control. IEEE Trans Autom Control 61(4):1063–1068
Tan KK, Zhao S, Huang S, Lee TH, Tay A (2009) A new repetitive control for LTI systems with input delay. J Process Control 19(4):711–716
Tao H, Paszke W, Yang H, Gałkowski K (2019) Finite frequency range robust iterative learning control of linear discrete system with multiple time-delays. J Franklin Inst 356(5):2690–2708
Tayebi A, Chien C (2007) A unified adaptive iterative learning control framework for uncertain nonlinear systems. IEEE Trans Autom Control 52(10):1907–1913
Tian D, Song X (2020) Iterative learning control for a type of modified Smith predictor. J Dyn Syst Meas Control 142(5):041009
Wang YQ, Gao F, Doyle FJ III (2009) Survey on iterative learning control, repetitive control, and run-to-run control. J Process Control 19(10):1589–1600
Wang L, Mo S, Qu H, Zhou D, Gao F (2013) $\mathcal{H}_\infty$ design of 2D controller for batch processes with uncertainties and interval time-varying delays. Control Eng Pract 21(10):1321–1333
Wang Y, Liu T, Zhao Z (2012) Advanced PI control with simple learning set-point design: application on batch processes and robust stability analysis. Chem Eng Sci 71(26):153–165
Wijdeven J, Donkers T, Bosgra O (2009) Iterative learning control for uncertain systems: robust monotonic convergence analysis. Automatica 45(10):2383–2391
Xu JX, Hu Q, Lee TH, Yamamoto S (2001) Iterative learning control with Smith time delay compensator for batch processes. J Process Control 11:321–328

Chapter 2
Proportional-Integral (PI) Based Iterative Learning Control

2.1 Introduction

Since the traditional proportional-integral-derivative (PID) controller has been extensively applied in various industrial batch systems, there are a large number of references focused on such a controller (including P-, PI-, PD-, PID-type)-based ILC designs in the past decades. There are mainly two manners for such ILC design: (1) the ILC law is designed to tune the control command of system input, for which the controller updating law or tuning rule is designed by adopting the existing PID control methods (e.g., Kim and Kim 1996; Bonvin et al. 2006; Ruan et al. 2008; Liu et al. 2010); (2) the inner-loop PID controller is adopted to tune the control command of system input, while an ILC law is established via the set-point of the inner loop for tracking the desired system output trajectory (Wu and Ding 2007; Tan et al. 2007; Wang et al. 2009, 2012; Liu et al. 2014).

Concerning the first manner of ILC design, the simplest control structure is referred to as P-type ILC, which has been widely recognized as the initial contribution for tracking performance optimization of industrial repetitive systems and batch processes (Uchiyama 1978; Arimoto et al. 1984; Bristow et al. 2006). Subsequently, PID-based ILC designs have been extensively studied over the past three decades. For instance, P-type ILC designs can be found in the references (Xiong and Zhang 2003; Hou et al. 2007; Tayebi and Chien 2007; Gao et al. 2025); the D-type ILC methods were presented in the references (Arimoto et al. 1984; Chen et al. 1998; Abidi and Xu 2011); the PI-type ILC designs were given in the references (Shi et al. 2005a, b; Liang et al. 2023); the PD-type ILC methods were proposed in the references (Mi et al. 2005; Li and Tso 2008; Zou et al. 2022); the PID-type ILC designs can be seen in the references (Kim and Kim 1996; Ruan et al. 2008; Li et al. 2023).

Concerning the second manner of ILC design, two control loops are typically configured in such an ILC scheme, where the inner loop includes a P-, PI-, or PID-type controller and the outer loop is tuned by an ILC law. The ILC law may be designed to regulate the system set-point (Wu and Ding 2007), or the control signal to the

T. Liu et al., *Robust Iterative Learning Control of Industrial Batch Systems*, Intelligent Control and Learning Systems 22, https://doi.org/10.1007/978-981-96-9778-6_2

system input (Xu and Yan 2004; Jiang et al. 2007), or even the inner-loop controller parameters (Bone 1995; Tayebi and Chien 2007). When the set-point command of such a system could be regarded as an adjustable parameter, the ILC design is named as a set-point-related (SPR) ILC scheme or alternatively an indirect-type ILC. It was surveyed in the literature (Wang et al. 2009) that no more than 10% of the existing ILC methods contributed to the above indirect-type ILC, which, however, has drawn more attention in recent years (Chi et al. 2023; Tao et al. 2023; Hao et al. 2025). Note that the indirect ILC framework was initially explored in the run-to-run (R2R) control schemes studied in the literature (Hankinson et al. 1997; Ganesan et al. 2007; Busch et al. 2007).

Considering that a PI controller has been overwhelmingly adopted in engineering practice owing to its capability of removing the steady-state output tracking error along the time direction, two alternative PI-based indirect-type ILC designs are, therefore, presented in this chapter. In each of these two ILC schemes, the inner loop is tuned by a PI controller while an ILC controller in the outer loop is designed to regulate the set-point command of the inner loop. The learning system stability is analyzed based on the 2D FM and 2D Roesser system descriptions, respectively. The advantages of both PI-based ILC schemes are demonstrated by illustrative examples.

2.2 Advanced PI Control Loop Design

2.2.1 Model-Based PI Tuning

Consider an industrial batch process involved with time-varying uncertainties and external disturbance described by

$$\begin{cases} x(t+1) = [A + \Delta A(t)]x(t) + [B + \Delta B(t)]u(t) + w(t), \\ y(t) = Cx(t), \end{cases} \tag{2.1}$$

where $u(t) \in \Re^m$, $x(t) \in \Re^n$ and $y(t) \in \Re^p$ denote the control input, system state and output at time t, respectively; the process disturbance is indicated by $w(t)$; the nominal system matrices are denoted by [A, B, C] with appropriate dimensions; the time-varying uncertainties are denoted by $\Delta A(t)$ and $\Delta B(t)$, complying with

$$\begin{bmatrix} \Delta A(t) & \Delta B(t) \end{bmatrix} = E\Delta(t)\begin{bmatrix} F_{\mathrm{A}} & F_{\mathrm{B}} \end{bmatrix}, \tag{2.2}$$

where F_{A}, F_{B}, and E are constant matrices pre-estimated in practice, and $\Delta(t)$ denotes the time-varying uncertainty satisfying $\Delta^{\mathrm{T}}(t)\Delta(t) \leq \boldsymbol{I}$.

To design a unity feedback control system for the above process, a widely used PI controller in practice is generally in the form of

$$u_{\mathrm{PI}}(t) = \tau_{\mathrm{P}} e(t) + \tau_{\mathrm{I}} \sum e(t), \tag{2.3}$$

where τ_P and τ_I denote the proportional and integral parameters of the PI controller, respectively, $e(t) = y_d(t) - y(t)$, and $y_d(t)$ is the desired output profile to be tracked.

Combining the sum of the time-direction output tracking error, i.e., $\sum e(t) = \sum_{i=0}^{t} e(i)$ with the process model in (2.1), an augmented plant description is given by

$$\begin{bmatrix} x(t+1) \\ \sum e(t) \end{bmatrix} = \begin{bmatrix} A + \Delta A(t) & 0 \\ -C & \boldsymbol{I} \end{bmatrix} \begin{bmatrix} x(t) \\ \sum e(t-1) \end{bmatrix} + \begin{bmatrix} B + \Delta B(t) \\ 0 \end{bmatrix} u(t) + \begin{bmatrix} \boldsymbol{I} \\ 0 \end{bmatrix} w(t). \tag{2.4}$$

By using the PI controller in (2.3) to the above augmented plant description in (2.4) gives the closed-loop system as below

$$\begin{cases} \hat{x}(t+1) = \tilde{A}\hat{x}(t) + \hat{D}w(t), \\ e(t) = \hat{C}\hat{x}(t), \end{cases} \tag{2.5}$$

where $\tilde{A} = \hat{A} + \hat{E}\Delta(t)\hat{F}$ and

$$\hat{A} = \begin{bmatrix} A - B(\tau_P + \tau_I)C & B\tau_I \\ -C & \boldsymbol{I} \end{bmatrix}, \ \hat{D} = \begin{bmatrix} \boldsymbol{I} \\ 0 \end{bmatrix}, \ \hat{x}(t) = \begin{bmatrix} x(t) \\ \sum e(t-1) \end{bmatrix}, \ \hat{E} = \begin{bmatrix} E^{\mathrm{T}} & 0 \end{bmatrix}^{\mathrm{T}},$$
$$\hat{C} = \begin{bmatrix} -C & 0 \end{bmatrix}, \ \hat{F} = \begin{bmatrix} F_A - F_B(\tau_P + \tau_I)C & B\tau_I \end{bmatrix}. \tag{2.6}$$

Considering that the above PI control system performance is primarily affected by the transfer function poles, it is expected to prescribe the closed-loop system poles to an ideal location so as to obtain the desired output response (Ge et al. 2002). In order to realize the desired system pole location by setting the PI controller parameters, the following constraints are imposed to address the time-varying process uncertainties and/or disturbance:

(i) The closed-loop system poles in (2.5) should be located in a specified region $D(\alpha, r)$ centered at $(\alpha, 0)$ with radius r satisfying $|\alpha| + r < 1$, namely

$$\lambda(\tilde{A}) \subset D(\alpha, r). \tag{2.7}$$

(ii) The $\mathcal{H}_\infty$ norm of the transfer function from the external disturbance $w(t)$ to the output tracking error $e(t)$, i.e., $H(z) = \hat{C}(z\boldsymbol{I} - \tilde{A})^{-1}\hat{D}$, should satisfy

$$\|H(z)\|_\infty < \gamma_{PI}, \tag{2.8}$$

where $\gamma_{PI} > 0$ denotes the $\mathcal{H}_\infty$ performance level.

For the purpose of determining the PI controller parameters, two useful lemmas are briefly introduced herein.

Lemma 2.1 (Gu et al. 2003) Concerning two matrices X and Y with proper dimensions, the following inequality is satisfied for any positive scalar $\varepsilon > 0$ and matrix Θ obeying $\Theta^{\mathrm{T}}\Theta \leq \boldsymbol{I}$ with proper dimension

$$X\Theta Y + Y^{\mathrm{T}}\Theta^{\mathrm{T}}X^{\mathrm{T}} \leq \varepsilon XX^{\mathrm{T}} + \varepsilon^{-1}Y^{\mathrm{T}}Y. \tag{2.9}$$

Lemma 2.2 (Xu et al. 2000) Given a circular region $D(\alpha, r)$ and a positive scalar γ, if a positive definite matrix $P > 0$ could comply with the following matrix inequalities

$$\begin{cases} \gamma^2\boldsymbol{I} - \hat{D}^{\mathrm{T}}P\hat{D} > 0, \\ A_{\mathrm{c}}^{\mathrm{T}}PA_{\mathrm{c}} - P + \beta\hat{C}^{\mathrm{T}}\hat{C} + \beta\tilde{A}^{\mathrm{T}}P\hat{D}(\gamma^2\boldsymbol{I} - \hat{D}^{\mathrm{T}}P\hat{D})^{-1}\hat{D}^{\mathrm{T}}P\tilde{A} < 0, \end{cases} \tag{2.10}$$

where $A_{\mathrm{c}} = (\tilde{A} - \alpha\boldsymbol{I})/r$ and $\beta = (1 - |\alpha|)/r^2$, the above constraints in (2.7) and (2.8) are satisfied for the closed-loop control system established by (2.5).

With the above two lemmas, the Theorem 2.1 presents a sufficient condition to assign the PI tuning parameters, while complying with the above two constraints.

Theorem 2.1 Given a user-specified region of $D(\alpha, r)$, the closed-loop control system in (2.5) could maintain robust D-stable and satisfy a $\mathcal{H}_\infty$ performance level γ_{PI}, if there are matrices $P_1 > 0$, $P_3 > 0$, $P_2 = P_2^{\mathrm{T}}$, R_1, R_2 and positive scalar ε satisfying

$$\begin{bmatrix} \Lambda_1 & 0 & \Lambda_2 & P\hat{C}^{\mathrm{T}} & 0 & P\hat{F}^{\mathrm{T}} \\ * & -\beta_1^{-1}\gamma_{\mathrm{PI}}^2\boldsymbol{I} & \hat{D}^{\mathrm{T}} & 0 & \hat{D}^{\mathrm{T}} & 0 \\ * & * & \Lambda_3 & 0 & 0 & 0 \\ * & * & * & -\beta_1^{-1}\boldsymbol{I} & 0 & 0 \\ * & * & * & * & -\beta_2 P & 0 \\ * & * & * & * & * & -\varepsilon\boldsymbol{I} \end{bmatrix} < 0, \tag{2.11}$$

where $\beta_1 = 1 - |\alpha|$, $\beta_2 = (\beta_1^{-1} - 1)^{-1}$, and

$$\Lambda_1 = -\alpha P\hat{A}^{\mathrm{T}} - \alpha\hat{A}P + (\alpha^2 - r^2)P + \varepsilon\alpha^2\hat{E}\hat{E}^{\mathrm{T}},\ \Lambda_2 = P\hat{A}^{\mathrm{T}} - \varepsilon\alpha\hat{E}\hat{E}^{\mathrm{T}},\ \Lambda_3 = -P + \varepsilon\hat{E}\hat{E}^{\mathrm{T}},$$

$$\hat{A}P = \begin{bmatrix} AP_1 - BR_1 & AP_2 - BR_2 \\ -CP_1 + P_2^{\mathrm{T}} & -CP_2 + P_3 \end{bmatrix},\ P\hat{F}^{\mathrm{T}} = \begin{bmatrix} P_1F_{\mathrm{A}}^{\mathrm{T}} - R_1^{\mathrm{T}}F_{\mathrm{B}}^{\mathrm{T}} \\ P_2^{\mathrm{T}}F_{\mathrm{A}}^{\mathrm{T}} - R_2^{\mathrm{T}}F_{\mathrm{B}}^{\mathrm{T}} \end{bmatrix},\ P = \begin{bmatrix} P_1 & P_2 \\ * & P_3 \end{bmatrix}. \tag{2.12}$$

Furthermore, when the above LMI in (2.11) could be feasible, the PI tuning parameters could be computed by

$$\begin{bmatrix} (\tau_{\mathrm{P}} + \tau_{\mathrm{I}})C & -\tau_{\mathrm{I}} \end{bmatrix} = \begin{bmatrix} R_1 & R_2 \end{bmatrix}P^{-1}. \tag{2.13}$$

Proof By multiplying r^2 on both sides of the second condition of (2.10) and using the well-known Schur complement, there stands

$$\begin{bmatrix} \tilde{A}^{\mathrm{T}}P\tilde{A} - \alpha\tilde{A}^{\mathrm{T}}P - \alpha P\tilde{A} + (\alpha^2 - r^2)P + \beta_1\hat{C}^{\mathrm{T}}\hat{C} & \tilde{A}^{\mathrm{T}}P\hat{D} \\ * & -\beta_1^{-1}(\gamma_{\mathrm{PI}}^2\mathbf{I} - \hat{D}^{\mathrm{T}}P\hat{D}) \end{bmatrix} < 0,$$

which can be written as

$$\begin{bmatrix} \tilde{A}^{\mathrm{T}} \\ \hat{D}^{\mathrm{T}} \end{bmatrix} P \begin{bmatrix} \tilde{A} & \hat{D} \end{bmatrix} + \begin{bmatrix} -\alpha\tilde{A}^{\mathrm{T}}P - \alpha P\tilde{A} + (\alpha^2 - r^2)P + \beta_1\hat{C}^{\mathrm{T}}\hat{C} & 0 \\ * & -\beta_1^{-1}\gamma_{\mathrm{PI}}^2 \boldsymbol{I} + \beta_2^{-1}\hat{D}^{\mathrm{T}}P\hat{D} \end{bmatrix} < 0.$$

Then it follows that by taking the Schur complement

$$\begin{bmatrix} -\alpha\tilde{A}^{\mathrm{T}}P - \alpha P\tilde{A} + (\alpha^2 - r^2)P & 0 & \tilde{A}^{\mathrm{T}} & \hat{C}^{\mathrm{T}} & 0 \\ * & -\beta_1^{-1}\gamma_{\mathrm{PI}}^2\mathbf{I} & \hat{D}^{\mathrm{T}} & 0 & \hat{D}^{\mathrm{T}} \\ * & * & -P^{-1} & 0 & 0 \\ * & * & * & -\beta_1^{-1}\mathbf{I} & 0 \\ * & * & * & * & -\beta_2 P^{-1} \end{bmatrix} < 0. \quad (2.14)$$

By defining $P = P^{-1}$ and multiplying diag$\{P^{-1}, \boldsymbol{I}, \boldsymbol{I}, \boldsymbol{I}, \boldsymbol{I}\}$ to both sides of (2.14), it follows from Lemma 2.1 that the above LMI in (2.11) holds true by denoting $R_1 = (\tau_{\mathrm{P}} + \tau_{\mathrm{I}})CP_1 - \tau_{\mathrm{I}}P_2^{\mathrm{T}}$ and $R_2 = (\tau_{\mathrm{P}} + \tau_{\mathrm{I}})CP_2 - \tau_{\mathrm{I}}P_3$. The proof is finished. □

By defining $(R_1\ R_2)P^{-1} = (\hat{\tau}_{\mathrm{P}}\ \hat{\tau}_{\mathrm{I}})$, the PI controller parameters could be computed as

$$\begin{aligned} \tau_{\mathrm{I}} &= -\hat{\tau}_{\mathrm{I}}, \\ \tau_{\mathrm{P}} &= \hat{\tau}_{\mathrm{P}}C^{\mathrm{T}}(CC^{\mathrm{T}})^{-1} + \hat{\tau}_{\mathrm{I}}. \end{aligned} \quad (2.15)$$

To minimize the $\mathcal{H}_\infty$ control performance level and thus the adverse impact from the external disturbance $w(t)$ to the output tracking error $e(t)$, the PI controller parameters could be derived from the following linear programming

$$\begin{aligned} &\min\ \gamma_{\mathrm{PI}} \\ &\text{s.t. } (2.11) \end{aligned} \quad (2.16)$$

In general, a smaller γ_{PI} will result in a faster output response but at the cost of a more sensitive control behavior, and vice versa. Hence, it aims to take a good compromise between the desirable closed-loop system performance and the controller behavior turned out by the above PI controller.

2.2.2 Model-Free PI Tuning

Consider an unknown nonlinear system generally expressed by

$$y(t+1) = f(y(t), y(t-1), \ldots, y(t-n_y), u(t), u(t-1), \ldots, u(t-n_u)) + \omega(t), \tag{2.17}$$

where $y(t) \in \Re$ and $u(t) \in \Re$ denote the system output and control input, respectively; $\omega(t)$ denotes the external disturbance; $f(\cdot)$ represents an unknown linear or nonlinear function; n_u and n_y are the unknown input and output orders.

Assumption 2.1 (Xiong and Hou 2022) With regard to $y(t)$, $y(t-1)$, and $u(t)$, the partial derivatives of $f(\cdot)$ exist and are all bounded.

Assumption 2.2 (Hou and Jin 2013) A nonlinear system described in (2.17) complies with the following *Lipschitz* condition

$$|\delta_t y(t+1)| \leq \beta_\Gamma \|\delta_t H(t)\| \tag{2.18}$$

for $\|\delta_t H(t)\| \neq 0$, where $\beta_\Gamma > 0$ is a finite constant and $\delta_t H(t) = \left[\, \delta_t y(t)\ \delta_t u(t) \,\right]^{\mathrm{T}}$.

Under Assumptions 2.1 and 2.2, the following time-domain dynamic linearization data model is established for the system described in (2.17)

$$\delta_t y(t+1) = \Gamma(t)\delta_t H(t), \tag{2.19}$$

where $\Gamma(t) = \left[\, \gamma_1(t),\ \gamma_2(t) \,\right]$ represents a time-varying vector named pseudo-gradient to be estimated. Further details about the above equivalent transformation in (2.19) can be found in Hou and Jin (2013).

To tune the time-domain PI controller parameters, a cost function is defined below

$$J_1(u(t)) = |y_d(t+1) - y(t+1)|^2 + \lambda_t |u(t) - u(t-1)|^2, \tag{2.20}$$

where $\lambda_t > 0$ is a weighting factor specified for assessing the effect of time-domain input variation in practice. By letting the derivative of $J_1(u(t))$ with regard to $u(t)$ be zero, it yields

$$\delta_t u(t) = \frac{\rho_t \gamma_2(t)}{\lambda_t + \gamma_2^2(t)} \left[y_d(t+1) - y(t) - \gamma_1(t)\delta_t y(t) \right], \tag{2.21}$$

where $\rho_t \in (0, 2)$ represents a user-specified step size. By taking a constant value of the system set-point command, the time-domain control increment in (2.21) can be described as

$$\begin{aligned}\delta_t u(t) &= \frac{\rho_t \gamma_2(t)}{\lambda_t + \gamma_2^2(t)} \big[e(t) - \gamma_1(t)\delta_t e(t)\big] \\ &= \tau_{\rm P}(t)\delta_t e(t) + \tau_{\rm I}(t)e(t),\end{aligned} \tag{2.22}$$

where

$$\tau_{\rm P}(t) \triangleq \frac{\rho_t \gamma_1(t)\gamma_2(t)}{\lambda_t + \gamma_2^2(t)}, \quad \tau_{\rm I}(t) \triangleq \frac{\rho_t \gamma_2(t)}{\lambda_t + \gamma_2^2(t)}.$$

It is straightforward to see that the above control increment in (2.22) corresponds to the PI controller described in the discrete-time domain. It should be pointed out that the above system with the discrete-time PI controller in (2.22) could maintain stability in the presence of a finite set-point variation, such as a harmonic-type signal with limited frequency, see Liu et al. (2014) for more details.

To determine the feasible tuning regions of the above PI controller parameters, a widely used parameter estimation algorithm is adopted to estimate the unknown $\gamma_1(t)$ and $\gamma_2(t)$

$$\begin{aligned}\hat{\Gamma}(t) &= \hat{\Gamma}(t-1) + \frac{\eta_t \delta_t H(t-1)}{\mu_t + \|\delta_t H(t-1)\|^2}\Big[\delta_t y(t) - \hat{\Gamma}^{\rm T}(t-1)\delta_t H(t-1)\Big], \\ \hat{\Gamma}(t) &= \hat{\Gamma}(0), \ \text{if } \big\|\hat{\Gamma}(t)\big\| \le \varepsilon_t, \ \text{or } \|\delta_t H(t-1)\| \le \varepsilon_t, \ \text{or } \operatorname{sign}(\hat{\Gamma}(t)) \ne \operatorname{sign}(\hat{\Gamma}(0)),\end{aligned} \tag{2.23}$$

where $\eta_t \in (0, 2)$ is another adjustable parameter, $\mu_t > 0$ is also a weighting factor, $\hat{\Gamma}(t) = \big[\,\hat{\gamma}_1(t)\ \hat{\gamma}_2(t)\,\big]$, $\hat{\gamma}_1(t)$, and $\hat{\gamma}_2(t)$ indicate the estimated values of $\gamma_1(t)$ and $\gamma_2(t)$, respectively; ε_t is a tiny constant used for initialization. Note that the convergence of the above system along with the boundedness of $\hat{\Gamma}(t)$ in the time domain, could be analyzed in a similar manner to those presented in the references (Hou and Jin 2013; Hou and Xiong 2019).

With the help of the above estimated values of $\hat{\gamma}_1(t)$ and $\hat{\gamma}_2(t)$, the feasible tuning regions of the above PI parameters could be evaluated as

$$\tau_{\rm P} \in \big[\tau_{\rm P}^{\min}, \tau_{\rm P}^{\max}\big], \quad \tau_{\rm I} \in \big[\tau_{\rm I}^{\min}, \tau_{\rm I}^{\max}\big], \tag{2.24}$$

where

$$\begin{aligned}\tau_{\rm P}^{\min} &= \min_{t\in\mathbb{Z}_{N-1}} \frac{\rho_t \hat{\gamma}_1(t)\hat{\gamma}_2(t)}{\lambda_t + \hat{\gamma}_2^2(t)}, \quad \tau_{\rm P}^{\max} = \max_{t\in\mathbb{Z}_{N-1}} \frac{\rho_t \hat{\gamma}_1(t)\hat{\gamma}_2(t)}{\lambda_t + \hat{\gamma}_2^2(t)}, \\ \tau_{\rm I}^{\min} &= \min_{t\in\mathbb{Z}_{N-1}} \frac{\rho_t \hat{\gamma}_2(t)}{\lambda_t + \hat{\gamma}_2^2(t)}, \quad \tau_{\rm I}^{\max} = \max_{t\in\mathbb{Z}_{N-1}} \frac{\rho_t \hat{\gamma}_2(t)}{\lambda_t + \hat{\gamma}_2^2(t)}.\end{aligned}$$

Based on the above regions in (2.24), a tuning procedure is given by the following Algorithm I to assign desirable PI controller parameters for implementation.

Algorithm 1 (Model-free PI controller tuning)

Step 1: The PI controller parameters are initialized as $\tau_P = \tau_P^{\min}$ and $\tau_I = \tau_I^{\min}$, while choosing the initial values of parameters ρ_t and η_t, two weighting factors λ_t and μ_t, user-specified thresholds ε_t and κ, the step sizes $\Delta\tau_P$ and $\Delta\tau_I$, and the time length N for iteration

Step 2: Assign the PI controller parameters τ_P and τ_I by following the procedure from (1) to (11):

(1) **while** ($\tau_P \le \tau_P^{\max}$) **do**

(2) **while** ($\tau_I \le \tau_I^{\max}$) **do**

(3) Use (2.21) and (2.23) for the nonlinear system described by (2.17) and compute $\text{ATE} = \sum_{t=1}^{N} |e(t)|/N$;

(4) **if** (ATE $\le \kappa$) **then**

(5) Assign the controller parameters τ_P and τ_I;

(6) **else if** (ATE $> \kappa$) **then**

(7) $\tau_I = \tau_I + \Delta\tau_I$;

(8) **end if**

(9) **end while**

(10) $\tau_P = \tau_P + \Delta\tau_P$, $\tau_I = \tau_I^{\min}$;

(11) **end while**

Remark 2.1 Concerning a time-invariant linear system with unknown dynamics commanded by a constant-type set-point command, the closed-loop PI parameters tuned by Algorithm I will surely converge to constant values, respectively, which can therefore be chosen as the desirable PI parameters. It should be mentioned that the steady-state PI tuning parameters may not exist for a nonlinear time-varying system. In this case, it is suggested to select a set of time-invariant PI controller parameters from the pre-estimated tuning regions in (2.24) in terms of some performance metric, in order to facilitate the subsequent set-point learning control design for batch performance optimization. ◊

Remark 2.2 When no desirable PI controller parameters could be acquired by the above Algorithm I under a prescribed κ, it is suggested to monotonically increase or decrease κ till the feasible PI controller parameters could be found. An alternative solution is to evaluate the corresponding ATE index for each of the PI parameters setting in the above tuning regions and then determine the optimal one with respect to the minimum of the above ATE. Note that other widely used performance criteria in engineering practice, such as mean-square error and maximum absolute error, could also be adopted to assess the system performance for the purpose of assigning proper PI controller parameters. It should be mentioned that all of the above indices related to the feasible PI tuning regions could be evaluated in advance by offline computation, and thus, the optimal PI parameter setting can be obtained before execution. ◊

2.3 Robust ILC Design by 2D Roesser System Description

2.3.1 PI-Based ILC Scheme

Consider an industrial batch process involved with time-varying uncertainties and disturbances described by

$$\begin{cases} x(t+1,k) = [A + \Delta A(t,k)]x(t,k) + [B + \Delta B(t,k)]u(t,k) + w(t,k), \\ y(t,k) = Cx(t,k), \\ x(0,k) = x_0, t = 0, 1, \ldots, N-1; k = 0, 1, \ldots \end{cases} \tag{2.25}$$

where $x(t,k) \in \Re^n$, $u(t,k) \in \Re^p$, and $y(t,k) \in \Re$ are the system state, control input, and process output at time t of the k-th batch run, respectively, t is the time step and k is the batch number, $w(t,k) \in \Re^n$ is the nonrepetitive load disturbance complying with $\|w(t,k) - w(t,k-1)\| \in l_2[0,+\infty)$; The nominal system matrices are denoted by A, B, C with proper dimensions; x_0 represents the identical initial condition for batch run; The uncertain perturbations of matrices A and B are denoted by $\Delta A(t,k)$ and $\Delta B(t,k)$, respectively, which can be represented as

$$\begin{bmatrix} \Delta A(t,k) & \Delta B(t,k) \end{bmatrix} = E\Delta(t,k)\begin{bmatrix} F_{\mathrm{A}} & F_{\mathrm{B}} \end{bmatrix}, \tag{2.26}$$

where E, F_{A}, and F_{B} are pre-estimated constant matrices characterizing the uncertainty structures, and $\Delta(t,k)$ is time-varying and satisfies

$$\Delta^{\mathrm{T}}(t,k)\Delta(t,k) \le \mathbf{I}, \quad \|\Delta(t,k) - \Delta(t,k-1)\| \in l_2[0,+\infty) \tag{2.27}$$

It is expected to design an ILC law to realize that the batch process output could track a desired target trajectory, $y_{\mathrm{d}}(t)$, as closely as possible, under time-varying process uncertainties and disturbance. The output tracking error is defined by

$$e(t,k) \triangleq y_{\mathrm{d}}(t) - y(t,k). \tag{2.28}$$

A PI-type control law usually takes the form of

$$u(t,k) = u_0 + \tau_{\mathrm{P}} e_s(t,k) + \tau_{\mathrm{I}} \sum e_s(t,k), \tag{2.29}$$

where u_0 is the initial control command to the batch process input, and

$$\begin{aligned} &e_s(t,k) \triangleq y_s(t,k) - y(t,k), \; \sum e_s(t,k) \triangleq \sum e_s(t-1,k) + e_s(t,k), \\ &e_s(0,k) \triangleq 0. \end{aligned} \tag{2.30}$$

Note that $e_s(t, k)$ and $e(t, k)$ are named as the set-point tracking error and the output tracking error, respectively. The set-point command denoted by $y_s(t, k)$ may be batch-varying, and updated by the following ILC law

$$y_s(t, k) = y_s(t, k-1) + L_1 \delta \sum e_s(t-1, k) + L_2 e(t+1, k-1), \tag{2.31}$$

where $\delta \sum e_s(t-1, k) = \sum e_s(t-1, k) - \sum e_s(t-1, k-1)$, L_1 and L_2 represent the set-point learning gains to be determined. The rationale behind (2.31) is to optimize the set-point command by the batch-wise variation of the set-point tracking error integral and the output tracking error in the last batch.

A schematic of the PI-based ILC scheme is shown in Fig. 2.1 for clarity. The advantages of the proposed scheme lie with: firstly, only an output-feedback-based control strategy is used with no need for the internal system state information; secondly, the inner-loop PI controller does not need to be modified, which may be tuned in advance in practical applications. That is to say, only an outer control loop is imposed to tune the set-point command of the inner control loop. Thirdly, both the inner-loop stability and the tracking convergence of the outer loop can be separately analyzed, such that the inner-loop PI controller tuning for holding the system stability and the outer-loop ILC design for tracking the desired output profile can be independently conducted.

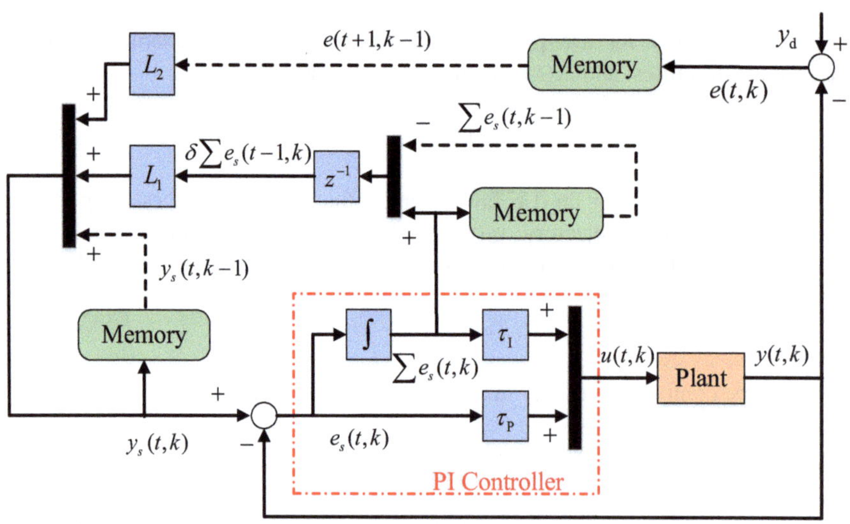

Fig. 2.1 Schematic of a PI-based ILC scheme in the 2D Roesser system framework

2.3.2 Robust Stability Analysis

A 2D Roesser system description is adopted herein to study the stability of the above PI-based ILC scheme. By using the PI-based ILC controller given in (2.29) and (2.31), the batch process in (2.25) can be transformed into

$$\begin{aligned}\delta x(t+1,k) &= \left[\overline{A}(t,k)-\overline{B}(t,k)(\tau_{\mathrm{P}}+\tau_{\mathrm{I}})C\right]\delta x(t,k)\\ &+\overline{B}(t,k)[\tau_{\mathrm{I}}+(\tau_{\mathrm{P}}+\tau_{\mathrm{I}})L_1]\delta\sum e_s(t-1,k)\\ &+\overline{B}(t,k)(\tau_{\mathrm{P}}+\tau_{\mathrm{I}})L_2 e(t+1,k-1)+\varpi(t,k),\end{aligned} \tag{2.32}$$

$$\delta\sum e_s(t,k) = -C\delta x(t,k)+(\boldsymbol{I}+L_1)\delta\sum e_s(t-1,k)+L_2 e(t+1,k-1) \tag{2.33}$$

$$\begin{aligned}e(t+1,k) &= -C\left[\overline{A}(t,k)-\overline{B}(t,k)(\tau_{\mathrm{P}}+\tau_{\mathrm{I}})C\right]\delta x(t,k)\\ &-C\overline{B}(t,k)[\tau_{\mathrm{I}}+(\tau_{\mathrm{P}}+\tau_{\mathrm{I}})L_1]\delta\sum e_s(t-1,k)\\ &+\left[\boldsymbol{I}-C\overline{B}(t,k)(\tau_{\mathrm{P}}+\tau_{\mathrm{I}})L_2\right]e(t+1,k-1)-C\varpi(t,k),\end{aligned} \tag{2.34}$$

where $\delta x(t,k)=x(t,k)-x(t,k-1)$, $\overline{A}(t,k)=A+\Delta A(t,k)$, $\overline{B}(t,k)=B+\Delta B(t,k)$ and

$$\begin{aligned}\varpi(t,k) &= [\Delta A(t,k)-\Delta A(t,k-1)]x(t,k)\\ &+[\Delta B(t,k)-\Delta B(t,k-1)]u(t,k)+\delta w(t,k).\end{aligned} \tag{2.35}$$

It can be easily verified that $\varpi(t,k)\in l_2[0,+\infty)$ could stand for any nonrepetitive process uncertainties or external disturbance described by (2.25), and thus is regarded as a non-repetitive disturbance to handle.

Correspondingly, a 2D Roesser system model for the above batch process is established by

$$\begin{cases}\begin{bmatrix}\delta x(t+1,k)\\ \delta\sum e_s(t,k)\\ e(t+1,k)\end{bmatrix}=\Pi\begin{bmatrix}\delta x(t,k)\\ \delta\sum e_s(t-1,k)\\ e(t+1,k-1)\end{bmatrix}+W\varpi(t,k),\\ z(t,k)=G\begin{bmatrix}\delta x(t,k)\\ \delta\sum e_s(t-1,k)\\ e(t+1,k-1)\end{bmatrix},\end{cases} \tag{2.36}$$

where

$$\Pi \triangleq \begin{bmatrix} \overline{A}(t,k) - \overline{B}(t,k)(\tau_P + \tau_I)C & \overline{B}(t,k)[\tau_I + (\tau_P + \tau_I)L_1] & \overline{B}(t,k)(\tau_P + \tau_I)L_2 \\ -C & \boldsymbol{I} + L_1 & L_2 \\ -C[\overline{A}(t,k) - \overline{B}(t,k)(\tau_P + \tau_I)C] & -C\overline{B}(t,k)[\tau_I + (\tau_P + \tau_I)L_1] & \boldsymbol{I} - C\overline{B}(t,k)(\tau_P + \tau_I)L_2 \end{bmatrix},$$

$$W \triangleq \begin{bmatrix} \boldsymbol{I} \\ 0 \\ -C \end{bmatrix}, G \triangleq \begin{bmatrix} 0 & 0 & \boldsymbol{I} \end{bmatrix}. \tag{2.37}$$

Denote

$$x^h(t,k) \triangleq \begin{bmatrix} \delta x(t,k) \\ \delta \sum e_s(t-1,k) \end{bmatrix}, x^v(t,k) \triangleq e(t+1,k-1).$$

Obviously, it is equivalent to analyze the stability of the Roesser system in (2.36) and the stability of the batch process in (2.1) regulated by the PI-based ILC controller in (2.29) and (2.31).

Note that

$$\begin{aligned} \Pi &= \begin{bmatrix} A - B(\tau_P + \tau_I)C & B[\tau_I + (\tau_P + \tau_I)L_1] & B(\tau_P + \tau_I)L_2 \\ -C & \boldsymbol{I} + L_1 & L_2 \\ -C[A - B(\tau_P + \tau_I)C] & -CB[\tau_I + (\tau_P + \tau_I)L_1] & \boldsymbol{I} - CB(\tau_P + \tau_I)L_2 \end{bmatrix} \\ &+ \begin{bmatrix} E \\ 0 \\ -CE \end{bmatrix} \Delta(t,k) \begin{bmatrix} F_A - F_B(\tau_P + \tau_I)C & F_B(\tau_I + (\tau_P + \tau_I)L_1) & F_B(\tau_P + \tau_I)L_2 \end{bmatrix} \\ &\triangleq \Pi_0 + \Phi\Delta(t,k)\Psi. \end{aligned} \tag{2.38}$$

Hence, $\varpi(t,k)$ indicates batch-wise variation of the process uncertainties. Meanwhile, $z(t,k)$ reflects the output tracking error $e(t+1,k-1)$ occurring in the last batch. The following theorem presents a sufficient condition to guarantee the asymptotic stability of the above 2D system described by (2.36).

Theorem 2.2 By taking matrices $Q_i > 0$, $i = 1, 2, 3$, Y_1, Y_2 and positive scalars $\varepsilon, \gamma > 0$ to satisfy the following LMI

$$\begin{bmatrix} -Q & 0 & \Xi^T & \Sigma^T & QG^T \\ * & -\gamma\mathbf{I} & W^T & 0 & 0 \\ * & * & -Q + \varepsilon\Phi\Phi^T & 0 & 0 \\ * & * & * & -\varepsilon\mathbf{I} & 0 \\ * & * & * & * & -\gamma\mathbf{I} \end{bmatrix} < 0, \tag{2.39}$$

where

$$
\begin{aligned}
&Q \triangleq \mathrm{diag}\{Q_1, Q_2, Q_3\}, \\
&\Xi \triangleq \begin{bmatrix} [A - B(\tau_P + \tau_I)C]Q_1 & B\tau_I Q_2 + B(\tau_P + \tau_I)Y_1 & B(\tau_P + \tau_I)Y_2 \\ -CQ_1 & Q_2 + Y_1 & Y_2 \\ -C[A - B(\tau_P + \tau_I)C]Q_1 & -CB\tau_I Q_2 - CB(\tau_P + \tau_I)Y_1 & Q_3 - CB(\tau_P + \tau_I)Y_2 \end{bmatrix}, \\
&\Sigma \triangleq \begin{bmatrix} F_A Q_1 - F_B(\tau_P + \tau_I)CQ_1 & F_B\tau_I Q_2 + F_B(\tau_P + \tau_I)Y_1 & F_B(\tau_P + \tau_I)Y_2 \end{bmatrix},
\end{aligned}
\tag{2.40}
$$

and the learning gains are taken as

$$
L_1 = Y_1 Q_2^{-1}, \; L_2 = Y_2 Q_3^{-1}, \tag{2.41}
$$

the 2D closed-loop system in (2.36) could maintain asymptotical stability with a specified robust $\mathcal{H}_\infty$ performance index γ.

Proof Define $P_1 = Q_1^{-1}$, $P_2 = Q_2^{-1}$, $P_3 = Q_3^{-1}$, and $P = \mathrm{diag}\{P_1, P_2, P_3\}$. The proof of this theorem is divided into two parts. The first part is to clarify that the satisfaction of the LMI in (2.39) implies

$$
\begin{bmatrix} \Pi^T \\ W^T \end{bmatrix} P \begin{bmatrix} \Pi & W \end{bmatrix} - \begin{bmatrix} P - \gamma^{-1}G^T G & 0 \\ * & \gamma \mathbf{I} \end{bmatrix} < 0. \tag{2.42}
$$

Then the second part is to prove the final conclusion by using (2.42).

Using the Schur complement and (2.42), one obtains

$$
\begin{bmatrix} -P + \gamma^{-1}G^T G & 0 & \Pi^T \\ * & -\gamma \boldsymbol{I} & W^T \\ * & * & -P^{-1} \end{bmatrix} < 0. \tag{2.43}
$$

It follows from (2.38) that

$$
\begin{bmatrix} -P + \gamma^{-1}G^T G & 0 & \Pi_0^T \\ * & -\gamma \boldsymbol{I} & W^T \\ * & * & -P^{-1} \end{bmatrix} + \begin{bmatrix} 0 \\ 0 \\ \Phi \end{bmatrix} \Delta(t,k) \begin{bmatrix} \Psi & 0 & 0 \end{bmatrix} + \begin{bmatrix} \Psi^T \\ 0 \\ 0 \end{bmatrix} \Delta^T(t,k) \begin{bmatrix} 0 & 0 & \Phi^T \end{bmatrix} < 0. \tag{2.44}
$$

By using Lemma 2.1, a sufficient condition for holding (2.44) can be established by

$$
\begin{bmatrix} -P + \gamma^{-1}G^T G & 0 & \Pi_0^T \\ * & -\gamma \boldsymbol{I} & W^T \\ * & * & -P^{-1} \end{bmatrix} + \varepsilon \begin{bmatrix} 0 \\ 0 \\ \Phi \end{bmatrix} \begin{bmatrix} 0 & 0 & \Phi^T \end{bmatrix} + \varepsilon^{-1} \begin{bmatrix} \Psi^T \\ 0 \\ 0 \end{bmatrix} \begin{bmatrix} \Psi & 0 & 0 \end{bmatrix} < 0, \tag{2.45}
$$

which, in fact, is equal to

$$\begin{bmatrix} -P+\gamma^{-1}G^{\mathrm{T}}G & 0 & \Pi_0^{\mathrm{T}} \\ * & -\gamma \boldsymbol{I} & W^{\mathrm{T}} \\ * & * & -P^{-1}+\varepsilon\Phi\Phi^{\mathrm{T}} \end{bmatrix} + \varepsilon^{-1}\begin{bmatrix} \Psi^{\mathrm{T}} \\ 0 \\ 0 \end{bmatrix}\begin{bmatrix} \Psi & 0 & 0 \end{bmatrix} < 0. \tag{2.46}$$

Applying the Schur complement to (2.46) yields

$$\begin{bmatrix} -P & 0 & \Pi_0^{\mathrm{T}} & \Psi^{\mathrm{T}} & G^{\mathrm{T}} \\ * & -\gamma \boldsymbol{I} & W^{\mathrm{T}} & 0 & 0 \\ * & * & -P^{-1}+\varepsilon\Phi\Phi^{\mathrm{T}} & 0 & 0 \\ * & * & * & -\varepsilon \boldsymbol{I} & 0 \\ * & * & * & * & -\gamma \boldsymbol{I} \end{bmatrix} < 0. \tag{2.47}$$

Multiplying diag$\{Q, \boldsymbol{I}, \boldsymbol{I}, \boldsymbol{I}, \boldsymbol{I}\}$ and its transpose on the left-hand side of the matrix inequality in (2.47) and using (2.41) with $Q = P^{-1}$gives the LMI in (2.39). Hence, it is concluded that (2.39) is a sufficient condition of (2.42). The first part is proved.

It follows from (2.42) that

$$\Pi^{\mathrm{T}}P\Pi - P < 0. \tag{2.48}$$

According to Theorem 2.1 in Wang et al. (2012), it is concluded that (2.48) is a sufficient condition ensuring the above 2D system stability.

According to (2.42), there exists an inequality satisfying for all $\left[\delta x^{\mathrm{T}}(t,k)\delta\sum e_s^{\mathrm{T}}(t-1,k)e^{\mathrm{T}}(t+1,k-1)\right]^{\mathrm{T}}$ and $\varpi(t,k)$,

$$\begin{bmatrix} \delta x(t,k) \\ \delta\sum e_s(t-1,k) \\ e(t+1,k-1) \\ \varpi(t,k) \end{bmatrix}^{\mathrm{T}} \left(\begin{bmatrix} \Pi^{\mathrm{T}} \\ W^{\mathrm{T}} \end{bmatrix} P \begin{bmatrix} \Pi & W \end{bmatrix} - \begin{bmatrix} P-\gamma^{-1}G^{\mathrm{T}}G & 0 \\ * & \gamma \boldsymbol{I} \end{bmatrix} \right) \begin{bmatrix} \delta x(t,k) \\ \delta\sum e_s(t-1,k) \\ e(t+1,k-1) \\ \varpi(t,k) \end{bmatrix} \le 0. \tag{2.49}$$

Using (2.36), one obtains

$$\begin{bmatrix} \delta x(t+1,k) \\ \delta\sum e_s(t,k) \\ e(t+1,k) \end{bmatrix}^{\mathrm{T}} P \begin{bmatrix} \delta x(t+1,k) \\ \delta\sum e_s(t,k) \\ e(t+1,k) \end{bmatrix} - \begin{bmatrix} \delta x(t,k) \\ \delta\sum e_s(t-1,k) \\ e(t+1,k-1) \end{bmatrix}^{\mathrm{T}} P \begin{bmatrix} \delta x(t,k) \\ \delta\sum e_s(t-1,k) \\ e(t,k-1) \end{bmatrix}$$
$$+\gamma^{-1}\|z(t,k)\|^2 - \gamma\|\varpi(t,k)\|^2 \le 0. \tag{2.50}$$

It can be easily derived that

$$
\begin{aligned}
&\sum_{t=0}^{N_1}\sum_{k=0}^{N_2}\left(\begin{bmatrix}\delta x(t+1,k)\\ \delta\sum e_s(t,k)\\ e(t+1,k)\end{bmatrix}^{\mathrm T}P\begin{bmatrix}\delta x(t+1,k)\\ \delta\sum e_s(t,k)\\ e(t+1,k)\end{bmatrix}-\begin{bmatrix}\delta x(t,k)\\ \delta\sum e_s(t-1,k)\\ e(t+1,k-1)\end{bmatrix}^{\mathrm T}P\begin{bmatrix}\delta x(t,k)\\ \delta\sum e_s(t-1,k)\\ e(t,k-1)\end{bmatrix}\right)\\
&=\sum_{t=0}^{N_1}\sum_{k=0}^{N_2}\Big\{\delta x^{\mathrm T}(t+1,k)P_1\delta x(t+1,k)-\delta x^{\mathrm T}(t,k)P_1\delta x(t,k)\\
&+\delta\sum e_s^{\mathrm T}(t,k)P_2\delta\sum e_s(t,k)-\delta\sum e_s^{\mathrm T}(t-1,k)P_2\delta\sum e_s(t-1,k)\\
&+e^{\mathrm T}(t+1,k)P_3e(t+1,k)-e^{\mathrm T}(t+1,k-1)P_3e(t+1,k-1)\Big\}\\
&=\sum_{k=0}^{N_2}\delta x^{\mathrm T}(N_1+1,k)P_1\delta x(N_1+1,k)+\sum_{k=0}^{N_2}\delta\sum e_s^{\mathrm T}(N_1,k)P_2\delta\sum e_s(N_1,k)\\
&+\sum_{t=1}^{N_1+1}e^{\mathrm T}(t,N_2)P_3e(t,N_2)\ge 0.
\end{aligned}
\tag{2.51}
$$

Note that the zero boundary conditions have been used in the proof of the second inequality in (2.51), namely, $\delta x(t,k)$, $\delta\sum e_s(t,k)$, $e(t,k)\equiv 0$, for $t\le 0$ or $k\le 0$.

Combining (2.50) with (2.51), it is concluded that $\sum_{t=0}^{N_1}\sum_{k=0}^{N_2}\|z(t,k)\|^2\le\gamma\sum_{t=0}^{N_1}\sum_{k=0}^{N_2}\|\varpi(t,k)\|^2$ and thus $\|z\|_2\le\gamma\|\varpi\|_2$.

According to the 2D control objective (1.20) in Chap. 1, it follows that the 2D system in (2.36) is asymptotically stable with a robust $\mathcal{H}_\infty$performance index γ. This completes the proof. □

Based on an extensive simulation study, it is found that the convergence rate of ILC is primarily tuned by L_2 in (2.31), namely, a larger L_2 leads to a faster convergence rate, and vice versa. Hence, the learning gain L_2 may be designed in advance, and meanwhile, the variable Y_2 in (2.40) is replaced by L_2Q_3. For the convenience of implementation, it is suggested to choose $L_2=\mathrm{diag}\{l_2^1,l_2^2,\ldots,l_2^m\}$, where $0<l_2^i<2$, $i=1,2,\ldots,m$.

By using Definition 1.1 in Chap. 1, one can draw the conclusion in Theorem 2.2 as t and k go to infinity. Since the time horizon of a batch process is finite, the following theorem is additionally presented for stability analysis.

Theorem 2.3 Under the sufficient conditions in Theorem 2.2 along with $\varpi\equiv 0$, the 2D system in (2.36) is exponentially stable with regard to Q_3^{-1}, satisfying the T-norm constraint along the batch direction, namely, there exists a positive scalar $0<\rho<1$ such that

$$\|e(t,k+1)\|_{Q_3^{-1},N}<\rho\|e(t,k)\|_{Q_3^{-1},N},\ \forall k>0. \tag{2.52}$$

Proof Define $P_1=Q_1^{-1}$, $P_2=Q_2^{-1}$, $P_3=Q_3^{-1}$, and $P=\mathrm{diag}\{P_1,P_2,P_3\}$. it follows from (2.42) that

$$\Pi^{\mathrm T}P\Pi-P<0.$$

Moreover, one obtains $\delta x(0, k) \equiv 0$, $\delta \sum e_s(-1, k) \equiv 0$. By Theorem 2.1 in Wang et al. (2012), one ensures that there is a scalar $0 < \rho < 1$ such that

$$\sum_{i=1}^{I_0} V_{P_3}(e(t, k+1)) < \rho^2 \sum_{i=1}^{I_0} V_{P_3}(e(t, k)),\ \forall k > 0,\ \forall I_0 > 0,\ \forall e(t, 0).$$

Using $P_3 = Q_3^{-1}$ and letting $I_0 = N - 1$, there follows

$$\|e(\cdot, k+1)\|^2_{Q_3^{-1},N} < \rho^2 \|e(\cdot, k)\|^2_{Q_3^{-1},N},$$

which means

$$\|e(\cdot, k+1)\|_{Q_3^{-1},N} < \rho \|e(\cdot, k)\|_{Q_3^{-1},N}.$$

This completes the proof. □

2.3.3 Simulation Tests on an Injection Molding Machine

Example 2.1 Consider an injection molding machine as shown in Fig. 2.2, which has been studied in the literature (Gao et al. 2001; Shi et al. 2006; Wang et al. 2012; Liu et al. 2014), typically composed of three main operation stages: filling, packing/holding, and cooling. A primarily controlled variable in the packing stage is the nozzle pressure, which should track a prespecified operation trajectory, in order to guarantee polymer product quality and cycle (batch) consistency. During the batch run, the shift from the raw material filling to packing is related to the state transfer from the injection velocity to nozzle pressure, which, however, is subject to the variation of injection molding operation, e.g., uneven raw material supply and nonlinear behavior of the hydraulic control valve. The resistance from the injection mold cavity pressure is usually regarded as a repetitive-type disturbance. Moreover, time-varying uncertainties are involved with the batch run, arising from unsteady initial values of the nozzle pressure, screw speed, and the molding load. These issues could cause the conventional ILC methods, e.g., P-type learning design, to be invalid for implementation. For illustration, the cyclic state transfer process from the injection velocity to the mold nozzle pressure for polymer molding is studied herein

$$P_\Delta : \begin{cases} x(t+1, k) = \left(\begin{bmatrix} 1.607 & 1 \\ -0.6086 & 0 \end{bmatrix} + \Delta A(t, k) \right) x(t, k) \\ \qquad + \left(\begin{bmatrix} 1.239 \\ -0.9282 \end{bmatrix} + \Delta B(t, k) \right) u(t, k) + \begin{bmatrix} w(t, k) \\ 0 \end{bmatrix}, \\ y(t, k) = \begin{bmatrix} 1 & 0 \end{bmatrix} x(t, k), \end{cases}$$

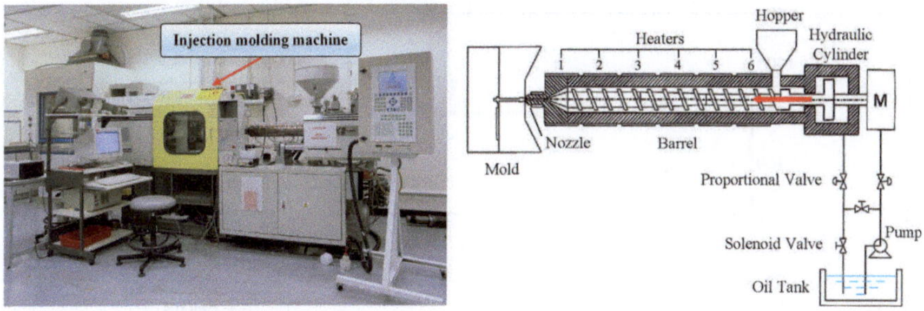

(a) (b)

Fig. 2.2 Illustration of an injection molding machine for batch run: **a** external view; **b** schematic diagram

where the process uncertainties are described by

$$\Delta A(t,k) = \begin{bmatrix} 0.0804\delta(t,k) & 0 \\ -0.0304\delta(t,k) & 0 \end{bmatrix} = \begin{bmatrix} 1 & 0 \\ 0 & 1 \end{bmatrix} \begin{bmatrix} \delta(t,k) & 0 \\ 0 & \delta(t,k) \end{bmatrix} \begin{bmatrix} 0.0804 & 0 \\ -0.0304 & 0 \end{bmatrix},$$
$$\Delta B(t,k) = \begin{bmatrix} 0.062\delta(t,k) \\ -0.0464\delta(t,k) \end{bmatrix} = \begin{bmatrix} 1 & 0 \\ 0 & 1 \end{bmatrix} \begin{bmatrix} \delta(t,k) & 0 \\ 0 & \delta(t,k) \end{bmatrix} \begin{bmatrix} 0.062 \\ -0.0464 \end{bmatrix},$$

and the scalar $|\delta(t,k)| \le 1$ denotes unknown perturbations.

The reference trajectory (y_d) takes a piece-wise linear form as studied in the previous references (Shi et al. 2005a; Wang et al. 2012; Liu et al. 2014)

$$y_d(t) = \begin{cases} 200, & 0 \le t \le 100; \\ 200 + 5(t-100), & 100 < t \le 120; \\ 300, & 120 < t \le N = 200. \end{cases}$$

Solving the linear program in (2.16) by setting $\alpha = 0.5$ and $r = 0.45$ gives the optimal $\mathcal{H}_\infty$ performance index $\gamma_{PI}^* = 2.08$. To prevent the control action from excessive ambition, one can resolve the LMI in (2.11) by taking $\gamma_{PI} = 15$, thus determining the PI tuning parameters $\tau_P = 0.6104$ and $\tau_I = 0.0037$.

By letting $L_2 = 1$, solving LMI in (2.39) and then using (2.41), the learning gains are computed as

$$L_1 = -0.0055, \; L_2 = 1. \tag{2.53}$$

In the absence of uncertainty and external disturbance, i.e., $\delta(t,k) = 0$ and $w(t,k) = 0$, the process output along with the control input in batches 1, 2, and 50 is shown in Fig. 2.3. It is seen that the output tracking performance is gradually enhanced from batch to batch. In particular, the process output in batch 50 nearly coincides with the desired trajectory.

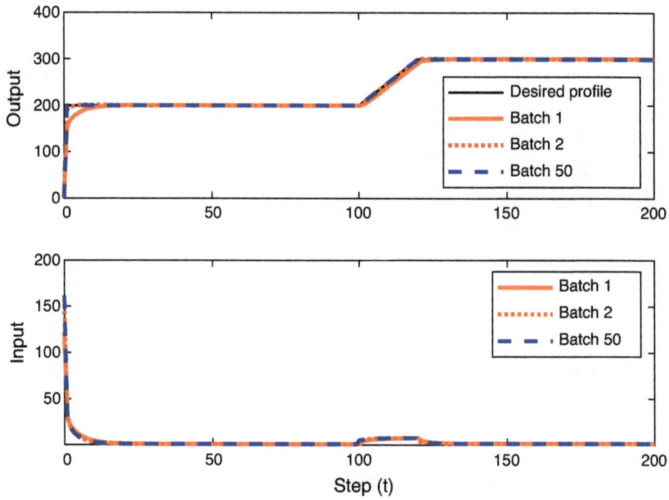

Fig. 2.3 Output and input curves of Example 2.1 under the PI-based ILC scheme in Sect. 2.3

For comparison, the following PI-type ILC algorithm given in Shi et al. (2005a) is performed

$$
\begin{aligned}
u(t,k) &= u(t,k-1) + K_L \begin{bmatrix} \delta x(t,k) \\ x_e(t+1,k) \\ e(t+1,k-1) \end{bmatrix}, \\
x_e(t+1,k) &= x_e(t,k) + e(t,k), x_e(0,k) = 0.
\end{aligned}
\tag{2.54}
$$

Based on the guidelines for controller tuning in Shi et al. (2005a), one obtains

$$
K_L = \begin{bmatrix} -1.3363 & -0.8565 & 0.1781 & 0.7266 \end{bmatrix}. \tag{2.55}
$$

In the above direct ILC scheme, u, x, x_e and e are stored in the control memory with a total dimension of $p + n + 2m$. By comparison, in the above control scheme in (2.29) and (2.31), y_s, $\sum e_s$ and e are stored in the control memory with only $3m$ dimensions. Besides, the direction type ILC law denoted by (2.54) and (2.55) depends on the exact system state information, which may be unavailable in many engineering applications.

To further assess the output tracking performance, the following system performance metric is adopted

$$
\text{ATE}(k) \triangleq \frac{1}{N} \sum_{t=1}^{N} |e(t,k)|, \tag{2.56}
$$

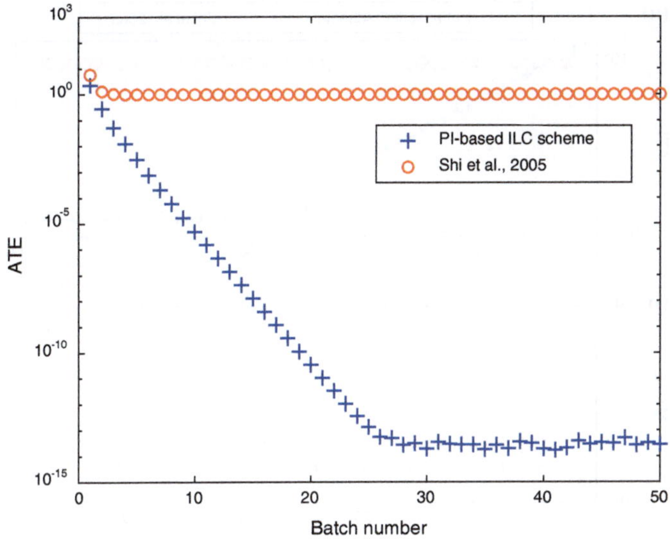

Fig. 2.4 ATE indices of Example 2.1 under two different methods

which is named as the averaged tracking error (ATE) for the k-th batch run. It is obvious that a smaller ATE(k) value corresponds to a better tracking performance.

The ATE values under the above PI-based ILC scheme and the ILC scheme in (2.54) and (2.55) are shown in Fig. 2.4. It is seen that the ATE values by both ILC schemes decrease along the batch direction, while the proposed scheme delivers better tracking performance.

In the following comparisons, both ILC schemes are considered for two cases: (a) there are repetitive external disturbances $w(t,k) = \sin(t)$ and the model uncertainties being fixed as their upper bounds; (b) there exist nonrepetitive uncertainties and nonrepetitive external disturbances, i.e., $\delta(t,k)$ varies randomly within $[-0.1, 0.1]$ along both the time and batch directions, and $w(t,k) = \sin(t + \theta(k))$ with $\theta(k)$ being randomly distributed within $[0, 2\pi]$. It is evident that the above system has stochastic properties in the second case. Hence, the Monte Carlo (MC) tests are adopted to compare the control performance of the above ILC methods. For each case, one hundred MC tests are performed. The ATE value for batch k is denoted by $\text{AET}^i(k)$ in the i-th test, the mean ATE is defined by

$$\text{MAET} = \frac{1}{100}\sum_{i=1}^{100} \text{AET}^i(k).$$

The corresponding ATE and mean ATE (MATE) values for both ILC schemes in two cases are depicted in Fig. 2.5. In all simulations, the proposed ILC scheme leads to better output tracking performance, demonstrating that the above PI-based

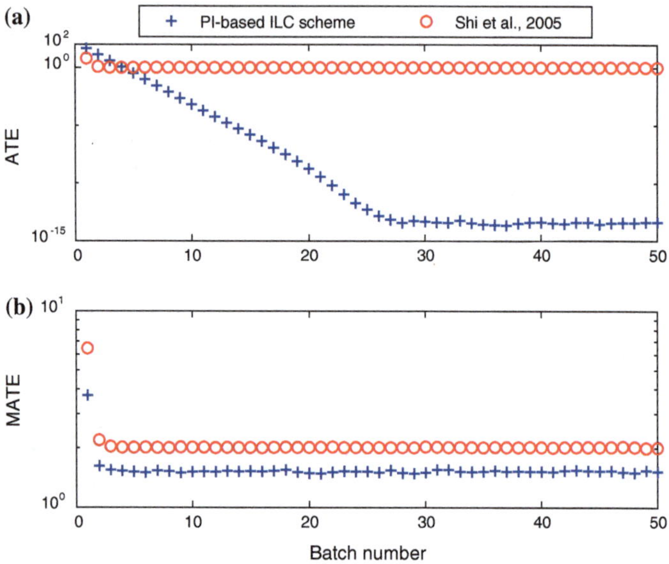

Fig. 2.5 ATE performance indices of Example 2.1 by two ILC methods in two cases: **a** with repetitive process uncertainties and disturbances; **b** with nonrepetitive process uncertainties and disturbances

ILC scheme acquires better robustness performance under nonrepetitive process uncertainties and external disturbance.

2.4 Robust ILC Design by 2D Fornasini-Marchesini System Description

2.4.1 PI-Based ILC Design

An alternative indirect-type ILC scheme depicted in Fig. 2.6, based on the widely-used PI controller in (2.29) is presented for practical application, where the set-point command is updated by the following ILC law

$$y_s(t,k) = y_s(t,k-1) + L_1\delta\sum e_s(t-1,k) + L_2e(t,k) + L_3e(t+1,k-1). \tag{2.57}$$

Unlike the indirect-type ILC scheme presented in Sect. 2.3, where only the output tracking error in the last batch run, $e(t+1,k-1)$, was adopted to update the system set-point, the current batch tracking error $e(t,k)$ is also considered to design the above ILC law.

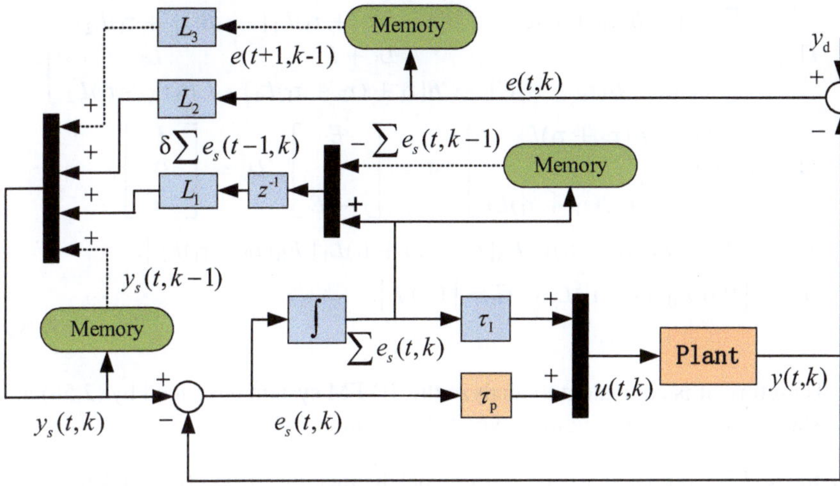

Fig. 2.6 Block diagram of another PI-based indirect-type ILC scheme in the 2D FM system framework

The set-point learning controllers are designed based on the robust stability analysis presented in the next section.

2.4.2 Robust Stability Analysis

Based on the ILC scheme shown in Fig. 2.6, a 2D FM system description is constructed from (2.32)–(2.34) as

$$\begin{cases} \begin{bmatrix} \delta x(t+1,k) \\ \delta \sum e_s(t,k) \\ e(t+1,k) \end{bmatrix} = \Pi_1 \begin{bmatrix} \delta x(t,k) \\ \delta \sum e_s(t-1,k) \\ e(t,k) \end{bmatrix} + \Pi_2 \begin{bmatrix} \delta x(t+1,k-1) \\ \delta \sum e_s(t,k-1) \\ e(t+1,k-1) \end{bmatrix} + D_{\mathrm{g}}\varpi(t,k), \\ z(t,k) = e(t,k) = G \begin{bmatrix} \delta x(t,k) \\ \delta \sum e_s(t-1,k) \\ e(t,k) \end{bmatrix}, \end{cases} \tag{2.58}$$

where $\Pi_1 = \Pi_{10} + \Phi\Delta(t,k)\Psi_{\mathrm{A}}$, $\Pi_2 = \Pi_{20} + \Phi\Delta(t,k)\Psi_{\mathrm{B}}$ and

$$\Pi_{10} = \begin{bmatrix} A - B(\tau_{\rm P} + \tau_{\rm I})C & B[\tau_{\rm I} + (\tau_{\rm P} + \tau_{\rm I})L_1] & B(\tau_{\rm P} + \tau_{\rm I})L_2 \\ -C & L_1 + \boldsymbol{I} & L_2 \\ -C[A - B(\tau_{\rm P} + \tau_{\rm I})C] & -CB[\tau_{\rm I} + (\tau_{\rm P} + \tau_{\rm I})L_1] & -CB(\tau_{\rm P} + \tau_{\rm I})L_2 \end{bmatrix},$$

$$\Pi_{20} = \begin{bmatrix} 0\ 0 & B(\tau_{\rm P} + \tau_{\rm I})L_3 \\ 0\ 0 & L_3 \\ 0\ 0 & \boldsymbol{I} - CB(\tau_{\rm P} + \tau_{\rm I})L_3 \end{bmatrix}, \ \Phi = \begin{bmatrix} E \\ 0 \\ -CE \end{bmatrix}, \ D_{\rm g} = \begin{bmatrix} \boldsymbol{I} \\ 0 \\ -C \end{bmatrix},$$

$$\Psi_{\rm A} = \left[F_{\rm A} - F_{\rm B}(\tau_{\rm P} + \tau_{\rm I})C \ F_{\rm B}[\tau_{\rm I} + (\tau_{\rm P} + \tau_{\rm I})L_1] \ F_{\rm B}(\tau_{\rm P} + \tau_{\rm I})L_2 \right],$$

$$\Psi_{\rm B} = \left[0\ 0\ F_{\rm B}(\tau_{\rm P} + \tau_{\rm I})L_3 \right], \ G = \left[0\ 0\ \boldsymbol{I} \right]. \tag{2.59}$$

Therefore, it is equivalent to analyze the 2D FM system described by (2.58) and the stability of the above indirect-type ILC scheme.

Remark 2.3 Compared to the current batch tracking error $e(t, k)$ taken as the controlled output for minimization in the above 2D FM system in (2.58), the previous batch tracking error $e(t + 1, k - 1)$ was adopted as the controlled output to obey the 2D Roesser model structure in the previous references (Wang et al. 2012, 2013; Liu et al. 2014), which, however, could not ensure the current batch tracking error to be the minimal.

For batch run optimization, it aims at minimizing the output tracking error along both the time and batch directions. Hence, a 2D $\mathcal{H}_\infty$control objective defined below is taken to design the ILC law

$$J_{\rm BP} = \sum_{t=0}^{N_1=N} \sum_{k=0}^{N_2 \to \infty} \left(\gamma_{\rm ILC}^{-1} \| e(t, k) \|_2^2 - \gamma_{\rm ILC} \| \varpi(t, k) \|_2^2 \right), \tag{2.60}$$

where $\gamma_{\rm ILC} > 0$ denotes a practically specified 2D $\mathcal{H}_\infty$performance index.

By establishing sufficient conditions for holding the above 2D FM system stability, the learning gains L_1, L_2, and L_3in the above ILC scheme could be derived. For this purpose, LMI-based conditions are presented in the following theorem to evaluate the system stability and meanwhile, determine these learning gains. ◊

Theorem 2.3 Given positive scalars $\sigma_t, \sigma_k > 1$, the 2D FM system described in (2.58) maintains asymptotically stable and satisfies a robust $\mathcal{H}_\infty$ performance level $\gamma_{\rm ILC}$, if there are matrices $Q_t > 0$, $Q_k > 0$, $Q_1 > 0$, $Q_2 > 0$, $Q_3 > 0$, $\hat{L}_1, \hat{L}_2, \hat{L}_3$ with appropriate dimensions, and a positive scalar ε, such that the following LMIs stand true

$$\begin{bmatrix} -Q+\varepsilon\Phi\Phi^{\mathrm{T}} & \Omega_1 & \Omega_2 & D_{\mathrm{g}} & 0 & 0 \\ * & -Q_t & 0 & 0 & \Omega_3 & \Omega_4 \\ * & * & -Q_k & 0 & 0 & \Omega_5 \\ * & * & * & -\gamma_{\mathrm{ILC}}\boldsymbol{I} & 0 & 0 \\ * & * & * & * & -\gamma_{\mathrm{ILC}}\boldsymbol{I} & 0 \\ * & * & * & * & * & -\varepsilon\boldsymbol{I} \end{bmatrix} < 0, \tag{2.61}$$

$$\sigma_t Q_t + \sigma_k Q_k < Q, \tag{2.62}$$

where $Q = \mathrm{diag}\{Q_1, Q_2, Q_3\}$ and

$$\Omega_1 = \begin{bmatrix} [A - B(\tau_{\mathrm{P}} + \tau_{\mathrm{I}})C]Q_1 & B\big[\tau_{\mathrm{I}}Q_2 + (\tau_{\mathrm{P}} + \tau_{\mathrm{I}})\hat{L}_1\big] & B(\tau_{\mathrm{P}} + \tau_{\mathrm{I}})\hat{L}_2 \\ -CQ_1 & \hat{L}_1 + Q_2 & \hat{L}_2 \\ -C[A - B(\tau_{\mathrm{P}} + \tau_{\mathrm{I}})C]Q_1 & -CB\big[\tau_{\mathrm{I}}Q_2 + (\tau_{\mathrm{P}} + \tau_{\mathrm{I}})\hat{L}_1\big] & -CB(\tau_{\mathrm{P}} + \tau_{\mathrm{I}})\hat{L}_2 \end{bmatrix},$$

$$\Omega_2 = \begin{bmatrix} 0\ 0 & B(\tau_{\mathrm{P}} + \tau_{\mathrm{I}})\hat{L}_3 \\ 0\ 0 & \hat{L}_3 \\ 0\ 0 & Q_3 - CB(\tau_{\mathrm{P}} + \tau_{\mathrm{I}})\hat{L}_3 \end{bmatrix}, \quad \Omega_3 = \begin{bmatrix} 0 \\ 0 \\ Q_3 \end{bmatrix},$$

$$\Omega_4 = \begin{bmatrix} Q_1[F_{\mathrm{A}} - F_{\mathrm{B}}(\tau_{\mathrm{P}} + \tau_{\mathrm{I}})C]^{\mathrm{T}} \\ \big[\tau_{\mathrm{I}}Q_2 + (\tau_{\mathrm{P}} + \tau_{\mathrm{I}})\hat{L}_1\big]^{\mathrm{T}} F_{\mathrm{B}}^{\mathrm{T}} \\ \hat{L}_2^{\mathrm{T}}(\tau_{\mathrm{P}} + \tau_{\mathrm{I}})^{\mathrm{T}} F_{\mathrm{B}}^{\mathrm{T}} \end{bmatrix}, \quad \Omega_5 = \begin{bmatrix} 0 \\ 0 \\ \hat{L}_3^{\mathrm{T}}(\tau_{\mathrm{P}} + \tau_{\mathrm{I}})^{\mathrm{T}} F_{\mathrm{B}}^{\mathrm{T}} \end{bmatrix}. \tag{2.63}$$

When the LMI conditions in (2.61) and (2.62) are feasible, the learning gains can be determined by

$$\begin{cases} L_1 = \hat{L}_1 Q_2^{-1}, \\ L_2 = \hat{L}_2 Q_3^{-1}, \\ L_3 = \hat{L}_3 Q_3^{-1}, \end{cases} \tag{2.64}$$

which ensures that the TCI and KCI of the above 2D system are no less than σ_t and σ_k, respectively, and the 2D convergence index is no less than $\rho = \min\{\sigma_t, \sigma_k\}$.

Proof For the convenience of analysis, take a 2D Lyapunov-Krasovskii function candidate (Kaczorek 1985) for the system stability analysis as below

$$V_{[\cdot]}(x_{\mathrm{g}}(t,k)) \triangleq x_{\mathrm{g}}^{\mathrm{T}}(t,k)[\cdot]x_{\mathrm{g}}(t,k), \tag{2.65}$$

where $x_{\mathrm{g}}(t,k) = \big[\delta x^{\mathrm{T}}(t,k)\ \delta \sum e_s^{\mathrm{T}}(t-1,k)\ e^{\mathrm{T}}(t,k)\big]^{\mathrm{T}}$ and $[\cdot]$ represents any positive definite matrix with appropriate dimensions.

Concerning the 2D FM system in (2.58) with $\varpi(t,k) = 0$, an increment of the above Lyapunov-Krasovskii function along both the time and batch directions is computed as

$$\begin{aligned}\Delta V(t,k) &= V_p(x_g(t+1,k)) - V_{p_t}(x_g(t,k)) - V_{p_k}(x_g(t+1,k-1)) \\ &= \begin{bmatrix} x_g(t,k) \\ x_g(t+1,k-1) \end{bmatrix}^{\mathrm{T}} \Omega \begin{bmatrix} x_g(t,k) \\ x_g(t+1,k-1) \end{bmatrix},\end{aligned} \tag{2.66}$$

where

$$\Omega = \begin{bmatrix} -P_t & 0 \\ * & -P_k \end{bmatrix} + \begin{bmatrix} \Pi_1^{\mathrm{T}} \\ \Pi_2^{\mathrm{T}} \end{bmatrix} P \begin{bmatrix} \Pi_1 & \Pi_2 \end{bmatrix}. \tag{2.67}$$

By letting $P_i = Q_i^{-1}$, $i = 1, 2, 3$, $P = \mathrm{diag}\{P_1, P_2, P_3\}$, $P_t = PQ_tP$, $P_k = PQ_kP$, and pre- and post-multiplying the left-hand side of (2.61) with $Q = \mathrm{diag}\{\boldsymbol{I}, P, P, \boldsymbol{I}, \boldsymbol{I}, \boldsymbol{I}\}$, it yields

$$\begin{bmatrix} -P^{-1} + \varepsilon\Phi\Phi^{\mathrm{T}} & \Pi_{10} & \Pi_{20} & D_g & 0 & 0 \\ * & -P_t & 0 & 0 & G^{\mathrm{T}} & \Psi_{\mathrm{A}}^{\mathrm{T}} \\ * & * & -P_k & 0 & 0 & \Psi_{\mathrm{B}}^{\mathrm{T}} \\ * & * & * & -\gamma_{\mathrm{ILC}}\boldsymbol{I} & 0 & 0 \\ * & * & * & * & -\gamma_{\mathrm{ILC}}\boldsymbol{I} & 0 \\ * & * & * & * & * & -\varepsilon\boldsymbol{I} \end{bmatrix} < 0. \tag{2.68}$$

By using Lemma 2.1, it follows that the inequality below is satisfied by the inequality in (2.68)

$$\Xi < 0, \tag{2.69}$$

where

$$\Xi \triangleq \begin{bmatrix} -P_t + \gamma_{\mathrm{ILC}}^{-1} G^{\mathrm{T}} G & 0 & 0 \\ * & -P_k & 0 \\ * & * & \gamma_{\mathrm{ILC}}\boldsymbol{I} \end{bmatrix} + \begin{bmatrix} \Pi_1^{\mathrm{T}} \\ \Pi_2^{\mathrm{T}} \\ D_g^{\mathrm{T}} \end{bmatrix} P \begin{bmatrix} \Pi_1 & \Pi_2 & D_g \end{bmatrix}.$$

Thus, $\Omega < 0$ is guaranteed by the matrix inequality in (2.69). Consequently, one obtains

$$V_P(x_g(t+1,k)) < V_{P_t}(x_g(t,k)) + V_{P_k}(x_g(t+1,k-1)). \tag{2.70}$$

It is assumed, without loss of generality, that $\rho = \sigma_t = \min\{\sigma_t, \sigma_k\}$. Multiplying P to both sides of the matrix inequality in (2.62) gives $P_t < \rho^{-1}P - P_k$. It follows that

$$V_P(x_g(t+1,k)) < \rho^{-1} V_P(x_g(t,k)) - V_{P_k}(x_g(t,k)) + V_{P_k}(x_g(t+1,k-1)). \tag{2.71}$$

Hence, the following inequalities stand true for any integers $T_0, K_0, i > 0$

$$\begin{aligned}
V_P(x_g(T_0+1, K_0+i)) &< \rho^{-1} V_P(x_g(T_0, K_0+i)) \\
&\quad - V_{P_k}(x_g(T_0, K_0+i)) + V_{P_k}(x_g(T_0+1, K_0+i-1)), \\
V_P(x_g(T_0+2, K_0+i-1)) &< \rho^{-1} V_P(x_g(T_0+1, K_0+i-1)) \\
&\quad - V_{P_k}(x_g(T_0+1, K_0+i-1)) + V_{P_k}(x_g(T_0+2, K_0+i-2)), \\
&\vdots \\
V_P(x_g(T_0+i, K_0+1)) &< \rho^{-1} V_P(x_g(T_0+i-1, K_0+1)) \\
&\quad - V_{P_k}(x_g(T_0+i-1, K_0+1)) + V_{P_k}(x_g(T_0+i, K_0)),
\end{aligned} \tag{2.72}$$

Summing up all inequalities in (2.72) gives

$$\begin{aligned}
&\sum_{\substack{t+k=T_0+K_0+i+1 \\ T_0 \le t < T_0+i \\ K_0 \le k \le K_0+i}} V_P(x_g(t,k)) \\
&< \rho^{-1} \sum_{\substack{t+k=T_0+K_0+i \\ T_0 \le t < T_0+i \\ K_0 \le k \le K_0+i}} V_P(x_g(t,k)) - \rho^{-1} V_P(x_g(T_0+i, K_0)), \\
&- V_{P_k}(x_g(T_0, K_0+i)) + V_{P_k}(x_g(T_0+i, K_0)), \\
&< \rho^{-1} \sum_{\substack{t+k=T_0+K_0+i \\ T_0 \le t < T_0+i \\ K_0 \le k \le K_0+i}} V_P(x_g(t,k)) - V_{P_k}(x_g(T_0, K_0+i)) - V_{P_t}(x_g(T_0+i, K_0)), \\
&< \rho^{-1} \sum_{\substack{t+k=T_0+K_0+i \\ T_0 \le t < T_0+i \\ K_0 \le k \le K_0+i}} V_P(x_g(t,k)).
\end{aligned} \tag{2.73}$$

By using Lemma 1.6, it is concluded that the LMI conditions in (2.61) and (2.62) guarantee the asymptotic stability of the above 2D FM system in (2.58).

By virtue of the inequality in (2.70) and $\sigma_t P_t + \sigma_k P_k < P$, one obtains

$$\begin{aligned}
&\sigma_t V_{P_t}(x_g(t+1, N)) + \sigma_k V_{P_k}(x_g(t+1, N)) \\
&\qquad < V_{P_t}(x_g(t, N)) + V_{P_k}(x_g(t+1, N-1)), \\
&\sigma_t V_{P_t}(x_g(t+1, N-1)) + \sigma_k V_{P_k}(x_g(t+1, N-1)) \\
&\qquad < V_{P_t}(x_g(t, N-1)) + V_{P_k}(x_g(t+1, N-2)), \\
&\qquad \vdots \\
&\sigma_t V_{P_t}(x_g(t+1, 0)) + \sigma_k V_{P_k}(x_g(t+1, 0)) \\
&\qquad < V_{P_t}(x_g(t, 0)) + V_{P_k}(x_g(t+1, -1)).
\end{aligned} \tag{2.74}$$

Summing up all inequalities in (2.74) under zero initial K-boundary condition gives

$$\begin{aligned}&\sigma_t \sum_{k=0}^{N} V_{P_t}(x_g(t+1,k)) + \sigma_k \sum_{k=0}^{N} V_{P_k}(x_g(t+1,k))\\&< \sum_{k=0}^{N} V_{P_t}(x_g(t,k)) + \sum_{k=0}^{N-1} V_{P_k}(x_g(t+1,k)).\end{aligned} \tag{2.75}$$

Thus, there stands

$$\begin{aligned}&\sigma_t \sum_{k=0}^{N} V_{P_t}(x_g(t+1,k)) < \sum_{k=0}^{N} V_{P_t}(x_g(t,k)) - \sigma_k V_{P_k}(x_g(t+1,N-1))\\&+(1-\sigma_k)\sum_{k=0}^{N-1} V_{P_k}(x_g(t+1,k)) < \sum_{k=0}^{N} V_{P_t}(x_g(t,k)).\end{aligned} \tag{2.76}$$

By using Lemma 1.6, it is obvious that the TCI of the 2D FM system guarantees no less than σ_t. Similarly, one can easily verify that the KCI of the above 2D system guarantees no less than σ_k.

Next, it should be clarified that the above 2D $\mathcal{H}_\infty$performance index is implementable when $\varpi(t,k) \neq 0$. It is therefore suggested to take the following control objective function

$$J(t,k) = \Delta V(t,k) + \gamma_{\text{ILC}}^{-1} z^{\text{T}}(t,k) z(t,k) - \gamma_{\text{ILC}} \varpi^{\text{T}}(t,k)\varpi(t,k). \tag{2.77}$$

By using (2.66) and (2.69), it follows for any $t, k = 1, 2, \ldots$

$$J(t,k) = \begin{bmatrix} x_g(t,k) \\ x_g(t+1,k-1) \\ \varpi(t,k) \end{bmatrix}^{\text{T}} \Xi \begin{bmatrix} x_g(t,k) \\ x_g(t+1,k-1) \\ \varpi(t,k) \end{bmatrix} < 0. \tag{2.78}$$

By assuming zero boundary conditions of the above 2D FM system in (2.58), one has

$$\begin{aligned}&\sum_{t=0}^{N_1}\sum_{k=0}^{N_2} \Delta V(t,k)\\&= \sum_{t=0}^{N_1}\sum_{k=0}^{N_2} \left[V_P(x_g(t+1,k)) - V_{P_t}(x_g(t,k)) - V_{P_k}(x_g(t+1,k-1))\right],\\&= \sum_{t=0}^{N_1}\sum_{k=0}^{N_2-1} V_{P-P_t-P_k}(x_g(t,k)) + \sum_{k=0}^{N_2-1} V_{P-P_k}(x_g(N_1+1,k)) + V_P(x_g(N_1+1,N_2)) \geq 0.\end{aligned} \tag{2.79}$$

Hence,

$$
\begin{aligned}
J_{\mathrm{BP}} &= \sum_{t=0}^{N_1=N} \sum_{k=0}^{N_2\to\infty} \left[\Delta V(t,k) + \gamma_{\mathrm{ILC}}^{-1}\|z(t,k)\|_2^2 - \gamma_{\mathrm{ILC}}\|\varpi(t,k)\|_2^2\right] \\
&\leq \sum_{t=0}^{N_1=N} \sum_{k=0}^{N_2\to\infty} J(t,k) < 0.
\end{aligned}
\tag{2.80}
$$

which means that the above 2D $\mathcal{H}_\infty$performance index is satisfied. The proof is finished. □

It should be mentioned that σ_t, σ_k, and γ_{ILC} in the above Theorem 2.3 are user-specified tuning parameters for implementation. Generally, it is recommended to take σ_t and σ_k as big as possible to achieve fast convergence rates along either the time- or batch-direction, and meanwhile, γ_{ILC} should be taken as a small value to keep low sensitivity to the time-varying process uncertainties and external disturbance.

In view of that a larger TCI (i.e., σ_t) could result in a faster convergence rate along the time direction but at a risk of more oscillatory transient response and higher sensitivity to the time-varying process uncertainties, it is suggested to take the TCI as $\sigma_t = 1$, so as to purely regard the KCI (σ_k) together with the robust $\mathcal{H}_\infty$performance index (γ_{ILC}) as the adjustable parameters in the performance optimization problem in order to satisfy different control objectives.

(i) (*2D $\mathcal{H}_\infty$ performance*) Solve the performance optimization problem under a specified $\sigma_k \geq 1$:

$$
\begin{aligned}
&\text{Minimize } \gamma_{\mathrm{ILC}} \\
&\text{s.t. (2.61) and (2.62),}
\end{aligned}
\tag{2.81}
$$

which implies that the optimal disturbance attenuation performance could be obtained under a specified batch-wise convergence rate.

(ii) (*Batch-wise convergence rate*) Solve the performance optimization problem under a specified $\gamma_{\mathrm{ILC}} > 0$:

$$
\begin{aligned}
&\text{Minimize } \eta \\
&\text{s.t. (2.61) and } \eta Q_t + Q_k < \eta Q
\end{aligned}
\tag{2.82}
$$

with $\eta = \sigma_k^{-1} > 0$, which means that the maximal batch-wise convergence rate is obtained according to a prespecified performance index γ_{ILC}.

It should be pointed out that the above performance optimization problem in Case (i) could be directly solved by using the LMI toolbox. In contrast, the performance optimization problem in Case (ii) is similar to the generalized eigenvalue problems (GEVPs) (Boyd et al. 1994), which can also be simply solved by the above toolbox.

2.4.3 An Illustrative Example

Example 2.2 Consider the above batch process of industrial injection molding and the target profile (y_d) as studied in Example 2.1. The inner-loop PI controller parameters are also taken as $\tau_P = 0.6104$ and $\tau_I = 0.0037$ by solving the above LMI condition in (2.11) under the same parameter settings.

By solving the optimization program in (2.81) under a prescribed $\sigma_k = 1$, the minimal robust $\mathcal{H}_\infty$performance index is computed as $\gamma^*_{ILC} = 17.96$, along with the corresponding ILC controller gains as $L_1 = -0.0060$, $L_2 = -2.8726 \times 10^{-21}$ and $L_3 = 0.6944$.

For comparison with the existing methods, the following three cases are tested.

Case 1. (Nominal case) Assume that $\Delta A(t, k)$ and $\Delta B(t, k)$ together with the disturbance $w(t, k)$ are all zeros. The output tracking results are plotted in Fig. 2.7, along with the corresponding ATE criterion depicted in Fig. 2.8. It is observed that the batch process output almost coincides with the desired trajectory after 5 cycles by using the proposed ILC scheme, in contrast to the existing indirect-type ILC methods (Wang et al. 2012; Liu et al. 2014), where almost 20 or 50 cycles are needed to achieve entire tracking.

Besides, to assess the impact of σ_k on the convergence speed of batch-wise tracking error, the established LMIs in (2.61) and (2.62) are resolved based on different σ_k with regard to a specified $\gamma_{ILC} = 110$ studied in Liu et al. (2014), the resulting ILC controller gains are shown in Table 2.1. Figure 2.9 shows the ATE performance index. It is observed that the convergence speed of batch-wise tracking error becomes faster as the KCI increases. The largest σ_k is calculated from the performance optimization

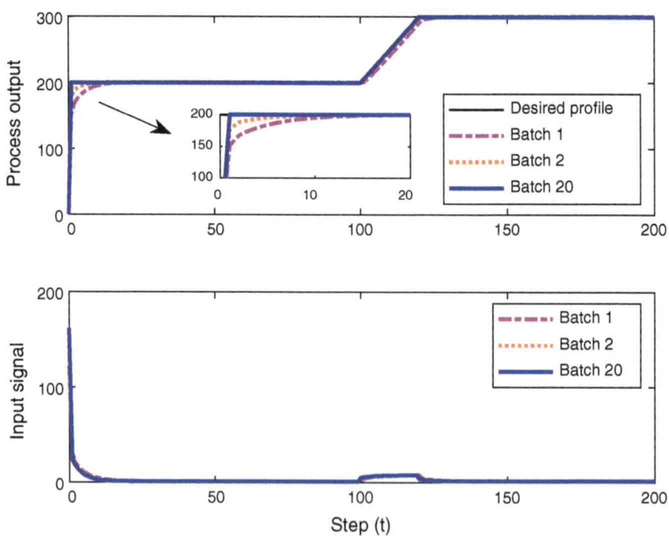

Fig. 2.7 Tracking performance for Case 1 in Example 2.2

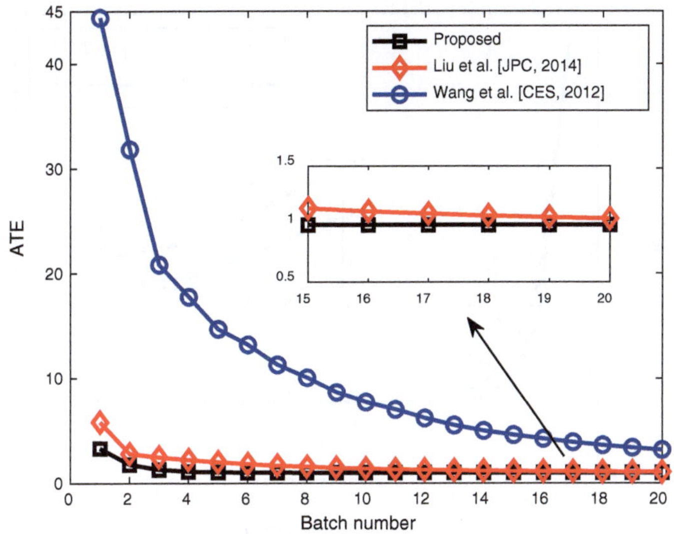

Fig. 2.8 ATE for Case 1 in Example 2.2

Table 2.1 Learning controller gains with respect to different values of K-RCI

Gains	$\sigma_k = 1$	$\sigma_k = 1.5$	$\sigma_k = 2.0$	$\sigma_k = 2.4$
L_1	−0.0064	−0.0057	−0.0053	−0.0059
L_2	2.8919×10^{-14}	4.1216×10^{-14}	-5.2355×10^{-10}	-4.9396×10^{-18}
L_3	0.3366	0.6934	0.7698	0.8178

problem in (2.82) as $\sigma_k^* = 2.4$. It is noted that the learning gain L_3 also tends to be larger by increasing the KCI, consistent with the fact that a larger L_3 could expedite a faster convergence speed, as concluded in Wang et al. (2013). However, L_3 was chosen manually in Wang et al. (2013) rather than by using a convergence index as proposed above.

Case 2. (Time-invariant process uncertainties and repetitive disturbance) Assume that $\Delta A(t, k)$ and $\Delta B(t, k)$ become their upper bounds. A repetitive-type external disturbance is mimicked by passing a unity step signal through a low-pass filter $G_d(z) = 150(z^{-1} + z^{-2})/(11 - 4z^{-1})$, which is inserted into the process output at $t =$ 60 (s). Figures 2.10 and 2.11 show the output tracking results and the corresponding ATE indices, respectively. It is observed that perfect tracking could be achieved by nearly 5 batches. The convergence speed under the proposed scheme is significantly improved, though the initial ATE value is larger than that in Liu et al. (2014). Note that the inflection point of ATE by the existing indirect-type ILC method (Liu et al. 2014) is caused by the PI retuning by using the designed feedforward controller.

Case 3. (Time-varying uncertainties and stochastic disturbance) Assume that the process uncertainties are time-varying with regard to $|\delta(t)| \leq 0.1$, and the external

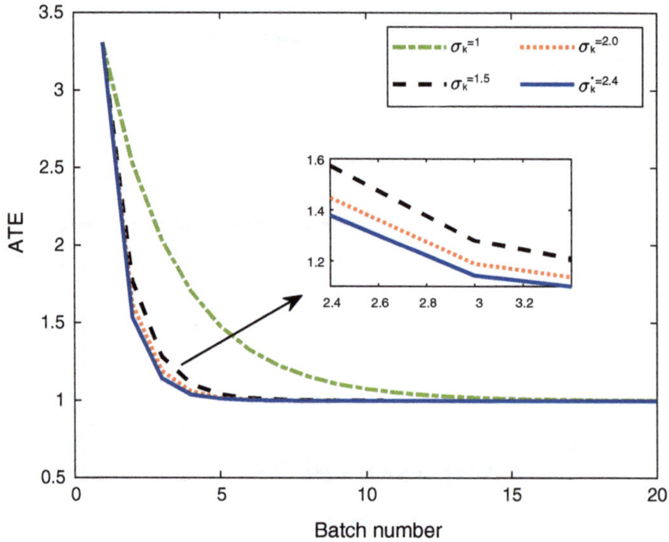

Fig. 2.9 Comparison of ATE with respect to different K-RCIs for Example 2.2

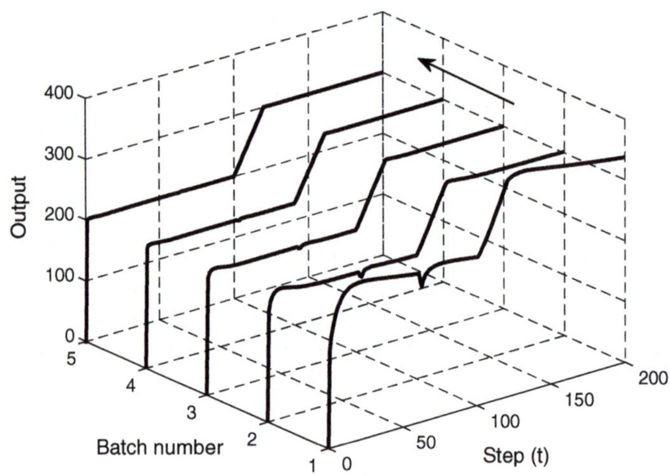

Fig. 2.10 Tracking performance for Case 2 of Example 2.2

disturbance denoted by $\omega(t, k) = sin(t+\theta(k))$is nonrepetitive, where $\theta(k)$ randomly varies in a range of $[0, 2\pi]$ as assumed in Wang et al. (2012) and Liu et al. (2014). In this case, $\gamma_{\text{ILC}} = 110$ is taken as studied in Liu et al. (2014) along with $\sigma_k = 1.5$, the ILC controller gains are computed as $L_1 = -0.0057$, $L_2 = 4.1216 \times 10^{-14}$, $L_3 = 0.6934$by solving the LMI conditions in (2.61) and (2.62). To further assess the output tracking performance and robustness of the proposed ILC scheme under

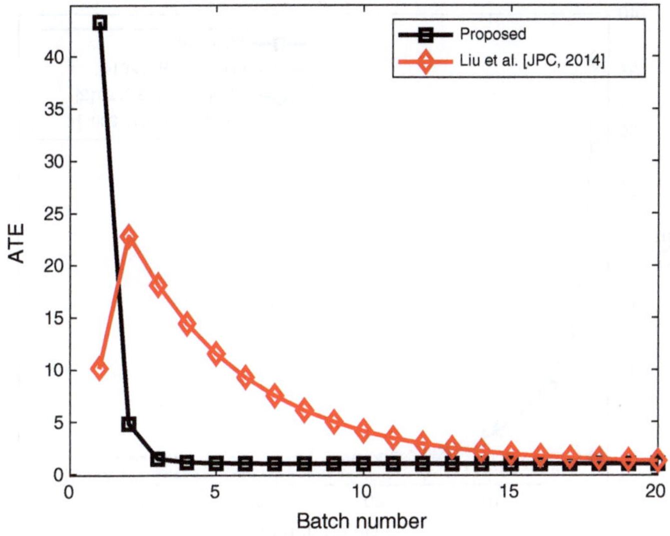

Fig. 2.11 ATE for Case 2 in Example 2.2

the stochastic disturbance, one hundred MC batch runs were conducted. The averaged ATEs are shown in Fig. 2.12, compared to those of Shi et al. (2005a), Wang et al. (2012), Liu et al. (2014). It is observed that the resulting control system well maintains robust stability along both the time and batch directions, by using the proposed indirect-type ILC method. The result further demonstrates that the proposed ILC scheme can be effectively adopted for robust output tracking under time-varying uncertainties and disturbance.

2.5 Summary

With the 2D Roesser and FM model frameworks, two alternative PI-based ILC schemes (Wang et al. 2012; Hao et al. 2019) have been presented in this chapter for industrial batch processes with time-varying uncertainties. A salient advantage of the proposed schemes lies with that the inner-loop PI controller design is independent of the outer-loop ILC law design. Therefore, any existing PI controller designed for controlling the batch process could be inherited for the indirect-type ILC design. To deal with time-varying batch run uncertainties and external disturbance, a robust PI tuning method has been offered by assigning the desired closed-loop system pole. In contrast, another model-free PI design is presented for application to real systems with unknown dynamics. For optimizing batch run performance, alternative indirect-type ILC laws have been developed to regulate the set-point command of the inner-loop control system. By virtue of a 2D Roesser or FM model description of the batch process, LMI-based sufficient conditions for holding robust stability of

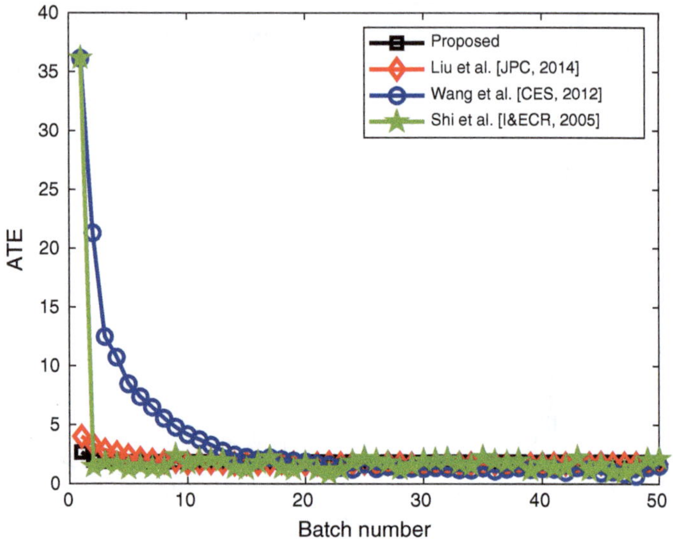

Fig. 2.12 ATE for Case 3 of Example 2.2

the resulting closed-loop learning system have been derived, respectively. Illustrative applications to the injection molding machine have well validated the effectiveness and superiority of the proposed ILC methods.

References

Abidi K, Xu J-X (2011) Iterative learning control for sampled-data systems: from theory to practice. IEEE Trans Ind Electron 58:3002–3015

Arimoto S, Kawamura S, Miyazaki F (1984) Bettering operation of dynamic systems by learning: a new control theory for servomechanism or mechatronic system. In: Proceedings of 23rd conference on decision and control, pp 1064–1069

Bone G (1995) A novel iterative learning control formulation of generalized predictive control. Automatica 31:1483–1487

Bonvin D, Srinivasan B, Hunkeler D (2006) Control and optimization of batch processes. IEEE Trans Control Syst Mag 26(1):34–45

Boyd S, Ghaoui LE, Feron E et al (1994) Linear Matrix inequalities in system and control theory. SIAM, Philadelphia, PA

Bristow DA, Tharayil M, Alleyne AG (2006) A survey of iterative learning control. IEEE Control Syst Mag 26:96–114

Busch J, Cruse A, Marquardt W (2007) Run-to-run control of membrane filtration processes. AIChE J 53:2316–2328

Chen Y, Gong Z, Wen C (1998) Analysis of a high-order iterative learning control algorithm for uncertain nonlinear systems with state delays. Automatica 34:345–353

Chi R, Li H, Shen D, Hou Z, Huang B (2023) Enhanced P-type control: indirect adaptive learning from set-point updates. IEEE Trans Autom Control 68(3):1600–1613

Ganesan R, Das T, Ramachandran K (2007) A multiresolution analysis-assisted reinforcement learning approach to run-by-run control. IEEE Trans Autom Sci Eng 4:182–193

Gao F, Yang Y, Shao C (2001) Robust iterative learning control with applications to injection molding process. Chem Eng Sci 56(24):7025–7034

Gao K, Zhou Y, Gao F, Lu J (2025) Optimally selected cycle based ILC for system with randomly varying initial state. IEEE Trans Autom Control 70(4):2714–2721

Ge M, Chiu M-S, Wang Q-G (2002) Robust PID controller design via LMI approach. J Process Control 12(1):3–13

Gu K, Chen J, Kharitonov VL (2003) Stability of time-delay systems. Birkhauser

Hankinson M, Vincent T, Irani K, Khargonekar P (1997) Integrated real-time and run-to-run control of etch depth in reactive ion etching. IEEE Trans Semicond Manuf 10:121–130

Hao S, Liu T, Gao F (2019) PI based indirect-type iterative learning control for batch processes with time-varying uncertainties: a 2D FM model based approach. J Process Control 78:57–67

Hao S, Liu T, Rogers E, Wang Y, Paszke W (2025) PI-based set-point learning control for batch processes with unknown dynamics and nonrepetitive uncertainties. IEEE Trans Autom Control 70(6):3649–3664

Hou Z, Jin S (2013) Model free adaptive control: theory and applications. CPC Press, New York, NY, USA

Hou Z, Xiong S (2019) On model-free adaptive control and its stability analysis. IEEE Trans Autom Control 64(11):4555–4569

Hou Z, Xu J-X, Zhong H (2007) Freeway traffic control using iterative learning control-based ramp metering and speed signaling. IEEE Trans Veh Technol 56:466–477

Jiang P, Bamforth L, Feng Z, Baruch J, Chen Y (2007) Indirect iterative learning control for a discrete visual servo without a camera-robot model. IEEE Trans Syst Man Cybern 37:863–876

Kaczorek T (1985) Two-dimensional linear systems. Springer, Berlin

Kim DI, Kim S (1996) An iterative learning control method with application for CNC machine tools. IEEE Trans Ind Appl 32(1):66–72

Li CH, Tso PL (2008) Experimental study on a hybrid-driven servo press using iterative learning control. Int J Mach Tools Manuf 48:209–219

Li H, Bai J, Wu F, Zou H (2023) A novel two-dimensional PID controller design using two-dimensional model predictive iterative learning control optimization for batch processes. Can J Chem Eng 101(12):6959–6969

Liang H, Yang C, Lv M, Sun B, Li Y (2023) An enhanced PI controller based on adaptive iterative learning control. Int J Robust Nonlinear Control 33(18):11200–11217

Liu T, Gao F, Wang YQ (2010) IMC-based iterative learning control for batch processes with uncertain time delay. J Process Control 20(2):173–180

Liu T, Wang X, Chen J (2014) Robust PID based indirect-type iterative learning control for batch processes with time-varying uncertainties. J Process Control 24(12):95–106

Mi C, Lin H, Zhang Y (2005) Iterative learning control of antilock braking of electric and hybrid vehicles. IEEE Trans Veh Technol 54:486–494

Ruan X, Bien Z, Park KH (2008) Decentralized iterative learning control to large-scale industrial processes for nonrepetitive trajectory tracking. IEEE Trans Syst Mans Cybern Part A: Syst Hum 38(1):238–252

Shi J, Gao F, Wu T-J (2005a) Integrated design and structure analysis of robust iterative learning control system based on a two-dimensional model. Ind Eng Chem Res 44:8095–8105

Shi J, Gao F, Wu T-J (2005b) Robust design of integrated feedback and iterative learning control of a batch process based on a 2D Roesser system. J Process Control 15(8):907–924

Shi J, Gao F, Wu T-J (2006) Robust iterative learning control design for batch processes with uncertain perturbations and initialization. AIChE J 52(6):2171–2187

Tan K, Zhao S, Xu J-X (2007) Online automatic tuning of a proportional integral derivative controller based on an iterative learning control approach. IET Control Theory Appl 1:90–96

Tao H, Zheng J, Wei J, Paszke W, Rogers E, Stojanovic V (2023) Repetitive process based indirect-type iterative learning control for batch processes with model uncertainty and input delay. J Process Control 132(103112):1–12

Tayebi A, Chien CJ (2007) A unified adaptive iterative learning control framework for uncertain nonlinear systems. IEEE Trans Autom Control 52:1907–1913

Uchiyama M (1978) Formulation of high-speed motion pattern of a mechanical arm by trial (in Janpanese). Trans SICE (Soc Instrum Control Eng) 14:706–712

Wang Y, Gao F, Doyle FJ III (2009) Survey on iterative learning control, repetitive control, and run-to-run control. J Process Control 19(10):1589–1600

Wang Y, Liu T, Zhao Z (2012) Advanced PI control with simple learning set-point design: application on batch processes and robust stability analysis. Chem Eng Sci 71:153–165

Wang Y, Yang Y, Zhao Z (2013) Robust stability analysis for an enhanced ILC-based PI controller. J Process Control 23(2):201–214

Wu J, Ding H (2007) Reference adjustment for a high-acceleration and high-precision platform via A-type of iterative learning control. Proc IMechE Part I: J Syst Control Eng 221:781–789

Xiong S, Hou Z (2022) Stabilizing regions of PID controller for a class of unknown nonlinear non-affine discrete-time systems. Int J Robust Nonlinear Control 32(18):9421–9437

Xiong Z, Zhang J (2003) Product quality trajectory tracking in batch processes using iterative learning control based on time-varying perturbation models. Ind Eng Chem Res 42:6802–6814

Xu J-X, Yan R (2004) Iterative learning control design without a priori knowledge of the control direction. Automatica 40:1803–1809

Xu S, Yang C, Zhou S (2000) Robust control for uncertain discrete-time systems with circular pole constraints. Syst Control Lett 39(1):13–18

Zou W, Shen Y, Paszke W, Tao H (2022) Robust feedback feed-forward PD-type iterative learning control for uncertain discrete systems over finite frequency ranges. Trans Inst Meas Control 44(14):2850–2862

Chapter 3
Proportional-Integral-Derivative (PID) Based Iterative Learning Control

3.1 Introduction

The developed PID-based ILC methods in the literature have been generally classified into two types (Bonvin et al. 2006; Ahn et al. 2007; Wang et al. 2009). One is called direct-type, which means the ILC design integrates the feedback control (primarily responsible for closed-loop stability) and the feedforward control (responsible for the setpoint tracking) through the identical PID controller or a combination of two or more P/PI/PID controllers. Another is called indirect-type, which implies that either the feedback or feedforward control could be implemented through different controllers (e.g., P, PI, PD, PID) that may be designed relatively independent.

For the design of direct-type ILC, the PID controller is mostly used to execute the integrated control for both the setpoint tracking and closed-loop stabilization, owing to its implemental simplicity (Kim and Kim 1996; Ruan et al. 2008). The achievable robustness and output tracking performance, however, have not yet been fully explored, in particular for the quantitative performance specifications (Tayebi 2007). Based on a two-dimensional (2D) state-space description of a batch process and using the linear quadratic optimal control criterion in combination with the robust control theory, a PI or PID controller were used in combination with full-order controller matrices (with respect to the process model order) to construct direct-type ILC schemes for injection molding processes (Shi et al. 2006a, b, c), but at the expense of control structure complexity along with considerable computation effort for implementation.

For the design of indirect-type ILC, the control structure is typically composed of two loops, the inner loop constructed with a PID or other type controller, and the outer loop used for adjusting the setpoint or the process input, similar to a feedforward control manner, by using a P or PD controller. Based on the internal model control (IMC) structure, a learning set-point design was proposed (Liu et al. 2010) to robustly track the set-point profile against the process input delay uncertainty. By comparison, a P-type learning algorithm was presented to adjust the set-point

T. Liu et al., *Robust Iterative Learning Control of Industrial Batch Systems*, Intelligent Control and Learning Systems 22, https://doi.org/10.1007/978-981-96-9778-6_3

in combination with the model predictive control (MPC) method for tracking the desired profile, which was successfully used for the control of artificial pancreatic beta-cell (Wang et al. 2010). Based on a PID control loop, a parallel learning-type PID was introduced to improve the set-point tracking performance without sacrificing the closed-loop stability (Tan et al. 2007). An alternative anticipatory-type ILC (A-ILC) was developed to adjust the set-point in terms of the PID control loop for robust tracking of the desired profile (Wu and Ding 2007). The robust stability condition of a learning-type set-point design in terms of a PI control loop was analyzed in Wang et al. (2012). A quadratic criterion was presented to analyze the ILC convergence in terms of an MPC structure for time-varying linear systems (Lee et al. 2000). The achievable tracking performance of an indirect-type ILC scheme was assessed by estimating the minimum output variance bound (Chen and Kong 2009). In Li et al. (2023), a two-dimensional PID controller design was developed to establish a 2D model predictive ILC scheme for batch processes subject to time-varying uncertainties. By taking a PID feedback control in the inner loop, an indirect-type ILC was presented in Chi et al. (2024) for a repetitive nonlinear system without modelling.

In this Chapter, an indirect-type ILC design based on the standard PID control loop (Liu et al. 2014) is presented for application to industrial batch processes subject to time-varying uncertainties. By taking a state-space model description of such a batch process together with norm-bounded uncertainties, a robust PID tuning algorithm is firstly offered in terms of the $\mathcal{H}_\infty$ control objective, which is primarily responsible for maintaining the closed-loop system robust stability and no steady output deviation. Then, an ILC scheme consisting of the learning controllers to adjust the set-point command and the feedforward controllers to adjust the process input is proposed to realize robust tracking against time-varying uncertainties and load disturbance. Accordingly, the PID tuning and the ILC design can be conducted relatively independent of each other, and therefore, more flexibility is allowed to prescribe the control system robust stability and tracking performance, respectively. By analyzing the sufficient conditions in terms of LMI constraints for maintaining robust stability of the PID control loop and the robust convergence of the ILC scheme, respectively, the PID and ILC controllers are derived along with an adjustable robust $\mathcal{H}_\infty$ performance level. The effectiveness of the proposed method is demonstrated through an illustrative example of injection model machine.

3.2 Robust PID Controller Tuning

A batch process with time-varying uncertainties can be generally described by the following observable canonical discrete-time state-space model

$$P_\Delta : \begin{cases} x(t+1,k) = [A_\mathrm{m} + \Delta\tilde{A}(t,k)]x(t,k) + [B_\mathrm{m} + \Delta\tilde{B}(t,k)]u(t,k) + w(t,k) \\ y(t,k) = Cx(t,k), 0 \le t \le N; \\ x(0,k) = x(0),\ k = 0, 1, \ldots, \end{cases} \tag{3.1}$$

where t and k denotes the time and batch indices, respectively, and k indicates the current batch (or cycle); $x(t,k) \in \Re^{n_x}$ denote the state variables, $u(t,k) \in \Re^{n_u}$ the control inputs, $y(t,k) \in \Re^{n_y}$ the process outputs, $w(t,k) \in \Re^{n_x}$ the external disturbances; Denote by A_{m} and B_{m} the nominal state matrices, and by $\Delta\tilde{A}(t,k)$ and $\Delta\tilde{B}(t,k)$ time-varying uncertainties that are not repetitive from cycle to cycle and practically specified as $\Delta\tilde{A}(t,k) = \Delta\overline{A}_1\tilde{\Theta}_1(t)\Delta\overline{A}_2$, $\Delta\tilde{B}(t,k) = \Delta\overline{B}_1\tilde{\Theta}_2(t)\Delta\overline{B}_2$, where $\Delta\overline{A}_1$, $\Delta\overline{A}_2$, $\Delta\overline{B}_1$, and $\Delta\overline{B}_2$ are constant matrices, and $\tilde{\Theta}_i^{\mathrm{T}}(t)\tilde{\Theta}_i(t) \leq \mathbf{I}, i = 1, 2$. Denote by N the time period of each batch, and $x(0)$ is the initial resetting condition of each batch. Note that other process uncertainties, such as from the input actuator and output measurement may also be lumped into $\Delta\tilde{A}(t,k)$ and $\Delta\tilde{B}(t,k)$ for analysis.

To design an indirect-type ILC scheme, the output error in the current batch is defined by

$$e(t,k) \triangleq y_{\mathrm{d}}(t) - y(t,k), \tag{3.2}$$

where $y_{\mathrm{d}}(t)$ denotes the desired output profile, and $y(t,k)$ the real output in the current batch. Correspondingly, the time integral of $e(t,k)$ is denoted by $\sum e(t,k)$, i.e.

$$\sum e(t,k) = \sum_{i=0}^{t} e(i,k), 0 \leq t \leq N. \tag{3.3}$$

According to the process state-space description in (3.1), by omitting the batch index for brevity, a PID control law is generally expressed by

$$u_{\mathrm{PID}}(t) = \tau_{\mathrm{P}}e(t) + \tau_{\mathrm{I}}\sum e(t) + \tau_{\mathrm{D}}[e(t+1) - e(t)], \tag{3.4}$$

where τ_{P}, τ_{I}, and τ_{D} are the proportional, integral, and derivative parameters of PID, respectively. Note that because $e(t+1)$ cannot be measured at the current moment for implementing $u(t)$, the differential term of $e(t+1) - e(t)$ is practically substituted by $e(t) - e(t-1)$, $[e(t) + e(t-2) - 2e(t-1)]/2$, or adding a low-pass filter for execution (Åström and Hägglund 1995), but at the cost of control performance degradation that should be addressed in the stability analysis.

By introducing an auxiliary state variable, $\sum e(t)$, an augmented control system is written as

$$\begin{cases} \begin{bmatrix} x(t+1) \\ \sum e(t) \end{bmatrix} = \begin{bmatrix} \tilde{A} & \mathbf{0} \\ -C & \mathbf{I} \end{bmatrix} \begin{bmatrix} x(t) \\ \sum e(t-1) \end{bmatrix} + \begin{bmatrix} \tilde{B} \\ \mathbf{0} \end{bmatrix} u(t) + \begin{bmatrix} \mathbf{I} \\ \mathbf{0} \end{bmatrix} w(t), \\ y(t) = \begin{bmatrix} C & \mathbf{0} \end{bmatrix} \begin{bmatrix} x(t) \\ \sum e(t-1) \end{bmatrix}, \end{cases} \tag{3.5}$$

where $\tilde{A} = A_{\mathrm{m}} + \Delta\tilde{A}(t)$ and $\tilde{B} = B_{\mathrm{m}} + \Delta\tilde{B}(t)$.

Substituting (3.2) with $y_{\mathrm{d}}(t) = 0$ (which has no influence on the closed-loop stability) into (3.4) in terms of the nominal process model described by (3.1) (i.e. $\Delta\tilde{A}(t) = 0$ and $\Delta\tilde{B}(t) = 0$) yields

$$\begin{aligned} u_{\mathrm{PID}}(t) = {} & (\mathbf{I} + \tau_{\mathrm{D}} C B_{\mathrm{m}})^{-1}(\tau_{\mathrm{P}} + \tau_{\mathrm{I}})e(t) + (\mathbf{I} + \tau_{\mathrm{D}} C B_{\mathrm{m}})^{-1}\tau_{\mathrm{I}} \sum e(t-1) \\ & + (\mathbf{I} + \tau_{\mathrm{D}} C B_{\mathrm{m}})^{-1}\tau_{\mathrm{D}}(C\mathbf{I} - A_{\mathrm{m}})x(t). \end{aligned} \tag{3.6}$$

Let

$$\hat{\tau}_{\mathrm{P}} = (\mathbf{I} + \tau_{\mathrm{D}} C B_{\mathrm{m}})^{-1}(\tau_{\mathrm{P}} + \tau_{\mathrm{I}}), \tag{3.7}$$

$$\hat{\tau}_{\mathrm{I}} = (\mathbf{I} + \tau_{\mathrm{D}} C B_{\mathrm{m}})^{-1}\tau_{\mathrm{I}}, \tag{3.8}$$

$$\hat{\tau}_{\mathrm{D}} = (\mathbf{I} + \tau_{\mathrm{D}} C B_{\mathrm{m}})^{-1}\tau_{\mathrm{D}}. \tag{3.9}$$

Then substituting (3.6)–(3.9) into (3.5) yields the closed-loop system

$$\begin{cases} \begin{bmatrix} x(t+1) \\ \sum e(t) \end{bmatrix} = \begin{bmatrix} \tilde{A} + \tilde{B}[\hat{\tau}_{\mathrm{D}}(C\mathbf{I} - A_{\mathrm{m}}) - \hat{\tau}_{\mathrm{P}} C] & \tilde{B}\hat{\tau}_{\mathrm{I}} \\ -C & \mathbf{I} \end{bmatrix} \begin{bmatrix} x(t) \\ \sum e(t-1) \end{bmatrix} + \begin{bmatrix} \mathbf{I} \\ \mathbf{0} \end{bmatrix} w(t), \\ y(t) = \begin{bmatrix} C & \mathbf{0} \end{bmatrix} \begin{bmatrix} x(t) \\ \sum e(t-1) \end{bmatrix}. \end{cases} \tag{3.10}$$

For tuning the PID controller to maintain the control system's robust stability, the $\mathcal{H}_\infty$ control objective is adopted here

$$\|e(t)\|_2 < \gamma_{\mathrm{PID}} \|w(t)\|_2, \tag{3.11}$$

where γ_{PID} denotes the robust performance level.

To achieve the $\mathcal{H}_\infty$ control objective, the following theorem is presented.

Theorem 3.1 *The PID control system in* (3.10) *subject to time-varying process uncertainties described by* (3.1) *is guaranteed robustly stable with a* $\mathcal{H}_\infty$ *control performance level,* γ_{PID}, *if there exist* $P_{11} > 0$, $P_{22} > 0$, *matrices* P_{12}, R_1, R_2 *with appropriate dimensions, and positive scalars* ε_1, ε_2, *such that the following LMI holds*

$$\begin{bmatrix} -P + \varepsilon_1 \Phi_{\mathrm{A1}} \Phi_{\mathrm{A1}}^{\mathrm{T}} + \varepsilon_2 \Phi_{\mathrm{B1}} \Phi_{\mathrm{B1}}^{\mathrm{T}} & \Gamma & D_{\mathrm{g}} & \mathbf{0} & \mathbf{0} & \mathbf{0} \\ * & -P & \mathbf{0} & P H^{\mathrm{T}} C^{\mathrm{T}} & P\Phi_{\mathrm{A2}}^{\mathrm{T}} & P\Phi_{\mathrm{B2}}^{\mathrm{T}} \\ * & * & -\gamma_{\mathrm{PID}}\mathbf{I} & \mathbf{0} & \mathbf{0} & \mathbf{0} \\ * & * & * & -\gamma_{\mathrm{PID}}\mathbf{I} & \mathbf{0} & \mathbf{0} \\ * & * & * & * & -\varepsilon_1 \mathbf{I} & \mathbf{0} \\ * & * & * & * & * & -\varepsilon_2 \mathbf{I} \end{bmatrix} < 0, \tag{3.12}$$

where $D_g = [\mathbf{I}\ \mathbf{0}]^T$, $H = [\mathbf{I}\ \mathbf{0}]$, $\Phi_{A1} = [\Delta\overline{A}_1^T,\ \mathbf{0}]^T$, $\Phi_{A2} = [\Delta\overline{A}_2 P_{11},\ \Delta\overline{A}_2 P_{12}]$, $\Phi_{B1} = [\Delta\overline{B}_1^T,\ \mathbf{0}]^T$, $\Phi_{B2} = [\Delta\overline{B}_2 R_1,\ \Delta\overline{B}_2 R_2]$,

$$P = \begin{bmatrix} P_{11} & P_{12} \\ * & P_{22} \end{bmatrix}, \Gamma = \begin{bmatrix} A_m P_{11} + B_m R_1 & A_m P_{12} + B_m R_2 \\ -CP_{11} + P_{12}^T & -CP_{12} + P_{22} \end{bmatrix}.$$

by parameterizing

$$[\ \hat{\tau}_D(C\mathbf{I} - A_m) - \hat{\tau}_P C\ \ \hat{\tau}_I\ ] = [R_1\ R_2]P^{-1}. \tag{3.13}$$

Proof Define the following Lyapunov-Krasinski inequality of state energy to guarantee the asymptotic stability of the closed-loop system described by (3.10)

$$V_P[\hat{x}(t+1)] - V_P[\hat{x}(t)] < -(\gamma_{PID}^{-1}\|e(t)\|_2^2 - \gamma_{PID}\|w(t)\|_2^2), \tag{3.14}$$

where $\hat{x}(t) = [x^T(t),\ \sum e(t-1)^T]^T$, and γ_{PID} is the robust performance level as shown in (3.11).

Considering that $e(t) = -Cx(t)$ by letting $y_d(t) = 0$, and $x(t) = [\mathbf{I},\mathbf{0}]\hat{x}(t)$, one gets

$$e(t) = -CH\hat{x}(t), \tag{3.15}$$

where H has been shown in (3.12).

By substituting (3.10) into (3.14), one obtains

$$\zeta^T \Xi_1 \zeta < 0, \tag{3.16}$$

where $\zeta = [\hat{x}^T(t),\ w^T(t)]^T$, $D_g = [\mathbf{I}\ 0]^T$, and

$$\tilde{A}_g = \begin{bmatrix} \tilde{A} + \tilde{B}[\hat{\tau}_D(C\mathbf{I} - A_m) - \hat{\tau}_P C] & \tilde{B}\hat{\tau}_I \\ -C & \mathbf{I} \end{bmatrix}, \tag{3.17}$$

$$\Xi_1 = \begin{bmatrix} \tilde{A}_g^T \\ D_g^T \end{bmatrix} P[\tilde{A}_g\ D_g] - \begin{bmatrix} P - \gamma_{PID}^{-1}H^T C^T CH & \mathbf{0} \\ * & \gamma_{PID}\mathbf{I} \end{bmatrix}. \tag{3.18}$$

By the Schur complement, it can be derived that (3.16) is guaranteed by

$$\begin{bmatrix} -P & \tilde{\Gamma} & D_g & \mathbf{0} \\ * & -P & \mathbf{0} & PH^T C^T \\ * & * & -\gamma_{PID}\mathbf{I} & \mathbf{0} \\ * & * & * & -\gamma_{PID}\mathbf{I} \end{bmatrix} < 0, \tag{3.19}$$

where

$$\tilde{\Gamma} = \begin{bmatrix} \tilde{A}P_{11} + \tilde{B}[\hat{\tau}_{\mathrm{D}}(C\mathbf{I} - A_{\mathrm{m}}) - \hat{\tau}_{\mathrm{P}}C]P_{11} + \tilde{B}\hat{\tau}_{\mathrm{I}}P_{12}^{\mathrm{T}} & \tilde{A}P_{12} + \tilde{B}[\hat{\tau}_{\mathrm{D}}(C\mathbf{I} - A_{\mathrm{m}}) - \hat{\tau}_{\mathrm{P}}C]P_{12} + \tilde{B}\hat{\tau}_{\mathrm{I}}P_{22} \\ -CP_{11} + P_{12}^{\mathrm{T}} & -CP_{12} + P_{22} \end{bmatrix}. \tag{3.20}$$

Note that $\tilde{\Gamma}$ can be reformulated as

$$\tilde{\Gamma} = \Gamma + \Phi_{\mathrm{A1}}\tilde{\Theta}_1(t)\Phi_{\mathrm{A2}} + \Phi_{\mathrm{B1}}\tilde{\Theta}_2(t)\Phi_{\mathrm{B2}}, \tag{3.21}$$

where Γ, Φ_{A1}, Φ_{A2}, Φ_{B1}, and Φ_{B2} have been shown in (3.12), $R_1 = [\hat{\tau}_{\mathrm{D}}(C\mathbf{I} - A_{\mathrm{m}}) - \hat{\tau}_{\mathrm{P}}C]P_{11} + \hat{\tau}_{\mathrm{I}}P_{12}^{\mathrm{T}}$ and $R_2 = [\hat{\tau}_{\mathrm{D}}(C\mathbf{I} - A_{\mathrm{m}}) - \hat{\tau}_{\mathrm{P}}C]P_{12} + \hat{\tau}_{\mathrm{I}}P_{22}$.

Using Lemma 2.1 in Chap. 2 and the Schur complement, it can be seen that (3.19) is guaranteed by (3.5) in Theorem 3.1. This completes the proof. □

It can be seen from (3.7)–(3.9) and (3.13) that the PID parameters can be retrieved by prescribing the derivative parameter(s), τ_{D}, that is, when τ_{D} is specified, the other parameters can be obtained by using (3.7)–(3.9) as

$$\tau_{\mathrm{I}} = \hat{\tau}_{\mathrm{I}}(\mathbf{I} + \tau_{\mathrm{D}}CB_{\mathrm{m}}), \tag{3.22}$$

$$\tau_{\mathrm{P}} = \hat{\tau}_{\mathrm{P}}(\mathbf{I} + \tau_{\mathrm{D}}CB_{\mathrm{m}}) - \tau_{\mathrm{I}}. \tag{3.23}$$

Note that letting $\tau_{\mathrm{D}} = 0$ leads to a PI controller, which may be preferred for practical application owing to its implemental simplicity.

To achieve good robust control performance, the PID or PI (by letting $\tau_{\mathrm{D}} = 0$) controller can be determined by the following optimization program

$$\underset{\Delta\tilde{A}(t),\, \Delta\tilde{B}(t)}{\text{Minimize}}\ \gamma_{\mathrm{PID}}. \tag{3.24}$$

In fact, a smaller value of γ_{PID} leads to a more aggressive control action and vice versa. Therefore, a trade-off should be made between the achievable control performance and the control action generated by the designed PID or PI controller.

3.3 Set-Point Learning Design Based on a PID Loop

In contrast to the output tracking error $e(t, k)$ in (3.2), the set-point tracking error in the current batch is defined by

$$e_{\mathrm{s}}(t, k) \triangleq y_{\mathrm{s}}(t, k) - y(t, k), \tag{3.25}$$

where $y_s(t, k)$ denotes the set-point command in the current batch, which is different with the desired system output profile, $y_d(t)$, in that it is timely adjusted in an indirect-type ILC scheme for tracking $y_d(t)$.

The time integral of $e_s(t, k)$ is denoted by $\sum e_s(t, k)$. It follows that

$$\sum e_s(t, k) = \sum e_s(t-1, k) + e_s(t, k), 0 \le t \le N. \tag{3.26}$$

Moreover, a batchwise error function is defined by

$$\delta f(t, k) \triangleq f(t, k) - f(t, k-1), \tag{3.27}$$

where f may denote x, y_s, u, e, e_s, or ω, respectively.

It follows from (3.1) using the definitions in (3.2) and (3.27) that

$$e(t, k) = e(t, k-1) - C\delta x(t, k), \tag{3.28}$$

$$\delta x(t+1, k) = [A_m + \Delta\tilde{A}(t, k)]\delta x(t, k) + [B_m + \Delta\tilde{B}(t, k)]\delta u(t, k) + \varpi(t, k), \tag{3.29}$$

where

$$\begin{aligned} \varpi(t, k) = {} & [\Delta\tilde{A}(t, k) - \Delta\tilde{A}(t, k-1)]x(t, k-1) \\ & + [\Delta\tilde{B}(t, k) - \Delta\tilde{B}(t, k-1)]u(t, k-1) + \delta w(t, k). \end{aligned} \tag{3.30}$$

It is obvious that $\varpi(t, k) \neq 0$ for any non-repeatable parameter uncertainties and varying initial process conditions from batch to batch, and therefore can be viewed as a nonrepetitive load disturbance to deal with.

Based on the classical PID control loop, the proposed indirect-type ILC scheme is shown in Fig. 3.1, as outlined by the dash-line box, where the learning controllers, L_1, L_2, L_3, are set to adjust the set-point command, i.e.

$$y_s(t, k) = y_s(t, k-1) + L_1 e(t+1, k-1) + L_2\delta e_s(t-1, k) + L_3\delta \sum e_s(t-1, k), \tag{3.31}$$

where $y_s(t, k-1)$ denotes the set-point input in the previous batch, and $e(t+1, k-1)$ the one-step ahead output error in the previous batch.

It follows from (3.25)–(3.27) that

$$\delta e_s(t-1, k) = e_s(t-1, k) - e_s(t-1, k-1), \tag{3.32}$$

$$\delta \sum e_s(t, k) = \delta \sum e_s(t-1, k) + \delta e_s(t, k). \tag{3.33}$$

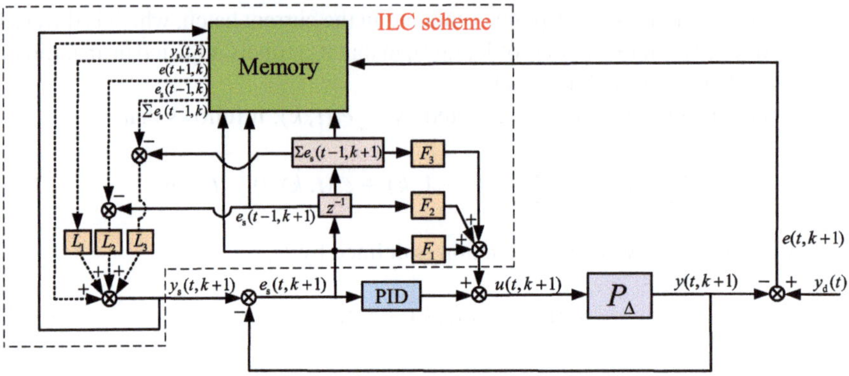

Fig. 3.1 Block diagram of the PID-based indirect-type ILC scheme

It is seen from (3.31)–(3.33) that the tracking errors of $e(t+1, k-1)$, $e_s(t-1, k-1)$, $e_s(t, k-1)$, and $\sum e_s(t-1, k-1)$ in the previous batch are used to construct the ILC updating law added to the set-point command in the current batch, relatively independent of the PID control loop shown in Fig. 3.1.

In Fig. 3.1, the feedforward controllers, F_1, F_2, F_3, are used to adjust the process input, i.e.

$$u(t, k) = u_{\text{PID}}(t, k) + F_1 e_s(t, k) + F_2 e_s(t-1, k) + F_3 \sum e_s(t-1, k), \quad (3.34)$$

where u_{PID} is the PID control output. The set-point tracking errors at the current moment and one-step ahead moment, and the error integral in the current batch are used to construct the feedforward control in the proposed scheme.

3.4 Robust Stability Analysis

According to the proposed ILC scheme shown in Fig. 3.1, it follows from (3.25), (3.27), and (3.28) that

$$y_s(t, k) - y_s(t, k-1) = \delta e_s(t, k) + C\delta x(t, k). \quad (3.35)$$

Substituting (3.35) into (3.31) yields

$$\delta e_s(t, k) = L_1 e(t+1, k-1) + L_2 \delta e_s(t-1, k) + L_3 \delta \sum e_s(t-1, k) - C\delta x(t, k). \quad (3.36)$$

Then by substituting (3.36) into (3.33), it yields

$$\delta \sum e_s(t,k) = L_1 e(t+1,k-1) + L_2 \delta e_s(t-1,k) + (\boldsymbol{I} + L_3)\delta \sum e_s(t-1,k) - C\delta x(t,k). \tag{3.37}$$

Substituting the PID control law of (3.4) with $e(t+1) - e(t)$ replaced by $e(t) - e(t-1)$ into (3.32), one obtains

$$\begin{aligned} u(t,k) &= \tau_P e_s(t,k) + \tau_I \sum e_s(t,k) + \tau_D[e_s(t,k) - e_s(t-1,k)] \\ &+ F_1 e_s(t,k) + F_2 e_s(t-1,k) + F_3 \sum e_s(t-1,k) \\ &= (\tau_P + \tau_I + \tau_D + F_1)e_s(t,k) + (F_2 - \tau_D)e_s(t-1,k) \\ &+ (\tau_I + F_3)\sum e_s(t-1,k). \end{aligned} \tag{3.38}$$

Note that the ideal derivative term in (3.4) is substituted by a practical form of $e(t) - e(t-1)$ for obtaining (3.38). Other similar terms may also be adopted to derive $u(t,k)$ and thus are omitted herein.

Correspondingly, it follows that

$$\begin{aligned} \delta u(t,k) &= (\tau_P + \tau_I + \tau_D + F_1)\delta e_s(t,k) + (F_2 - \tau_D)\delta e_s(t-1,k) \\ &+ (\tau_I + F_3)\delta \sum e_s(t-1,k). \end{aligned} \tag{3.39}$$

Based on the robust PID design given in Sect. 3.2, by substituting (3.36), (3.37), and (3.39) into (3.29), one obtains

$$\begin{aligned} \delta x(t+1,k) &= [\tilde{A} - \tilde{B}(\tau_P + \tau_I + \tau_D + F_1)C]\delta x(t,k) \\ &+ \tilde{B}(\tau_P + \tau_I + \tau_D + F_1)L_1 e(t+1,k-1) \\ &+ \tilde{B}[(\tau_P + \tau_I + \tau_D + F_1)L_2 + F_2 - \tau_D]\delta e_s(t-1,k) \\ &+ \tilde{B}[(\tau_P + \tau_I + \tau_D + F_1)L_3 + F_3 + \tau_I]\delta \sum e_s(t-1,k) + \varpi(t,k), \end{aligned} \tag{3.40}$$

where $\tilde{A} = A_m + \Delta\tilde{A}(t,k)$ and $\tilde{B} = B_m + \Delta\tilde{B}(t,k)$.

Consequently, the predicted output error can be derived in terms of (3.28) as

$$\begin{aligned} e(t+1,k) &= e(t+1,k-1) - C\delta x(t+1,k) \\ &= [-C\tilde{A} + C\tilde{B}(\tau_P + \tau_I + \tau_D + F_1)C]\delta x(t,k) \\ &+ [\boldsymbol{I} - C\tilde{B}(\tau_P + \tau_I + \tau_D + F_1)L_1]e(t+1,k-1) \\ &- C\tilde{B}[(\tau_P + \tau_I + \tau_D + F_1)L_2 + F_2 - \tau_D]\delta e_s(t-1,k) \\ &- C\tilde{B}[(\tau_P + \tau_I + \tau_D + F_1)L_3 + F_3 + \tau_I]\delta \sum e_s(t-1,k) - C\varpi(t,k). \end{aligned} \tag{3.41}$$

Therefore, a 2D system description of the above ILC scheme is written as

$$\begin{cases} \begin{bmatrix} \delta x(t+1,k) \\ \delta e_s(t,k) \\ \delta \sum e_s(t,k) \\ e(t+1,k) \end{bmatrix} = \tilde{\Psi} \begin{bmatrix} \delta x(t,k) \\ \delta e_s(t-1,k) \\ \delta \sum e_s(t-1,k) \\ e(t+1,k-1) \end{bmatrix} + D_w \varpi(t), \\ \varsigma(t,k) = G \begin{bmatrix} \delta x(t,k) \\ \delta e_s(t-1,k) \\ \delta \sum e_s(t-1,k) \\ e(t+1,k-1) \end{bmatrix}. \end{cases} \tag{3.42}$$

where $G = [\mathbf{0}\ \mathbf{0}\ \mathbf{0}\ \mathbf{I}]$, $D_w = [\mathbf{I}\ \mathbf{0}\ \mathbf{0}\ -C^T]^T$,

$$\tilde{\Psi} = \begin{bmatrix} \tilde{A} - \tilde{B}(\tau_P + \tau_I + \tau_D + F_1)C & \tilde{B}[(\tau_P + \tau_I + \tau_D + F_1)L_2 + F_2 - \tau_D] \\ -C & L_2 \\ -C & L_2 \\ -C\tilde{A} + C\tilde{B}(\tau_P + \tau_I + \tau_D + F_1)C & -C\tilde{B}[(\tau_P + \tau_I + \tau_D + F_1)L_2 + F_2 - \tau_D] \end{bmatrix.$$
$$\begin{matrix} \tilde{B}[(\tau_P + \tau_I + \tau_D + F_1)L_3 + F_3 + \tau_I] & \tilde{B}(\tau_P + \tau_I + \tau_D + F_1)L_1 \\ L_3 & L_1 \\ \mathbf{I} + L_3 & L_1 \\ -C\tilde{B}[(\tau_P + \tau_I + \tau_D + F_1)L_3 + F_3 + \tau_I] & \mathbf{I} - C\tilde{B}(\tau_P + \tau_I + \tau_D + F_1)L_1 \end{matrix}\Bigg].$$

Note that $\varsigma(t,k) = e(t+1,k-1)$ is regarded as the controlled variable to be minimized under process uncertainties, varying initial process conditions from batch to batch, and load disturbance. That is to say, the robust 2D control objective can be determined in terms of a batch process control specification (Liu and Wang 2012) as

$$J_{BP} = \sum_{t=0}^{N_1=N} \sum_{k=0}^{N_2 \to \infty} \left(\gamma_{ILC}^{-1} \|\varsigma(t,k)\|_2^2 - \gamma_{ILC} \|\varpi(t,k)\|_2^2 \right) < 0. \tag{3.43}$$

By defining

$$x^h(t,k) = \begin{bmatrix} \delta x(t,k) \\ \delta e_s(t-1,k) \\ \delta \sum e_s(t-1,k) \end{bmatrix}, \ x^v(t,k) = e(t+1,k-1), \tag{3.44}$$

the 2D system in (3.42) can be viewed as a typical Roesser's system in the form of (1.1).

Hence, analyzing the robust stability of the above ILC scheme is equivalent to that of the 2D system in (3.42). The following theorem is given to assess the robust stability and determine the ILC controller parameters.

Theorem 3.2 *The 2D control system in* (3.42) *subject to time-varying process uncertainties described by* (3.1) *is guaranteed robustly stable with a* $\mathcal{H}_\infty$ *control performance level,* γ_{ILC}*, if there exist* $Q_1 > 0$*,* $Q_2 > 0$*,* $Q_3 > 0$*,* $Q_4 > 0$*, matrices* $\hat{F}_2$*,* $\hat{F}_3$*,* $\hat{L}_1$*,* $\hat{L}_2$*,* $\hat{L}_3$ *with appropriate dimensions, and positive scalars* ε_1*,* ε_2*, such that the following LMI holds*

$$\begin{bmatrix} -Q+\varepsilon_1\Omega_{\text{A1}}\Omega_{\text{A1}}^{\text{T}}+\varepsilon_2\Omega_{\text{B1}}\Omega_{\text{B1}}^{\text{T}} & \Pi & D_{\text{w}} & \mathbf{0} & \mathbf{0} & \mathbf{0} \\ * & -Q & \mathbf{0} & QG^{\text{T}} & P\Omega_{\text{A2}}^{\text{T}} & P\Omega_{\text{B2}}^{\text{T}} \\ * & * & -\gamma_{\text{ILC}}\mathbf{I} & \mathbf{0} & \mathbf{0} & \mathbf{0} \\ * & * & * & -\gamma_{\text{ILC}}\mathbf{I} & \mathbf{0} & \mathbf{0} \\ * & * & * & * & -\varepsilon_1\mathbf{I} & \mathbf{0} \\ * & * & * & * & * & -\varepsilon_2\mathbf{I} \end{bmatrix} < 0, \quad (3.45)$$

where $Q = \text{diag}\{Q_1, Q_2, Q_3, Q_4\}, D_{\text{g}} = [\mathbf{I}\ \mathbf{0}]^{\text{T}}$, $H = [\mathbf{I}\ \mathbf{0}]$, $\Omega_{\text{A1}} = [\Delta\overline{A}_1^{\text{T}}, \mathbf{0}, \mathbf{0}, -\Delta\overline{A}_1^{\text{T}}C^{\text{T}}]^{\text{T}}$, $\Omega_{\text{A2}} = [\Delta\overline{A}_2, \mathbf{0}, \mathbf{0}, \mathbf{0}]$, $\Omega_{\text{B1}} = [\Delta\overline{B}_1^{\text{T}}, \mathbf{0}, \mathbf{0}, -\Delta\overline{B}_1^{\text{T}}C^{\text{T}}]^{\text{T}}$,

$$\begin{aligned} \Omega_{\text{B2}} = \big[&-\Delta\overline{B}_2(\tau_{\text{P}}+\tau_{\text{I}}+\tau_{\text{D}}+F_1)C,\ \Delta\overline{B}_2[(\tau_{\text{P}}+\tau_{\text{I}}+\tau_{\text{D}}+F_1)\hat{L}_2+\hat{F}_2-\tau_{\text{D}}], \\ &\Delta\overline{B}_2[(\tau_{\text{P}}+\tau_{\text{I}}+\tau_{\text{D}}+F_1)\hat{L}_3+\hat{F}_3+\tau_{\text{I}}],\ \Delta\overline{B}_2(\tau_{\text{P}}+\tau_{\text{I}}+\tau_{\text{D}}+F_1)\hat{L}_1\big]. \end{aligned}$$

$$\Pi = \begin{bmatrix} A_{\text{m}}Q_1-B_{\text{m}}(\tau_{\text{P}}+\tau_{\text{I}}+\tau_{\text{D}}+F_1)CQ_1 & B_{\text{m}}(\tau_{\text{P}}+\tau_{\text{I}}+\tau_{\text{D}}+F_1)\hat{L}_2+B_{\text{m}}\hat{F}_2-B_{\text{m}}\tau_{\text{D}}Q_2 \\ -CQ_1 & \hat{L}_2 \\ -CQ_1 & \hat{L}_2 \\ -CA_{\text{m}}Q_1+CB_{\text{m}}(\tau_{\text{P}}+\tau_{\text{I}}+\tau_{\text{D}}+F_1)CQ_1 & -CB_{\text{m}}(\tau_{\text{P}}+\tau_{\text{I}}+\tau_{\text{D}}+F_1)\hat{L}_2-CB_{\text{m}}\hat{F}_2+CB_{\text{m}}\tau_{\text{D}}Q_2 \end{bmatrix}$$
$$\begin{bmatrix} B_{\text{m}}(\tau_{\text{P}}+\tau_{\text{I}}+\tau_{\text{D}}+F_1)\hat{L}_3+B_{\text{m}}\hat{F}_3+B_{\text{m}}\tau_{\text{I}}Q_3 & B_{\text{m}}(\tau_{\text{P}}+\tau_{\text{I}}+\tau_{\text{D}}+F_1)\hat{L}_1 \\ \hat{L}_3 & \hat{L}_1 \\ Q_3+\hat{L}_3 & \hat{L}_1 \\ -CB_{\text{m}}(\tau_{\text{P}}+\tau_{\text{I}}+\tau_{\text{D}}+F_1)\hat{L}_3-CB_{\text{m}}\hat{F}_3-CB_{\text{m}}\tau_{\text{I}}Q_3 & Q_4-CB_{\text{m}}(\tau_{\text{P}}+\tau_{\text{I}}+\tau_{\text{D}}+F_1)\hat{L}_1 \end{bmatrix}, \quad (3.46)$$

by parameterizing

$$\begin{cases} L_1 = \hat{L}_1 Q_4^{-1}, \\ L_2 = \hat{L}_2 Q_2^{-1}, \\ L_3 = \hat{L}_3 Q_3^{-1}, \end{cases} \quad (3.47)$$

$$\begin{cases} F_2 = \hat{F}_2 Q_2^{-1}, \\ F_3 = \hat{F}_3 Q_3^{-1}. \end{cases} \quad (3.48)$$

Proof The robust 2D control objective in (3.43) can be rewritten as

$$J_{\text{BP}} = \sum_{t=0}^{N_1=N}\sum_{k=0}^{N_2\to\infty}\left(\gamma_{\text{ILC}}^{-1}\|\varsigma(t,k)\|_2^2-\gamma_{\text{ILC}}\|\varpi(t,k)\|_2^2+\Delta V\right)-\sum_{t=0}^{N_1=N}\sum_{k=0}^{N_2\to\infty}\Delta V < 0, \quad (3.49)$$

where ΔV is the difference of a Lyapunov–Krasovskii function used for analysis of 2D asymptotic stability, i.e.

$$\Delta V = V_Q\begin{bmatrix} x^h(t+1,k-1) \\ x^v(t,k) \end{bmatrix} - V_Q\begin{bmatrix} x^h(t,k-1) \\ x^v(t,k-1) \end{bmatrix}. \tag{3.50}$$

Using the boundary conditions from an initial resetting of batch process operation, i.e.

$$\begin{cases} \delta x(0,0) = \delta x(0,1) = \delta x(1,0) = 0, \\ \delta e_s(0,0) = \delta e_s(0,1) = \delta e_s(1,0) = 0, \\ \delta \sum e_s(0,0) = \delta \sum e_s(0,1) = \delta \sum e_s(1,0) = 0, \\ e(0,0) = e(0,1) = e(1,0) = 0, \end{cases} \tag{3.51}$$

it can be easily verified by using $Q = \text{diag}\{Q_1, Q_2, Q_3, Q_4\}$ that

$$\begin{aligned} \sum_{t=0}^{N_1=N}\sum_{k=0}^{N_2\to\infty} \Delta V = & \sum_{t=0}^{N_1=N}\sum_{k=0}^{N_2\to\infty} \{V_{Q_1}[\delta x(t+1,k)] - V_{Q_1}[\delta x(t,k)] + V_{Q_2}[\delta e_s(t,k)] \\ & - V_{Q_2}[\delta e_s(t-1,k)] + V_{Q_3}[\delta \sum e_s(t,k)] - V_{Q_3}[\delta \sum e_s(t-1,k)] \\ & + V_{Q_4}[e(t+1,k)] - V_{Q_4}[e(t+1,k-1)], \\ = & \sum_{k=0}^{N_2\to\infty} V_{Q_1}[\delta x(N_1+1,k)] + \sum_{k=0}^{N_2\to\infty} V_{Q_2}[\delta e_s(N_1,k)] \\ & + \sum_{k=0}^{N_2\to\infty} V_{Q_3}[\delta \sum e_s(N_1,k)] + \sum_{t=0}^{N_1} V_{Q_4}[e(t+1,N_2+1)] > 0. \end{aligned} \tag{3.52}$$

Therefore, a sufficient condition to ensure the control objective in (3.49) is that

$$\gamma_{\text{ILC}}^{-1}\|\varsigma(t,k)\|_2^2 - \gamma_{\text{ILC}}\|\varpi(t,k)\|_2^2 + \Delta V < 0. \tag{3.53}$$

By substituting the 2D system description in (3.42) and (3.50) into (3.53), one obtains

$$\xi^{\text{T}}\Xi_2\xi < 0, \tag{3.54}$$

where $\xi = \left[[x^h(t,k-1)]^{\text{T}}, [x^v(t,k-1)]^{\text{T}}, \varpi^{\text{T}}(t,k)\right]^{\text{T}}$, and

$$\Xi_2 = \begin{bmatrix} \tilde{\Psi}^{\text{T}} \\ D_{\text{w}}^T \end{bmatrix} Q[\,\tilde{\Psi}\; D_{\text{w}}\,] - \begin{bmatrix} Q - \gamma_{\text{ILC}}^{-1}G^{\text{T}}G & \mathbf{0} \\ * & \gamma_{\text{ILC}}\mathbf{I} \end{bmatrix}. \tag{3.55}$$

By the Schur complement, it can be derived that (3.54) is guaranteed by

$$\begin{bmatrix} -Q & \tilde{\Pi} & D_{\mathrm{w}} & \mathbf{0} \\ * & -Q & \mathbf{0} & QG^{\mathrm{T}} \\ * & * & -\gamma_{\mathrm{ILC}}\mathbf{I} & \mathbf{0} \\ * & * & * & -\gamma_{\mathrm{ILC}}\mathbf{I} \end{bmatrix} < 0, \tag{3.56}$$

where

$$\tilde{\Pi} = \Pi + \Omega_{\mathrm{A1}}\tilde{\Theta}_1(t)\Omega_{\mathrm{A2}} + \Omega_{\mathrm{B1}}\tilde{\Theta}_2(t)\Omega_{\mathrm{B2}}, \tag{3.57}$$

where Π, Ω_{A1}, Ω_{A2}, Ω_{B1}, and Ω_{B2} have been shown in (3.45).

Using Lemma 2.1 in Chap. 2 and the Schur complement, it can be seen that (3.56) is guaranteed by (3.45) in Theorem 3.2. This completes the proof. □

Note that the feedforward controller, F_1, is prescribed for solving the LMI condition in (3.45). To facilitate the feasibility of the LMI condition in (3.45), the choice of F_1 should be made to keep all the eigenvalues of $A_{\mathrm{m}} - B_{\mathrm{m}}(\tau_{\mathrm{P}} + \tau_{\mathrm{I}} + \tau_{\mathrm{D}} + F_1)C$ in the unit circle in the z-transfer plane, i.e.

$$|\lambda_i| = |\mathrm{eig}[A_{\mathrm{m}} - B_{\mathrm{m}}(\tau_{\mathrm{P}} + \tau_{\mathrm{I}} + \tau_{\mathrm{D}} + F_1)C]| < 1, \ i = 1, 2, \ldots, n_x. \tag{3.58}$$

In fact, all the feedforward controllers, F_1, F_2, F_3, corresponding to F_1, $\hat{F}_2$, $\hat{F}_3$ in (3.45) taken as slack variables to facilitate the LMI feasibility, are used to increase the flexibility of the indirect-type ILC in the proposed control scheme in Fig. 3.1, for the purpose of robustly tracking the desired output profile under process uncertainties and load disturbance.

To obtain the optimal robust tracking performance, the ILC controller parameters can be determined by the following optimization program

$$\underset{\Delta\tilde{A}(t),\, \Delta\tilde{B}(t)}{\text{Minimize}}\ \gamma_{\mathrm{ILC}}. \tag{3.59}$$

Similarly, by specifying the learning controllers, L_1, L_2, L_3, which determine the convergence rate of the ILC scheme, the achievable robust performance can be assessed through the LMI condition in (3.45), and so is for the admissible process uncertainty bounds denoted by $\Delta\tilde{A}(t, k)$ and $\Delta\tilde{B}(t, k)$. Note that the admissible variation of initial process conditions from batch to batch can also be assessed through the LMI condition in (3.45) by lumping the variation bound into the magnitude (D_{w}) of the disturbance as shown in (3.42).

3.5 An Illustrative Example

Consider the batch process of injection molding and the target profile as studied in Example 2.1. By performing the optimization program in (3.24), the minimal $\mathcal{H}_\infty$ robust performance level is determined as $\gamma^*_{\mathrm{PID}} = 1.28$. To avoid over aggressive control signal, $\gamma_{\mathrm{PID}} = 5$ is taken here to solve the LMI condition in (3.12), obtaining the PI controller parameters, $\tau_{\mathrm{P}} = 1.2889$ and $\tau_{\mathrm{I}} = 0.0336$. For the ILC design, it can be easily verified that the range of $F_1 \in [-1.3, 0.1]$ can ensure all the eigenvalues of $A_{\mathrm{m}} - B_{\mathrm{m}}(\tau_{\mathrm{P}} + \tau_{\mathrm{I}} + \tau_{\mathrm{D}} + F_1)C$ located in the unit circle in the z-transfer plane. Herein, choose $F_1 = -0.5$ to perform the optimization program in (3.59), obtaining the minimal $\mathcal{H}_\infty$ robust performance level, $\gamma^*_{\mathrm{ILC}} = 110$, and correspondingly, $L_1 = 0.1776$, $L_2 = 0$, $L_3 = -0.029$, $F_2 = 0$, and $F_3 = -0.0097$.

For illustration, the following cases of process uncertainties are tested.

Case 1. Time-invariant process uncertainties. In this case, $\Delta\tilde{A}(t)$ and $\Delta\tilde{B}(t)$ are assumed to be fixed as their upper bounds. The tracking results are shown in Fig. 3.2a, b, while the output tracking error in terms of the following criterion is plotted in Fig. 3.3

$$\mathrm{ATE}(k) = \sum_{t=1}^{N} |e(t,k)|/N.$$

It is seen that perfect tracking is realized through 20 cycles by the proposed ILC scheme after an initial run of the PID tuning, compared to the reference (Wang et al. 2012), which needed almost 50 cycles to reach perfect tracking. Moreover, there exists no steady output tracking error in each cycle, owing to using the output tracking error in the current cycle for 2D ILC design as shown in (3.42).

Case 2. Time-varying uncertainties and disturbance. Assume that the process state transfer matrices become time-varying with $|\delta(t)| \le 0.1$, together with non-repetitive load disturbance, $w(t,k) = \sin(t+\theta(k))$ where $\theta(k)$ is a random variable uniformly distributed in the range of $[0, 2\pi]$ as assumed in the reference (Wang et al. 2012). Since the closed-loop system becomes a stochastic process, here 100 Monte Carlo tests are performed, each of which includes 100 cycles. The averaged results of ATE are plotted in Fig. 3.4, in comparison with those of Refs. (Shi et al. 2005; Wang et al. 2012). It is seen that the closed-loop system maintains robust stability well along both the time and batch directions by the proposed ILC scheme, thus demonstrating that it can be reliably used for robust tracking of the desired profile and on-line optimization against batch-to-batch process uncertainties and load disturbance.

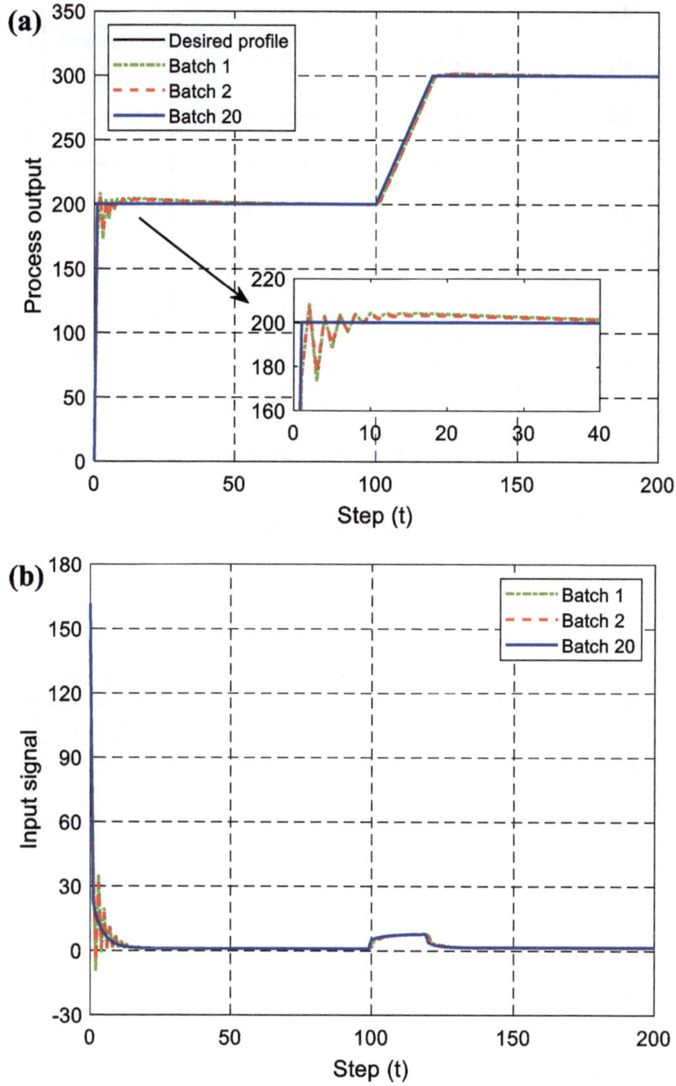

Fig. 3.2 Tracking performance for Case 1

3.6 Summary

For industrial batch processes subject to time-varying uncertainties from batch to batch, a robust indirect-type ILC method (Liu et al. 2014) has been presented in this Chapter based on the widely used PID control loop. In the proposed control scheme, either the closed-loop PID controller or the ILC updating law can be designed relatively independent, along with the feedforward controllers added to increase the

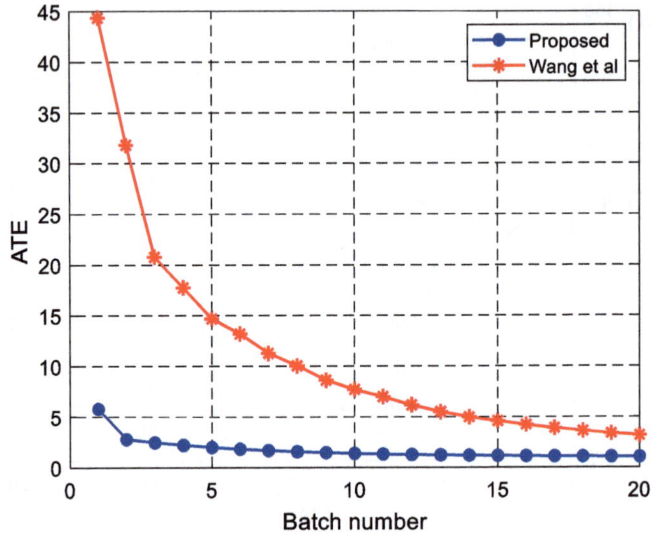

Fig. 3.3 Plot of ATE for Case 1

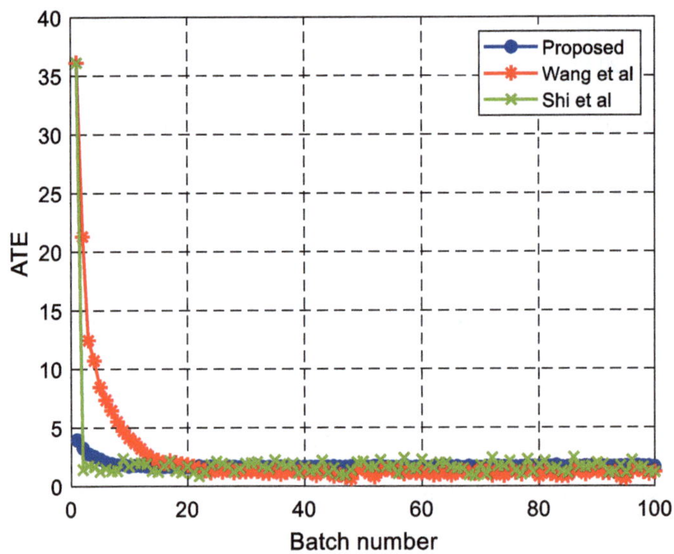

Fig. 3.4 Plot of ATE for Case 2

control flexibility. To accommodate time-varying process uncertainties, a robust PID design has been given based on the robust $\mathcal{H}_\infty$ control objective. For implementational simplicity, it is preferred to use a PI controller if such a controller can be derived from the LMI conditions established for maintaining the closed-loop robust stability.

For the batchwise direction, an ILC scheme consisting of a learning set-point strategy and a feedforward control added to the process input has been developed based on an equivalent 2D system description of the batch process and the LMI condition formulated in terms of the robust $\mathcal{H}_\infty$ control objective for robust convergence. Only measured output errors of current and previous cycles are used to implement the proposed ILC scheme for the convenience of practical application. The application of an illustrative example from the literature has well demonstrated the effectiveness and merits of the proposed indirect-type ILC scheme.

References

Ahn H-S, Chen YQ, Moore KL (2007) Iterative learning control: brief survey and categorization. IEEE Trans Syst Man Cybern Part C Appl Rev 37(6):1099–1121

Åström KJ, Hägglund T (1995) PID controller: theory, design, and tuning, 2nd edn. ISA Society of America, Research Triangle Park, NC

Bonvin D, Srinivasan B, Hunkeler D (2006) Control and optimization of batch processes. IEEE Trans Control Syst Mag 26(6):34–45

Chen J, Kong C-K (2009) Performance assessment for iterative learning control of batch units. J Process Control 19:1043–1053

Chi R, Li H, Lin N, Huang B (2024) Data-driven indirect iterative learning control. IEEE Trans Cybern 54(3):1650–1660

Kim DI, Kim S (1996) An iterative learning control method with application for CNC machine tools. IEEE Tran Ind Appl 32:66–72

Lee JH, Lee KS, Kim WC (2000) Model-based iterative learning control with a quadratic criterion for time-varying linear systems. Automatica 36:641–657

Li H, Bai J, Wu F, Zou H (2023) A novel two-dimensional PID controller design using two-dimensional model predictive iterative learning control optimization for batch processes. Can J Chem Eng 101(12):6959–6976

Liu T, Gao F, Wang Y (2010) IMC-based iterative learning control for batch processes with uncertain time delay. J Process Control 20(2):173–180

Liu T, Wang Y (2012) A synthetic approach for robust constrained iterative learning control of piecewise affine batch processes. Automatica 48(11):2762–2775

Liu T, Wang XZ, Junghui C (2014) Robust PID based indirect-type iterative learning control for batch processes with time-varying uncertainties. J Process Control 24(12):95–106

Ruan X, Bien Z, Park KH (2008) Decentralized iterative learning control to large-scale industrial processes for nonrepetitive trajectory tracking. IEEE Tran Syst Man Cybern Part A Syst Hum 38:238–252

Shi J, Gao F, Wu T-J (2005) Integrated design and structure analysis of robust iterative learning control system based on a two-dimensional model. Ind Eng Chem Res 44:8095–8105

Shi J, Gao F, Wu T-J (2006a) From two-dimensional linear quadratic optimal control to iterative learning control. Paper 1. Two-dimensional linear quadratic optimal controls and system analysis. Ind Eng Chem Res 45:4603–4616

Shi J, Gao F, Wu T-J (2006b) From two-dimensional linear quadratic optimal control to iterative learning control. Paper 2. Iterative learning controls for batch processes. Ind Eng Chem Res 45:4617–4628

Shi J, Gao F, Wu T-J (2006c) Robust iterative learning control design for batch processes with uncertain perturbations and initialization. AIChE J 52(6):2171–2187

Tan K, Zhao S, Xu JX (2007) Online automatic tuning of a proportional integral derivative controller based on an iterative learning control approach. IET Control Theor Appl 1(1):90–96

Tayebi A (2007) Analysis of two particular iterative learning control schemes in frequency and time domains. Automatica 43:1565–1572
Wang Y, Gao F, Doyle F (2009) Survey on iterative learning control, repetitive control, and run-to-run control. J Process Control 19(10):1589–1600
Wang Y, Liu T, Zhao Z (2012) Advanced PI control with simple learning set-point design: application on batch processes and robust stability analysis. Chem Eng Sci 71(1):153–165
Wang Y, Zisser H, Dassau E, Jovanovič L, Doyle F (2010) Model predictive control with learning-type reference: application in artificial pancreatic beta-cell. AIChE J 20:173–180
Wu J, Ding H (2007) Reference adjustment for a high-acceleration and high-precision platform via A-type of iterative learning control. Proc IMechE, Part I: J Syst Control Eng 221:781–789

Chapter 4
Closed-Loop ILC Scheme with State Feedback

4.1 Introduction

Industrial and chemical batch processes with a wide range of operating conditions are generally nonlinear and therefore have been widely studied for advanced control design to obtain the desired system performance, such as perfect tracking of the set-point profile to guarantee good product quality and consistency. Based on the system state measurement or estimation, closed-loop ILC methods with state feedback have been explored for linear or nonlinear batch processes to progressively improve tracking performance from batch to batch (or named cycle to cycle) (Ahn et al. 2007; Bonvin et al. 2006; Wang et al. 2009). In fact, many batch processes, e.g., the membrane filtration process and pharmaceutical crystallization, are slowly time-varying from cycle to cycle, while repeating fundamental dynamic response characteristics or subject to repetitive and/or nonrepetitive load disturbance (Busch et al. 2007; Nagy et al. 2008). The early ILC methods based on using time-invariant process information to design the state feedback control law could not maintain robust stability for these batch processes (Rogers et al. 2007). For uncertain nonlinear systems with specific structural properties or uncertainty types, adaptive ILC schemes have been developed to ensure the boundedness or asymptotic convergence of the set-point tracking error (Tayebi and Chien 2007; Qu and Xu 2002; Li et al. 2024; Li and Huang 2024). Besides, robust ILC schemes based on specific nonlinear model structures were proposed to guarantee batch-to-batch stability while complying with the process input constraints (Nagy et al. 2007; Hermanto et al. 2011).

It has been widely recognized that many industrial nonlinear processes, including batch processes, can be decomposed into a series of linearized operating regions for system operation. Accordingly, piecewise affine control strategies, also called linear parameter varying (LPV) control or gain-scheduling control (using multiple linear models), have been developed for robust control of these nonlinear processes (Rugh and Shamma 2000; Lu and Arkun 2002; Özkan et al. 2003; Park and Jeong 2004; Wan and Kothare 2003, 2008; Forni and Galeani 2010; Spiegel et al. 2024). However, it is

T. Liu et al., *Robust Iterative Learning Control of Industrial Batch Systems*, Intelligent Control and Learning Systems 22, https://doi.org/10.1007/978-981-96-9778-6_4

not trivial to extend these piecewise affine control methods to batch process operation while guaranteeing the control system robust stability along both the time (during a cycle) and batch (from cycle to cycle) directions (Galkowski et al. 2002; Rogers et al. 2007; Pakshin et al. 2024). To cope with cycle-to-cycle varying process uncertainties, the references (Liu and Gao 2010; Shi et al. 2005, 2006) developed robust 2D ILC methods based on the norm-bounded description of process uncertainties. To enhance system performance against nonrepetitive load disturbance, alternative on-line adaptive ILC schemes were also developed for practical implementation and performance improvement (Chin et al. 2004; Zhang 2008).

In this Chapter, a robust 2D ILC method based on the state feedback structure (Liu and Wang 2012) is presented for linear or nonlinear batch processes that can be practically divided into a series of piecewise affine operating regions. By using a 2D system description of the process for batch control design, both time-varying process uncertainties and nonrepetitive load disturbance are taken into account, together with the process input and state constraints for implementation. By introducing the desired performance objective or robust control objective in combination with the 2D Lyapunov–Krasovskii functions that can guarantee monotonic state energy (or output error) decrease along both the time and batch directions, LMI conditions are correspondingly established for the ILC controller design and performance optimization. In these LMI conditions, there are adjustable convergence indices for the time and batch directions, and an adjustable closed-loop robust control performance level. The effectiveness of the proposed ILC method is demonstrated through the application to a highly nonlinear continuous stirred tank reactor (CSTR) subject to cycle-to-cycle uncertainties and load disturbance.

4.2 State Feedback Based ILC Design

Consider a nonlinear batch process that can be practically divided into a series of piecewise-affine operating regions

$$\begin{cases} \dot{x}(t,k) = f[x(t,k), u(t,k)], \\ y(t,k) = g[x(t,k)], \quad 0 \le t \le N, \\ y(t,k) \in \Omega = \cup_{i=1}^{m} \Omega_i, \\ x(0,k) = \overline{x}(0), \quad k = 0, 1, \ldots \end{cases} \tag{4.1}$$

where $x(t,k) \in \Re^{n_x}$, $y(t,k) \in \Re^{n_y}$, $u(t,k) \in \Re^{n_u}$, t and k denote the time and cycle indices, respectively; $\Omega \subset \Re^{n_y}$ denotes the real space of the output response, which includes $m$piecewise affine operating regions denoted by Ω_i $(i = 1, 2, \ldots, m)$; N is the cycling time, and $\overline{x}(0)$ denotes the initial resetting condition of each cycle that may be reset to zero with respect to each affine operating region for the convenience of control design.

A typical scenario of linearizing the nonlinear batch process in (4.1) with regard to a number of equilibrium operating points $\left(\left[y_{\text{eq}}^{[i]}, x_{\text{eq}}^{[i]}, u_{\text{eq}}^{[i]}\right], i = 1, 2, \ldots, m\right)$ is

$$\begin{cases} \dot{x}^{[i]}(t,k) = f\left(x_{\text{eq}}^{[i]}, u_{\text{eq}}^{[i]}\right) + \frac{\partial f}{\partial x}\left(x_{\text{eq}}^{[i]}, u_{\text{eq}}^{[i]}\right)\Delta x^{[i]}(t,k) + \frac{\partial f}{\partial u}\left(x_{\text{eq}}^{[i]}, u_{\text{eq}}^{[i]}\right)\Delta u^{[i]}(t,k) \\ y^{[i]}(t,k) = g\left(x_{\text{eq}}^{[i]}\right) + \frac{\partial g}{\partial x}\left(x_{\text{eq}}^{[i]}\right)\Delta x^{[i]}(t,k) \\ \Delta x^{[i]}(t,k) = x(t,k) - x_{\text{eq}}^{[i]}(t,k) \\ \Delta u^{[i]}(t,k) = u(t,k) - u_{\text{eq}}^{[i]}(t,k) \end{cases} \tag{4.2}$$

Alternatively, such linear affine models can be derived using a piecewise model identification method (Liu and Gao 2012; Ljung 1999).

To describe model uncertainties during a cycle or from cycle to cycle, the polyhedral uncertainty type and the norm-bounded uncertainty type are typically considered for analysis. Correspondingly, a discrete-time model structure of the observable canonical form with the polyhedral uncertainties can be written as

$$\begin{cases} x^{[i]}(t+1,k) = A_j^{[i]}(t,k)x^{[i]}(t,k) + B_j^{[i]}(t,k)u^{[i]}(t,k) + w^{[i]}(t,k), \\ y^{[i]}(t,k) = C^{[i]}x^{[i]}(t,k), \quad 0 \le t \le N^{[i]}, \\ x^{[i]}(0,k) = x^{[i]}(0), \quad k = 0, 1, \ldots \\ \left[A_j^{[i]}(t,k), B_j^{[i]}(t,k)\right] \in \text{Co}\left\{\left[A_j^{[i]}(t,k), B_j^{[i]}(t,k)\right]\right\}, \\ i = 1, 2, \ldots, mj = 1, 2, \ldots, l \end{cases} \tag{4.3}$$

where t denotes the current time step and k the current cycle index, and $w^{[i]}(t,k) \in \Re^{n_x}$ denotes unmodeled process dynamics and load disturbance. $\left[A_j^{[i]}(t,k), B_j^{[i]}(t,k)\right](j = 1, 2, \ldots, l)$ are the vertices of a convex hull that are generally specified as the extreme cases ('worst-cases') of the process response in cycling.

In contrast, a model with the norm-bounded uncertainties is expressed by

$$\begin{cases} x^{[i]}(t+1,k) = \left[A^{[i]} + \Delta\tilde{A}^{[i]}(t,k)\right]x^{[i]}(t,k) + \left[B^{[i]} + \Delta\tilde{B}^{[i]}(t,k)\right]u^{[i]}(t,k) + w^{[i]}(t,k), \\ y^{[i]}(t,k) = C^{[i]}x^{[i]}(t,k), \quad 0 \le t \le N^{[i]}, \\ x^{[i]}(0,k) = x^{[i]}(0), \quad k = 0, 1, \ldots; i = 1, 2, \ldots, m \end{cases} \tag{4.4}$$

where $\Delta\tilde{A}^{[i]}(t,k) = \overline{F}_{1,1}^{[i]}\tilde{\Theta}_1(t)\overline{F}_{1,2}^{[i]}$, $\Delta\tilde{B}^{[i]}(t,k) = \overline{F}_{2,1}^{[i]}\tilde{\Theta}_2(t)\overline{F}_{2,2}^{[i]}$, $\overline{F}_{i,1}^{[i]}$ and $\overline{F}_{i,2}^{[i]}$ are constant matrices of the upper or lower bounds estimated in practice, and $\tilde{\Theta}_i^{\text{T}}(t)\tilde{\Theta}_i(t) \le \mathbf{I}, i = 1, 2$ are time-varying factors.

For batch run operation, the process input constraint is usually specified by

$$-\overline{u}^{[i]} \leq u(t+1,k) - u_{\text{eq}}^{[i]} \leq \overline{u}^{[i]}, \tag{4.5}$$

where $\overline{u}^{[i]} = \left[\overline{u}_1^{[i]}, \overline{u}_2^{[i]}, \ldots, \overline{u}_{n_u}^{[i]}\right]^{\mathrm{T}}, \overline{u}_j^{[i]} > 0, j = 1, 2, \ldots, n_u$.

When the process state variables can be measured, the following state constraint may be considered

$$-\overline{\chi}^{[i]} \leq \psi\left[x(t+1,k) - x_{\text{eq}}^{[i]}\right] \leq \overline{\chi}^{[i]}, \tag{4.6}$$

where $\psi \in \Re^{q \times n_x}$ is a constant matrix specified in terms of the process operation, $\overline{\chi}^{[i]} = \left[\overline{\chi}_1^{[i]}, \overline{\chi}_2^{[i]}, \ldots, \overline{\chi}_q^{[i]}\right]^{\mathrm{T}}, \overline{\chi}_j^{[i]} > 0, j = 1, 2, \ldots, q$, denote the upper bounds relating to the admissible ranges of individual states, respectively.

For implementation, it is usually adopted to switch the piecewise control laws in terms of the cycling time of each operating region. An unsteady initial process state in each operating region can therefore be viewed as a nonrepetitive load disturbance from cycle to cycle, requiring to be suppressed for consistent batch operation. This requirement will be treated in the following control design.

For a piecewise affine batch process as described above, when an ILC strategy is applied, the ILC updating control law should have the following general form

$$\begin{cases} u^{[i]}(t,k) = u^{[i]}(t,k-1) + \delta u^{[i]}(t,k), \\ u^{[i]}(0,1) = u^{[i]}(0), 0 \leq t \leq N^{[i]}, \quad k = 0, 1, \ldots, i = 1, 2, \ldots, m. \end{cases} \tag{4.7}$$

where $\delta u^{[i]}(t,k)$ is the control output to be determined in the current cycle (k) for the i-th operating region, and $u^{[i]}(0)$ denotes the initial value of iteration that is generally reset to zero (or a constant value with respect to $u_{\text{eq}}^{[i]}$) for implementation.

To facilitate a closed-loop ILC design that allows for time-varying process uncertainties from cycle to cycle and nonrepetitive load disturbance, the output tracking error in the current cycle for each operating region is defined by

$$e^{[i]}(t,k) \triangleq y^{[i]}(t,k) - y_{\text{d}}^{[i]}(t), \quad i = 1, 2, \ldots, m, \tag{4.8}$$

where $y_{\text{d}}^{[i]}(t)$ denotes the set-point profile (or the desired output trajectory) in each of these operating regions for the cycling operation.

When the process uncertainties are described in the polyhedral form of (4.3), it follows from (4.3) and (4.7) along with the definition in (4.8) that

$$e^{[i]}(t,k) = e^{[i]}(t,k-1) + C^{[i]}\delta x^{[i]}(t,k), \tag{4.9}$$

$$\delta x^{[i]}(t+1,k) = A_j^{[i]}(t,k)\delta x^{[i]}(t,k) + B_j^{[i]}(t,k)\delta u^{[i]}(t,k) + \omega^{[i]}(t,k), \tag{4.10}$$

$$\omega^{[i]}(t,k) = \left[A_j^{[i]}(t,k) - A_j^{[i]}(t,k-1)\right]x^{[i]}(t,k-1)$$

$$+\left[B_j^{[i]}(t,k)-B_j^{[i]}(t,k-1)\right]u^{[i]}(t,k-1)+\delta w^{[i]}(t,k). \tag{4.11}$$

Obviously, there exists $\omega^{[i]}(t,k)\neq 0$ in the presence of time-varying process uncertainties. Hereafter, it will be used to denote various sources of load disturbances.

When the process uncertainties are described in the norm-bounded form of (4.4), the output and state error expressions can be similarly derived and thus are omitted.

To meet with a practical requirement of no steady-state tracking error for each cycle, the information of tracking error integral is introduced as an additional state variable for the ILC design, which is defined by

$$\sum e^{[i]}(t,k)\triangleq\sum_{i=0}^{t}e^{[i]}(i,k),\ 1\leq t\leq N^{[i]}, i=1,2,\ldots,m. \tag{4.12}$$

It follows that

$$\sum e^{[i]}(t,k)=\sum e^{[i]}(t-1,k)+e^{[i]}(t,k). \tag{4.13}$$

Therefore, an equivalent 2D system description of the above batch process shown in (4.3) or (4.4), together with two augmented state variables of the tracking error information, can be expressed regarding the polyhedral uncertainties by

$$\begin{cases}\hat{x}^{[i]}(t+1,k)=A_{1,j}^{[i]}\hat{x}^{[i]}(t,k)+A_2^{[i]}\hat{x}^{[i]}(t+1,k-1)\\ \qquad\qquad +B_{g,j}^{[i]}\delta u^{[i]}(t,k)+D_g^{[i]}\omega^{[i]}(t,k),\\ y_e^{[i]}(t,k)=\begin{bmatrix}e^{[i]}(t,k-1)\\ e^{[i]}(t,k)\\ \sum e^{[i]}(t,k)\end{bmatrix}=G^{[i]}\hat{x}^{[i]}(t,k),\quad 1\leq t\leq N^{[i]},\\ z^{[i]}(t,k)=e^{[i]}(t,k)=L^{[i]}\hat{x}^{[i]}(t,k),\\ \hat{x}_e^{[i]}(0,k)=\mathbf{0},\quad k=0,1,\ldots,\quad i=1,2,\ldots,m,\end{cases} \tag{4.14}$$

or the norm-bounded uncertainties by

$$\begin{cases}\hat{x}^{[i]}(t+1,k)=\left(A_1^{[i]}+\Delta A_1^{[i]}\right)\hat{x}^{[i]}(t,k)+A_2^{[i]}\hat{x}^{[i]}(t+1,k-1)\\ \qquad\qquad +\left(B_g^{[i]}+\Delta B_g^{[i]}\right)\delta u^{[i]}(t,k)+D_g^{[i]}\omega^{[i]}(t,k),\\ y_e^{[i]}(t,k)=\begin{bmatrix}e^{[i]}(t,k-1)\\ e^{[i]}(t,k)\\ \sum e^{[i]}(t,k)\end{bmatrix}=G^{[i]}\hat{x}^{[i]}(t,k),\ 1\leq t\leq N^{[i]},\\ z^{[i]}(t,k)=e^{[i]}(t,k)=L\hat{x}^{[i]}(t,k),\\ \hat{x}_e^{[i]}(0,k)=\mathbf{0},\quad k=0,1,\ldots;i=1,2,\ldots,m,\end{cases} \tag{4.15}$$

where $L = [\mathbf{0}, \mathbf{I}, \mathbf{0}]$,

$$\begin{aligned}
\hat{x}^{[i]}(t,k) &= \Big[\delta x^{[i]}(t,k)^{\mathrm{T}}, e^{[i]}(t,k), \sum e^{[i]}(t-1,k)\Big]^{\mathrm{T}},\\
A_{1,j}^{[i]} &= \begin{bmatrix} A_j^{[i]} & \mathbf{0}\,\mathbf{0} \\ C^{[i]}A_j^{[i]} & \mathbf{0}\,\mathbf{0} \\ \mathbf{0} & \mathbf{I}\,\mathbf{I} \end{bmatrix}, B_{g,j}^{[i]} = \begin{bmatrix} B_j^{[i]} \\ C^{[i]}B_j^{[i]} \\ \mathbf{0} \end{bmatrix}, \quad j \in \{1, 2, \ldots, l\},\\
A_1^{[i]} &= \begin{bmatrix} A^{[i]} & \mathbf{0}\,\mathbf{0} \\ C^{[i]}A^{[i]} & \mathbf{0}\,\mathbf{0} \\ \mathbf{0} & \mathbf{I}\,\mathbf{I} \end{bmatrix}, A_2^{[i]} = \begin{bmatrix} \mathbf{0}\,\mathbf{0}\,\mathbf{0} \\ \mathbf{0}\,\mathbf{I}\,\mathbf{0} \\ \mathbf{0}\,\mathbf{0}\,\mathbf{0} \end{bmatrix}, B_g^{[i]} = \begin{bmatrix} B^{[i]} \\ C^{[i]}B^{[i]} \\ \mathbf{0} \end{bmatrix},\\
D_g^{[i]} &= \begin{bmatrix} \mathbf{I} \\ C^{[i]} \\ \mathbf{0} \end{bmatrix}, G^{[i]} = \begin{bmatrix} -C^{[i]} & \mathbf{I}\,\mathbf{0} \\ \mathbf{0} & \mathbf{I}\,\mathbf{0} \\ \mathbf{0} & \mathbf{I}\,\mathbf{I} \end{bmatrix},\\
\Delta A_1^{[i]} &= F_{1,1}^{[i]}\Theta_1(t)F_{1,2}^{[i]}, \ \Delta B_g^{[i]} = F_{2,1}^{[i]}\Theta_2(t)F_{2,2}^{[i]},\\
F_{1,1}^{[i]} &= \Big[\big(\bar{F}_{1,1}^{[i]}\big)^{\mathrm{T}}, \big(C^{[i]}\bar{F}_{1,1}^{[i]}\big)^{\mathrm{T}}, \mathbf{0}\Big]^{\mathrm{T}}, \ F_{2,1}^{[i]} = \Big[\big(\bar{F}_{2,1}^{[i]}\big)^{\mathrm{T}}, \big(C^{[i]}\bar{F}_{2,1}^{[i]}\big)^{\mathrm{T}}, \mathbf{0}\Big]^{\mathrm{T}},\\
F_{1,2}^{[i]} &= \Big[\bar{F}_{1,2}^{[i]}, \mathbf{0}, \mathbf{0}\Big], \quad F_{2,2}^{[i]} = \bar{F}_{2,2}^{[i]}, \Theta_i^{\mathrm{T}}(t)\Theta_i(t) \le \mathbf{I}, \quad i = 1, 2, \ldots, m.
\end{aligned}$$

From a view of control design, all the deviation states in (4.14) or (4.15), denoted by $\hat{x}^{[i]}(t,k)$, in terms of the equilibrium point of each operating region, are to be stabilized to the normalized origin in the state space.

Since the ILC scheme could be applied to each operating region, the superscript indicating the region number will be omitted hereafter for simplicity when this does not lead to any confusion.

When the process state variables are measured for the control design, a closed-loop ILC scheme is proposed as shown in Fig. 4.1, where the solid lines denote the real-time information in the current cycle and the dashed lines denote the stored information of the previous cycle. Correspondingly, a 2D ILC updating law is constructed as

$$u_{\mathrm{e}}(t,k) = C_{\mathrm{c}1}\hat{x}(t,k) + C_{\mathrm{c}2}\hat{x}(t+1,k-1). \tag{4.16}$$

For the polyhedral uncertainties described in (4.14), substituting (4.16) into (4.14) yields the closed-loop system

$$\hat{x}(t+1,k) = (A_{1,j} + B_{\mathrm{g},j}C_{\mathrm{c}1})\hat{x}(t,k) + (A_2 + B_{\mathrm{g},j}C_{\mathrm{c}2})\hat{x}(t+1,k-1) + D_{\mathrm{g}}\omega(t,k). \tag{4.17}$$

For the norm-bounded uncertainties described in (4.15), substituting (4.16) into (4.15) yields the closed-loop system

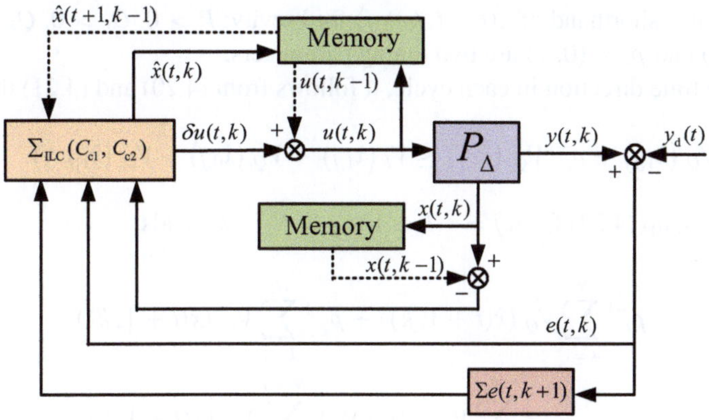

Fig. 4.1 Diagram of the 2D state feedback ILC scheme

$$\hat{x}(t+1,k) = (\hat{A}_1 + \Delta\hat{A}_1)\hat{x}(t,k) + (\hat{A}_2 + \Delta\hat{A}_2)\hat{x}(t+1,k-1) + D_g\omega(t,k), \tag{4.18}$$

where $\hat{A}_i = A_i + B_g C_c,\ i = 1, 2,\ \Delta\hat{A}_1 = \Delta A_1 + \Delta B_g C_{c1},\ \Delta\hat{A}_2 = \Delta B_g C_{c2}$.

4.3 Robust Stability Analysis

When the process uncertainties are time-invariant or time-varying but cycle-to-cycle invariant, together with repetitive load disturbance, the control objective for realizing perfect tracking of the set-point profile along both the time and batch directions is proposed as

$$\min J_{BP} = \sum_{t=0}^{N_1=N} \sum_{k=0}^{N_2\to\infty} \left(\|z(t,k)\|_2^2 - \|\omega(t,k)\|_2^2\right), \tag{4.19}$$

where $z(t,k)$ and $\omega(t,k)$ have been defined in (4.14) or (4.15).

For this purpose, the following Lyapunov–Krasovskii functions and inequalities of state energy is defined

$$V_P(\hat{x}_{i+1,j}) - V_{Q_t}(\hat{x}_{i,j}) - V_{Q_k}(\hat{x}_{i+1,j-1}) \leq \\ -\left(\|z(t+i,k+j)\|_2^2 - \|\omega(t+i,k+j)\|_2^2\right), \tag{4.20}$$

$$\frac{Q_t}{\rho_t} + \frac{Q_k}{\rho_k} \leq P, \tag{4.21}$$

where $\hat{x}_{i,j}$ is a shorthand of $\hat{x}(t+i, k+j)$ for brevity; $P > 0$, $Q_t > 0$, $Q_k > 0$, and $\rho_t \in (0, 1)$ and $\rho_k \in (0, 1)$ are two tuning parameters.

For the time direction in each cycle, it follows from (4.20) and (4.21) that

$$\rho_t^{-1} V_{Q_t}(\hat{x}_{1,j}) + \rho_k^{-1} V_{Q_k}(\hat{x}_{1,j}) \leq V_P(\hat{x}_{1,j}) < V_{Q_t}(\hat{x}_{0,j}) + V_{Q_k}(\hat{x}_{1,j-1}). \tag{4.22}$$

Summing up (4.22) from $j = -k + 1$ to $j = N_2 - k$ yields

$$\begin{aligned} &\rho_t^{-1} \sum_{k=1}^{N_2} V_{Q_t}(\hat{x}(t+1, k)) + \rho_k^{-1} \sum_{k=1}^{N_2} V_{Q_k}(\hat{x}(t+1, k)) \\ &\qquad < \sum_{k=1}^{N_2} V_{Q_t}(\hat{x}(t, k)) + \sum_{k=0}^{N_2-1} V_{Q_k}(\hat{x}(t+1, k)). \end{aligned} \tag{4.23}$$

With a resetting condition of $\hat{x}(t, 0) = \mathbf{0}$, it follows that

$$\begin{aligned} \rho_t^{-1} \sum_{k=1}^{N_2} V_{Q_t}(\hat{x}(t+1, k)) &< \sum_{k=1}^{N_2} V_{Q_t}(\hat{x}(t, k)) \\ &\quad - \left(\rho_k^{-1} - 1\right) \sum_{k=1}^{N_2-1} V_{Q_k}(\hat{x}(t+1, k)) - \rho_k^{-1} V_{Q_k}(\hat{x}(t+1, N_2)) \end{aligned} \tag{4.24}$$

and therefore,

$$\sum_{k=1}^{N_2} V_{Q_t}(\hat{x}(t+1, k)) < \rho_t \sum_{k=1}^{N_2} V_{Q_t}(\hat{x}(t, k)), \quad \forall t > 0. \tag{4.25}$$

For the batch direction from cycle to cycle, following a similar derivation as above, one obtains

$$\sum_{t=1}^{N_1} V_{Q_k}(\hat{x}(t, k+1)) < \rho_k \sum_{t=1}^{N_1} V_{Q_k}(\hat{x}(t, k)), \quad \forall k > 0. \tag{4.26}$$

Assume that $\rho = \rho_t = \max\{\rho_t, \rho_k\}$ without loss of generality, it can be seen from (4.21) that $Q_t \leq \rho P - Q_k$. Correspondingly, it follows from (4.20) that

$$V_P(\hat{x}(t+1, k+1)) < \rho V_P(\hat{x}(t, k+1)) - V_{Q_k}(\hat{x}(t, k+1)) + V_{Q_k}(\hat{x}(t+1, k)). \tag{4.27}$$

Therefore, the following inequalities hold for any positive integers T_0, K_0, and i,

$$V_P(\hat{x}(T_0+1, K_0+i)) < \rho V_P(\hat{x}(T_0, K_0+i)) - V_{Q_k}(\hat{x}(T_0, K_0+i)) + V_{Q_k}(\hat{x}(T_0+1, K_0+i-1)). \tag{4.28}$$

$$V_P(\hat{x}(T_0+2, K_0+i-1)) < \rho V_P(\hat{x}(T_0+1, K_0+i-1)) - V_{Q_k}(\hat{x}(T_0+1, K_0+i-1)) + V_{Q_k}(\hat{x}(T_0+2, K_0+i-2)), \tag{4.29}$$

$$\vdots$$

$$V_P(\hat{x}(T_0+i, K_0+1)) < \rho V_P(\hat{x}(T_0+i-1, K_0+1)) - V_{Q_k}(\hat{x}(T_0+i-1, K_0+1)) + V_{Q_k}(\hat{x}(T_0+i, K_0)). \tag{4.30}$$

Summing up the above inequalities yields

$$\begin{aligned}
&\sum_{\substack{t+k=T_0+K_0+i+1\\ T_0\le t\le I_0+i\\ K_0\le k\le K_0+i}} V_P(\hat{x}(t,k)) < \rho\left[-V_P(\hat{x}(T_0+i, K_0)) + \sum_{\substack{t+k=T_0+K_0+i\\ T_0\le t\le T_0+i\\ K_0\le k\le K_0+i}} V_P(\hat{x}(t,k))\right]\\
&- V_{Q_k}(\hat{x}(T_0, K_0+i)) + V_{Q_k}(\hat{x}(T_0+i, K_0)) <\\
&- V_{Q_t}(\hat{x}(T_0+i, K_0)) - V_{Q_k}(\hat{x}(T_0, K_0+i))\\
&+ \rho \sum_{\substack{t+k=T_0+K_0+i\\ T_0\le t\le T_0+i\\ K_0\le k\le K_0+i}} V_P(\hat{x}(t,k)) < \rho \sum_{\substack{t+k=T_0+K_0+i\\ T_0\le t\le T_0+i\\ K_0\le k\le K_0+i}} V_P(\hat{x}(t,k)).
\end{aligned} \tag{4.31}$$

According to Lemma 1.6, the closed-loop system described in (4.17) or (4.18) is guaranteed 2D stable with regard to (4.20) and (4.21), and has a 2D convergence index no larger than $\rho = \max\{\rho_t, \rho_k\}$.

Summing up (4.20) from $i = 0$ to $N_1 - t$ yields

$$\begin{aligned}
&V_P\left(\hat{x}_{N_1-t+1,j}\right) + \sum_{i=1}^{N_1-t} V_{P-Q_t}\left(\hat{x}_{i,j}\right) - V_{Q_t}\left(\hat{x}_{0,j}\right) - \sum_{i=1}^{N_1-t+1} V_{Q_k}\left(\hat{x}_{i,j-1}\right)\\
&\le -\sum_{i=0}^{N_1-t}\left(\|z(t+i, k+j)\|_2^2 - \|\omega(t+i, k+j)\|_2^2\right),
\end{aligned} \tag{4.32}$$

where $V_P\left(\hat{x}_{N_1-t+1,j+1}\right)$ goes to zero when $t \to N_1 + 1$.

Then, summing up (4.32) from $j = 0$ to $N_2 - k$ yields

$$\sum_{i=1}^{N_1-t} V_{P-Q_t}\left(\hat{x}_{i,N_2-k}\right) + \sum_{i=1}^{N_1-t}\sum_{j=0}^{N_2-1-k} V_{P-Q_t-Q_k}\left(\hat{x}_{i,j}\right) - \sum_{j=0}^{N_2-k} V_{Q_t}\left(\hat{x}_{0,j}\right) - \sum_{i=1}^{N_1-t-1} V_{Q_k}\left(\hat{x}_{i,-1}\right)$$
$$\leq -\sum_{t=1}^{N_1}\sum_{k=1}^{N_2}\left(\|z(t,k)\|_2^2 - \|\omega(t,k)\|_2^2\right), \tag{4.33}$$

where $V_{P-Q_t}\left(\hat{x}_{i,N_2-k}\right)$ goes to zero when $N_2 \to \infty$.

Substituting the objective function in (4.19) into (4.33) and using (4.25) and (4.26) yields

$$\begin{aligned} J_{\mathrm{BP}} &\leq \sum_{j=0}^{N_2-k} V_{Q_t}\left(\hat{x}_{0,j}\right) + \sum_{i=1}^{N_1-t-1} V_{Q_k}\left(\hat{x}_{i,-1}\right) - \sum_{i=1}^{N_1-t}\sum_{j=0}^{N_2-1-k} V_{P-Q_t-Q_k}\left(\hat{x}_{i,j}\right) \\ &< \frac{1-\rho_k^{N_2-k+1}}{1-\rho_k} V_{Q_t}\left(\hat{x}_{0,0}\right) + \frac{1-\rho_t^{N_1-t-1}}{1-\rho_t} V_{Q_k}\left(\hat{x}_{1,-1}\right) \\ &\quad - \frac{\left(1-\rho_k^{N_2-k}\right)\left(\rho_t-\rho_t^{N_1-t+1}\right)}{(1-\rho_k)(1-\rho_t)} V_{P-Q_t-Q_k}\left(\hat{x}_{0,0}\right), \end{aligned} \tag{4.34}$$

which gives an upper bound on the optimal performance objective.

By denoting $\sup(J_{\mathrm{BP}}) = \gamma_{\mathrm{opt}}$, the optimal control objective in (4.19) can be implemented by

$$\begin{aligned} &\text{Minimize } \gamma_{\mathrm{opt}} \\ &(\text{subject to (4.34)}, P > 0, Q_t > 0, \text{ and } Q_k > 0). \end{aligned} \tag{4.35}$$

By the Schur complement, the inequality in (4.34) can be transformed into

$$\begin{bmatrix} \gamma_{\mathrm{opt}} & \hat{x}(t,k)\left[\frac{1-\rho_k^{N_2-k+1}}{1-\rho_k}Q_t - \frac{(1-\rho_k^{N_2-k})(\rho_t-\rho_t^{N_1-t+1})}{(1-\rho_k)(1-\rho_t)}(P-Q_t-Q_k)\right] & \hat{x}(t,k)Q_k \\ * & \frac{1-\rho_k^{N_2-k+1}}{1-\rho_k}Q_t - \frac{(1-\rho_k^{N_2-k})(\rho_t-\rho_t^{N_1-t+1})}{(1-\rho_k)(1-\rho_t)}(P-Q_t-Q_k) & \mathbf{0} \\ * & * & \frac{1-\rho_t^{N_1-t-1}}{1-\rho_t}Q_k \end{bmatrix} \geq 0. \tag{4.36}$$

In the presence of time-varying process uncertainties from cycle to cycle, a robust control objective is proposed to minimize the maximum of (4.19) in the worst case, i.e.

$$\min_{\gamma_{\mathrm{opt}}} \max_{[A_j B_j] \in \mathrm{Co}} J_{\mathrm{BP}} = \sum_{t=0}^{N_1=N}\sum_{k=0}^{N_2\to\infty}\left(\|z(t,k)\|_2^2 - \|\omega(t,k)\|_2^2\right) \tag{4.37}$$

for the polyhedral uncertainties described by (4.14), or

$$\min_{\gamma_{\text{opt}}} \max_{\Delta\tilde{A},\Delta\tilde{B}} J_{\text{BP}} = \sum_{t=0}^{N_1=N} \sum_{k=0}^{N_2\to\infty} \left(\|z(t,k)\|_2^2 - \|\omega(t,k)\|_2^2\right) \tag{4.38}$$

for the norm-bounded uncertainties described by (4.15).

Correspondingly, the following theorem is given to derive a robust ILC updating law with state feedback under the polyhedral uncertainties.

Theorem 4.1 *The closed-loop 2D control system in (4.17) with the polyhedral uncertainties described by (4.14) is guaranteed robustly stable with a TCI no larger than $\rho_t \in (0,1)$ and a KCI no larger than $\rho_k \in (0,1)$, if there exist $\widehat{P}_j > 0$, $\widehat{Q}_{t,j} > 0$, $\widehat{Q}_{k,j} > 0$, nonsingular matrix R, matrices $\widehat{C}_{c1}$, $\widehat{C}_{c2}$ with approximate dimensions, such that the LMI conditions below hold true*

$$\begin{bmatrix} -\widehat{P}_j + D_g D_g^{\mathrm{T}} & A_{1,j}R + B_{g,j}\widehat{C}_{c1} & A_2 R + B_{g,j}\widehat{C}_{c2} & \mathbf{0} \\ * & -\widehat{Q}_{t,j} & \mathbf{0} & R^{\mathrm{T}}L^{\mathrm{T}} \\ * & * & -\widehat{Q}_{k,j} & \mathbf{0} \\ * & * & * & -\mathbf{I} \end{bmatrix} < 0, \tag{4.39}$$

$$\frac{\widehat{Q}_{t,j}}{\rho_t} + \frac{\widehat{Q}_{k,j}}{\rho_k} \le R + R^{\mathrm{T}} - \widehat{P}_j, \tag{4.40}$$

$j \in \{1, 2, \ldots, l\}$, by parameterizing $\widehat{C}_{ci} = C_{ci}R$, $i = 1, 2$.

Proof Substituting the 2D system in (4.17) along with the process described in (4.14) into (4.20) yields

$$\zeta^{\mathrm{T}} \Xi_1 \zeta < 0, \tag{4.41}$$

where $\zeta = \left[\hat{x}_{0,1}^{\mathrm{T}}, \hat{x}_{1,0}^{\mathrm{T}}, \omega^{\mathrm{T}}(t,k)\right]^{\mathrm{T}}$, and

$$\Xi_1 = \begin{bmatrix} A_{1,j}^{\mathrm{T}} + C_{c1}^{\mathrm{T}}\tilde{B}_{g,j}^{\mathrm{T}} \\ A_2^{\mathrm{T}} + C_{c2}^{\mathrm{T}}\tilde{B}_{g,j}^{\mathrm{T}} \\ D_g^{\mathrm{T}} \end{bmatrix} P\left[A_{1,j} + \tilde{B}_{g,j}C_{c1}\ A_2 + \tilde{B}_{g,j}C_{c2}\ D_g\right] - \begin{bmatrix} Q_t - L^{\mathrm{T}}L & \mathbf{0} & \mathbf{0} \\ * & Q_k & \mathbf{0} \\ * & * & \mathbf{I} \end{bmatrix}. \tag{4.42}$$

By the Schur complement, it can be seen that (4.41) is guaranteed by

$$\begin{bmatrix} -P^{-1} & A_{1,j}+B_{g,j}C_{c1} & A_2+B_{g,j}C_{c2} & D_g & \mathbf{0} \\ * & -Q_t & \mathbf{0} & \mathbf{0} & L^{\mathrm{T}} \\ * & * & -Q_k & \mathbf{0} & \mathbf{0} \\ * & * & * & -\mathbf{I} & \mathbf{0} \\ * & * & * & * & -\mathbf{I} \end{bmatrix} < 0. \tag{4.43}$$

A slack matrix R is introduced into (4.43) by pre- and post-multiplying both sides with $\hat{R}^{\mathrm{T}} = \mathrm{diag}\big(\mathbf{I}, R^{\mathrm{T}}, R^{\mathrm{T}}, \mathbf{I}, \mathbf{I}\big)$ and $\hat{R}$, thus obtaining

$$\begin{bmatrix} -\widehat{P} & A_{1,j}R+B_{g,j}\widehat{C}_{c1} & A_2R+B_{g,j}\widehat{C}_{c2} & D_g & \mathbf{0} \\ * & -\widehat{Q}_t & \mathbf{0} & \mathbf{0} & R^{\mathrm{T}}L^{\mathrm{T}} \\ * & * & -\widehat{Q}_k & \mathbf{0} & \mathbf{0} \\ * & * & * & -\mathbf{I} & \mathbf{0} \\ * & * & * & * & -\mathbf{I} \end{bmatrix} < 0, \tag{4.44}$$

where $\widehat{P} = P^{-1}$, $\widehat{Q}_t = R^{\mathrm{T}}Q_tR$, and $\hat{Q}_k = R^{\mathrm{T}}Q_kR$.

Since there exists $\left(R^{\mathrm{T}} - \widehat{P}^{-1}\right)\hat{P}\left(R - \widehat{P}^{-1}\right) \geq 0, \forall \widehat{P} > 0$, it follows that

$$R + R^{\mathrm{T}} - \widehat{P} \leq R^{\mathrm{T}}\widehat{P}^{-1}R. \tag{4.45}$$

By substituting (4.45) into (4.40) while temporarily leaving out the subscript j, it can be seen that the resulting inequality ensures (4.21) by using $\widehat{P} = P^{-1}$ and a pre- and post-multiplication of R^{T} and R. In fact, it is obvious that (4.44) is affine in $\left[A_{1,j}, B_{g,j}\right]$ and $\widehat{P}$, $\widehat{Q}_t$ and $\widehat{Q}_k$. Hence, both (4.40) and (4.44) (equivalent to (4.39) by the Schur complement) are satisfied for $[A_j^{[i]}, B_j^{[i]}] \in \mathrm{Co}\left\{[A_j^{[i]}, B_j^{[i]}], j = 1, 2, \ldots, l\right\}$. Robust stability is therefore guaranteed for the closed-loop 2D control system in (4.17). Note that the convergence indices of TCI and KCI follow from the derivation of (4.25) and (4.26). This completes the proof. □

Remark 4.1 By introducing a slack matrix R into (4.39) and (4.40) in Theorem 4.1, the Lyapunov matrices $\hat{P}_j (j \in \{1, 2, \ldots, l\})$ are not involved in any product with the state or controller matrices. This merit enables more flexibility in the design of robust performance and controller matrices, compared with the previously developed Lyapunov-function-based 2D control methods (Du et al. 2001; Galkowski et al. 2002; Rogers et al. 2007; Shi et al. 2005, 2006). ◊

For the norm-bounded uncertainties described by (4.15), the following theorem is given accordingly:

Theorem 4.2 *The closed-loop 2D control system in* (4.18) *with the norm-bounded uncertainties described by* (4.15) *is guaranteed robustly stable with a TCI no larger* *than $\rho_t \in (0, 1)$ and a KCI no larger than $\rho_k \in (0, 1)$, if there exist $\widehat{P} > 0$, $\widehat{Q}_t > 0$, $\widehat{Q}_k > 0$, nonsingular matrix R, matrices $\widehat{C}_{c1}$, $\widehat{C}_{c2}$, positive scalars, ε_1, ε_2 and ε_3, such that the following LMI conditions hold true*

$$\begin{bmatrix} \Gamma & A_1R + B_g\widehat{C}_{c1} & A_2R + B_g\widehat{C}_{c2} & \mathbf{0} & \mathbf{0} & \mathbf{0} & \mathbf{0} \\ * & -\widehat{Q}_t & \mathbf{0} & R^{\mathrm{T}}L^{\mathrm{T}} & R^{\mathrm{T}}F_{1,2}^{\mathrm{T}} & \widehat{C}_{c1}^{\mathrm{T}}F_{2,2}^{\mathrm{T}} & \mathbf{0} \\ * & * & -\widehat{Q}_k & \mathbf{0} & \mathbf{0} & \mathbf{0} & \widehat{C}_{c2}^{\mathrm{T}}F_{2,2}^{\mathrm{T}} \\ * & * & * & -\mathbf{I} & \mathbf{0} & \mathbf{0} & \mathbf{0} \\ * & * & * & * & -\varepsilon_1\mathbf{I} & \mathbf{0} & \mathbf{0} \\ * & * & * & * & * & -\varepsilon_2\mathbf{I} & \mathbf{0} \\ * & * & * & * & * & * & -\varepsilon_3\mathbf{I} \end{bmatrix} < 0, \tag{4.46}$$

$$\frac{\widehat{Q}_t}{\rho_t} + \frac{\widehat{Q}_k}{\rho_k} \leq R + R^{\mathrm{T}} - \widehat{P}, \tag{4.47}$$

$\Gamma = -\widehat{P} + D_gD_g^{\mathrm{T}} + \varepsilon_1F_{11}F_{11}^{\mathrm{T}} + \varepsilon_2F_{21}F_{21}^{\mathrm{T}} + \varepsilon_3F_{21}F_{21}^{\mathrm{T}}$, by parameterizing $\widehat{C}_{ci} = C_{ci}R$, $i = 1, 2$.

Proof The asymptotic stability of the nominal 2D system (i.e. $\Delta A_1 = \Delta B_g = \mathbf{0}$) can be proved similar to that for Theorem 4.1, by replacing $A_{1,j}$ and $B_{g,j}$ with A_1 and B_g.

Given ΔA_1 shown in (4.15), it follows from Lemma 2.1 that

$$\begin{bmatrix} F_{1,1} \\ \mathbf{0} \\ \mathbf{0} \\ \mathbf{0} \end{bmatrix} \Theta_1 \begin{bmatrix} \mathbf{0} \\ F_{1,2}^{\mathrm{T}} \\ \mathbf{0} \\ \mathbf{0} \end{bmatrix}^{\mathrm{T}} + \begin{bmatrix} \mathbf{0} \\ F_{1,2}^{\mathrm{T}} \\ \mathbf{0} \\ \mathbf{0} \end{bmatrix} \Theta_1^{\mathrm{T}} \begin{bmatrix} F_{1,1} \\ \mathbf{0} \\ \mathbf{0} \\ \mathbf{0} \end{bmatrix}^{\mathrm{T}} \leq \varepsilon_1 \begin{bmatrix} F_{1,1} \\ \mathbf{0} \\ \mathbf{0} \\ \mathbf{0} \end{bmatrix} \begin{bmatrix} F_{1,1} \\ \mathbf{0} \\ \mathbf{0} \\ \mathbf{0} \end{bmatrix}^{\mathrm{T}} + \varepsilon_1^{-1} \begin{bmatrix} \mathbf{0} \\ F_{1,2}^{\mathrm{T}} \\ \mathbf{0} \\ \mathbf{0} \end{bmatrix} \begin{bmatrix} \mathbf{0} \\ F_{1,2}^{\mathrm{T}} \\ \mathbf{0} \\ \mathbf{0} \end{bmatrix}^{\mathrm{T}}. \tag{4.48}$$

Substituting (4.48) into (4.43) where $A_{1,j}$ should be replaced by $A_1 + \Delta A_1$, together with a similar substitution of $B_g + \Delta B_g$ for $B_{g,j}$, and using the Schur complement, one can obtain (4.46) in Theorem 4.2. The remaining proof is similar to that for Theorem 4.1 and thus is omitted. □

Given the state feedback shown in Fig. 4.1, the process input constraints can be determined by substituting the 2D ILC updating law in (4.16) into (4.5) as

$$\begin{aligned}&-\bar{u}^{[i]} \le u^{[i]}(t,k) + C_{c1}^{[i]}\hat{x}^{[i]}(t,k) + C_{c2}^{[i]}\hat{x}^{[i]}(t+1,k-1)\\&\quad - u_{eq}^{[i]} \le \bar{u}^{[i]}, \quad i = 1,2,\ldots,m.\end{aligned} \tag{4.49}$$

For the polyhedral uncertainties described in (4.14), the state constraints can be determined by substituting (4.17) with $\omega^{[i]}(t,k) = 0$ into (4.6) as

$$\begin{aligned}&-\overline{\chi}^{[i]} \le \psi\left\{x^{[i]}(t+1,k-1) + \vec{\mathbf{I}}\left[\left(A_{1,j}^{[i]} + B_{g,j}^{[i]}C_{c1}^{[i]}\right)\hat{x}^{[i]}(t,k)\right.\right.\\&\left.\left.+\left(A_2^{[i]} + B_{g,j}^{[i]}C_{c2}^{[i]}\right)\hat{x}^{[i]}(t+1,k-1)\right] - x_{eq}^{[i]}\right\} \le \overline{\chi}^{[i]},\end{aligned} \tag{4.50}$$

where $\vec{\mathbf{I}} = \left[\mathbf{I}_{n_x\times n_x}, \mathbf{0}\right]$, $i = 1,2,\ldots,m.$, $j \in \{1,2,\cdots,l\}$.

For the norm-bounded uncertainties in (4.15), it is practical to let $\left\|\overline{\chi}_{\max}^{[i]}\right\|_2 = \left\|\left(\psi^{\mathrm{T}}\psi\right)^{-1}\psi^{\mathrm{T}}\overline{\chi}^{[i]} + x_{eq}^{[i]} - x^{[i]}(t+1,k-1)\right\|_2$ to determine the state constraints for implementation. In case $\psi^{\mathrm{T}}\psi$ becomes singular, one may specify $\psi = \mathbf{I}$ in (4.6) and let $\left\|\overline{\chi}_{\max}^{[i]}\right\|_2 = \left\|\overline{\chi}^{[i]} + x_{eq}^{[i]} - x^{[i]}(t+1,k-1)\right\|_2$. It follows that

$$\left\|\hat{x}^{[i]}(t+1,k)\right\|_2 \le \left\|\overline{\chi}_{\max}^{[i]}\right\|_2. \tag{4.51}$$

Note that according to Lemma 2.1, there exists

$$\begin{aligned}&(\hat{A} + [\Delta\hat{A}_1, \Delta\hat{A}_2])^{\mathrm{T}}(\hat{A} + [\Delta\hat{A}_1, \Delta\hat{A}_2])\\&\le \hat{A}^{\mathrm{T}}\left(\mathbf{I} - \delta_1 F_{1,1}^{[i]}\left(F_{1,1}^{[i]}\right)^{\mathrm{T}} - \delta_2 F_{2,1}^{[i]}\left(F_{2,1}^{[i]}\right)^{\mathrm{T}}\right)^{-1}\hat{A}\\&+\delta_1^{-1}\left[F_{1,2}^{[i]},\mathbf{0}\right]^{\mathrm{T}}\left[F_{1,2}^{[i]},\mathbf{0}\right] + \delta_2^{-1}\left[C_{c1}^{[i]}, C_{c2}^{[i]}\right]^{\mathrm{T}}\left(F_{2,2}^{[i]}\right)^{\mathrm{T}}F_{2,2}^{[i]}\left[C_{c1}^{[i]}, C_{c2}^{[i]}\right],\end{aligned} \tag{4.52}$$

where $\hat{A} = \left[\hat{A}_1, \hat{A}_2\right]$, δ_1 and δ_2 may be practically specified from $\mathbf{I} - \delta_1\left(F_{1,1}^{[i]}\right)^{\mathrm{T}}F_{1,1}^{[i]} > 0$ and $\mathbf{I} - \delta_2\left(F_{2,1}^{[i]}\right)^{\mathrm{T}}F_{2,1}^{[i]} > 0$.

Substituting (4.18) with $\omega^{[i]}(t,k) = 0$ and (4.52) into (4.51), the state constraints below are obtained for evaluation

$$\begin{aligned}&\left\|\left\{\hat{A}^{\mathrm{T}}\left(\mathbf{I} - \delta_1 F_{1,1}^{[i]}\left(F_{1,1}^{[i]}\right)^{\mathrm{T}} - \delta_2 F_{2,1}^{[i]}\left(F_{2,1}^{[i]}\right)^{\mathrm{T}}\right)^{-1}\hat{A}\right.\right.\\&\left.\left.+\delta_1^{-1}\left[F_{1,2}^{[i]},\mathbf{0}\right]^{\mathrm{T}}\left[F_{1,2}^{[i]},\mathbf{0}\right] + \delta_2^{-1}\left[C_{c1}^{[i]}, C_{c2}^{[i]}\right]^{\mathrm{T}}\left(F_{2,2}^{[i]}\right)^{\mathrm{T}}F_{2,2}^{[i]}\left[C_{c1}^{[i]}, C_{c2}^{[i]}\right]\right\}\begin{bmatrix}\hat{\chi}^{[i]}(t,k)\\ \hat{\chi}^{[i]}(t+1,k-1)\end{bmatrix}\right\|_2\\&\le \left\|\overline{\chi}_{\max}^{[i]}\right\|_2.\end{aligned} \tag{4.53}$$

With the above quantitative input constraints, off- and on-line control algorithms can be programmed for implementation. For instance, in the case of time-invariant process uncertainties and repetitive load disturbance, an on-line optimization algorithm with state feedback may be programmed in terms of

$$\underset{\rho_t\in(0,1),\rho_k\in(0,1)}{\text{Minimize}}\ \gamma_{\text{opt}}. \tag{4.54}$$

(subject to (4.36), (4.39), (4.40), (4.49) and (4.50) for the polyhedral uncertainties, or alternatively, (4.36), (4.46), (4.47), (4.49) and (4.53) for the norm-bounded uncertainties).

Generally, a smaller value of γ_{rob} corresponds to a larger controller gain and vice versa. Note that the convergence indices of TCI and KCI are user specified in practice, which may be prescribed to determine the cycling periods for all the piecewise affine operating regions and the on- or off-line control algorithms, in consideration of the process input constraints.

4.4 An Illustrative Example

Consider an industrial continuous stirred tank reactor (CSTR) studied by Wan and Kothare (2003)

$$\dot{C}_{\text{A}} = \frac{q}{V}(C_{\text{Af}} - C_{\text{A}}) - k_0 \exp\left(-\frac{E}{RT}\right)C_{\text{A}}, \tag{4.55}$$

$$\dot{T} = \frac{q}{V}(T_{\text{f}} - T) - \frac{\Delta H}{\rho C_{\text{p}}} k_0 \exp\left(-\frac{E}{RT}\right)C_{\text{A}} + \frac{UA}{\rho V C_{\text{p}}}(T_{\text{c}} - T), \tag{4.56}$$

where C_{A} is the concentration of A for an exothermic irreversible reaction ($A \rightarrow B$), T is the temperature of the reactor, T_{C} is the temperature of the coolant stream, which is taken as the manipulated variable, $q = 100(L/\text{min})$, $V = 100(\text{L})$, $C_{\text{Af}} = 1(\text{mol/L})$, $T_{\text{f}} = 400(\text{K})$, $\rho = 1000(\text{g/L})$, $C_{\text{p}} = 1(\text{J/gK})$, $k_0 = 4.71 \times 10^8(\text{min}^{-1})$, $E/R = 8000(\text{K})$, $\Delta H = -2 \times 10^5(\text{J/mol})$, and $UA = 1 \times 10^5(\text{J/minK})$. The operating constraints are $200 \leq T_{\text{c}} \leq 450(\text{K})$, $0.01 \leq C_{\text{A}} \leq 1(\text{mol/L})$, and $250 \leq T \leq 500(\text{K})$. More details about the process dynamics and chemical properties can be found in the referred reference and thus are omitted for brevity.

Figure 4.2 shows the static equilibrium curves of this CSTR. Under the desired operating condition, $T_{\text{c}} = 302(\text{K})$, the reactor has three equilibrium points as shown in Fig. 4.2. The operating objective is to transfer the system from the two undesired equilibrium points of $\left(T^{\text{eq}}, C_{\text{A}}^{\text{eq}}\right) = (361.141, 0.8986)$ (denoted by 1#) and $(432.083, 0.1892)$ (denoted by 6#) to the desired equilibrium point, $(398.972, 0.52)$,

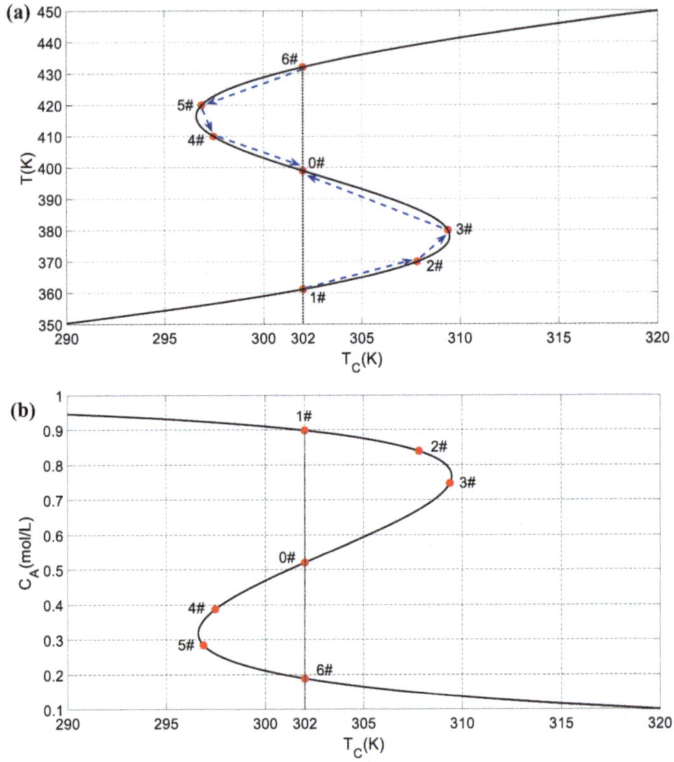

Fig. 4.2 Equilibrium curve of CSTR with a linear operating division

denoted by 0#, when the system response is moved to either of these two undesired equilibrium points in the cycling operation.

To realize the above batch control objective, six equilibrium points are chosen for a piecewise affine control design for the cycling operation, 1# to 3# for the transition from 1# point to 0# point, and 4# to 6# for the transition from 6# to 0#, as shown in Fig. 4.2. Correspondingly, six piecewise affine operating regions are divided for the control design. For illustration, the above state feedback control scheme is adopted for the transition from 1# to 0#. The sampling time is taken as 0.02 (min), and the first-order Euler's approximation is used to discretize the continuous-time system for simulation tests.

For the state feedback control design, linearized state-space models of these operating regions are simply obtained by using the Jacobian matrix of the original system with respect to the equilibrium points of 1# to 3#, as listed in Table 4.1. It can be verified that these linear models are indeed very rough to represent the dynamic response characteristics in these regions, and therefore, may be viewed as the 'worst-case' for the control design. The corresponding 2D ILC controller can be derived using the proposed control algorithms, such as an off-line type for simplicity. By specifying the convergence indices for the time and batch directions as $\rho_t = 0.99$ and $\rho_k = 0.98$, and taking $\gamma_{\mathrm{rob}} = 50$ to comply with the operating constraints, the controllers are derived as listed in Table 4.2.

The set-point profiles for the two transition tasks are specified in terms of the measurable reactor temperature as

$$T_{\mathrm{d},1} = \begin{cases} 361.141 + t \times (370 - 361.141)/2, \\ \quad 0 \le t \le 2(\mathrm{min}); \\ 370 + (t-2) \times (380 - 370)/3, \\ \quad 2 < t \le 5(\mathrm{min}); \\ 380 + (t-5) \times (398.972 - 380)/3, \\ \quad 5 < t \le 8(\mathrm{min}); \\ 398.972, \\ \quad 8 < t \le N = 10(\mathrm{min}). \end{cases}$$

Table 4.1 Piecewise linear models for a CSTR

$(T^{\mathrm{eq}}, C_{\mathrm{A}}^{\mathrm{eq}}, T_{\mathrm{c}}^{\mathrm{eq}})$	A	B	C
$1\#: (361.141, 0.8986, 302)$	$\begin{bmatrix} 0.9849 & 0.4514 \\ -0.0001 & 0.9777 \end{bmatrix}$	$\begin{bmatrix} 0.02 \\ 0 \end{bmatrix}$	$\begin{bmatrix} 1 & 0 \end{bmatrix}$
$2\#: (370, 0.8391, 307.8121)$	$\begin{bmatrix} 0.9976 & 0.7672 \\ -0.0002 & 0.9762 \end{bmatrix}$	$\begin{bmatrix} 0.02 \\ 0 \end{bmatrix}$	$\begin{bmatrix} 1 & 0 \end{bmatrix}$
$3\#: (380, 0.7469, 309.3844)$	$\begin{bmatrix} 1.0161 & 1.3553 \\ -0.0003 & 0.9732 \end{bmatrix}$	$\begin{bmatrix} 0.02 \\ 0 \end{bmatrix}$	$\begin{bmatrix} 1 & 0 \end{bmatrix}$
$4\#: (410, 0.3874, 297.4838)$	$\begin{bmatrix} 1.0433 & 0.0184 \\ -0.0299 & 0.9834 \end{bmatrix}$	$\begin{bmatrix} -0.1587 \\ -0.0748 \end{bmatrix}$	$\begin{bmatrix} -0.1551 & 0.0471 \end{bmatrix}$
$5\#: (420, 0.2844, 296.8796)$	$\begin{bmatrix} 1.035 & -0.0437 \\ 0.0175 & 0.984 \end{bmatrix}$	$\begin{bmatrix} -0.0643 \\ 0.0475 \end{bmatrix}$	$\begin{bmatrix} -0.3747 & -0.09 \end{bmatrix}$
$6\#: (432.083, 0.1892, 302)$	$\begin{bmatrix} 1.0176 & 0.0298 \\ -0.0625 & 0.9753 \end{bmatrix}$	$\begin{bmatrix} -0.1599 \\ -0.1453 \end{bmatrix}$	$\begin{bmatrix} -0.1562 & 0.0637 \end{bmatrix}$

Table 4.2 State feedback 2D ILC controller matrices

	C_{c1}	C_{c2}
1#	[−49.1277, −22.5835, 3.592, −4.1909]	[−0.0303, 0.0003, −54.8823, −0.0087]
2#	[−49.8934, −38.3497, −5.1705, −2.9933]	[0.1238, 0.0012, −49.5643, 0.0536]
3#	[−50.7606, −67.7705, −11.2626, −4.3211]	[0.1941, 0.0016, −54.0561, 0.011]

$$T_{\mathrm{d},2} = \begin{cases} 432.083 + t \times (420 - 432.083)/3, \\ \quad 0 \le t \le 3(\mathrm{min}); \\ 420 + (t-3) \times (410 - 420)/2, \\ \quad 3 < t \le 5(\mathrm{min}); \\ 410 + (t-5) \times (398.972 - 410)/2, \\ \quad 5 < t \le 7(\mathrm{min}); \\ 398.972, \\ \quad 7 < t \le N = 9(\mathrm{min}). \end{cases}$$

For illustration, the following two cases are tested.

Case 1: (Time-invariant uncertainties and repetitive load disturbance). Assume that the reactor dynamics are invariant from cycle to cycle, so that the model mismatches may be viewed as time-invariant uncertainties. A repetitive load disturbance such as periodically feeding the raw materials to the reactor is introduced at $t = 2\,(\mathrm{min})$ in the first operating region, which causes T to drop down with a cooling speed of $-5(\mathrm{K/min})$ to the maximal deviation of 2(K) together with the corresponding variation of C_A by -5% (or 10%) with regard to the initial equilibrium point of 1# (or 6#), and then the disturbance is removed at $t = 5\,(\mathrm{min})$ in a reverse manner to test the control stability. The results are shown in Fig. 4.3. Note that the tracking errors for T plotted in Fig. 4.3d are computed in terms of the root-sum-squared-error (RSSE) criterion. It is seen that perfect tracking is achieved through almost 5 cycles by either of the above ILC scheme. Moreover, there is no steady-state output tracking error from the first cycle.

Case 2: (Time-varying uncertainties and nonrepetitive load disturbance). Assume that besides the above time-invariant uncertainties and repetitive load disturbance, time-varying process uncertainties and nonrepetitive load disturbance further cause T to randomly vary with a magnitude of 1.5(K) in each cycle, together with a random variation of C_A by 2% (or 5%) with regard to the initial equilibrium point of 1# (or 6#). The RSSE plots for 100 cycles are shown in Fig. 4.4. It is seen that the proposed control scheme maintains good robust stability along both the time and batch directions. Note that when the fluctuation magnitude of T is reduced to 1 (K), as illustrated from the 101st to the 180th cycle in Fig. 4.4, the tracking error is correspondingly reduced. If the influence to T and C_A then becomes repetitive from cycle to cycle, which is illustrated from the 181st to the 200th cycle in Fig. 4.4, it is seen that the corresponding RSSE goes to zero, indicating that perfect tracking is reached thereafter. It is thus demonstrated that the proposed state feedback based method is capable of robust tracking and online optimization under cycle-to-cycle time-invariant or varying uncertainties and repetitive or nonrepetitive load disturbances.

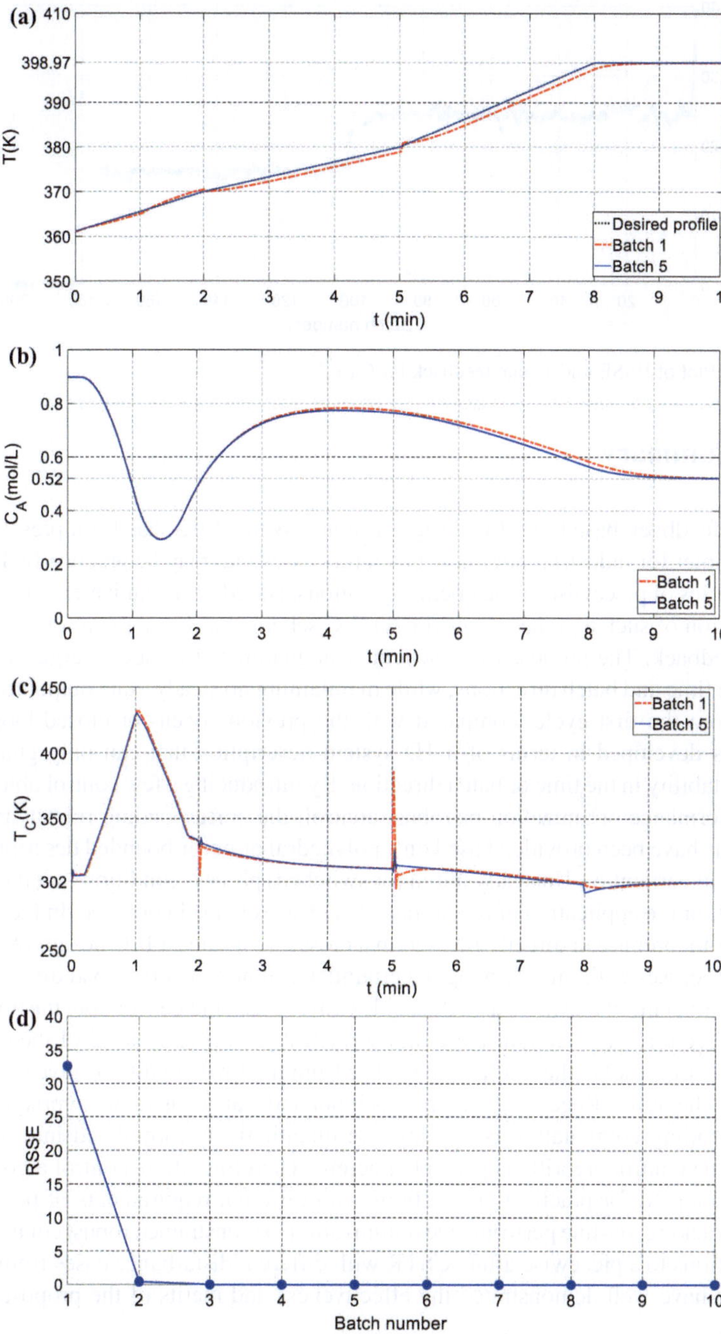

Fig. 4.3 Tracking performance under state feedback for Case 1

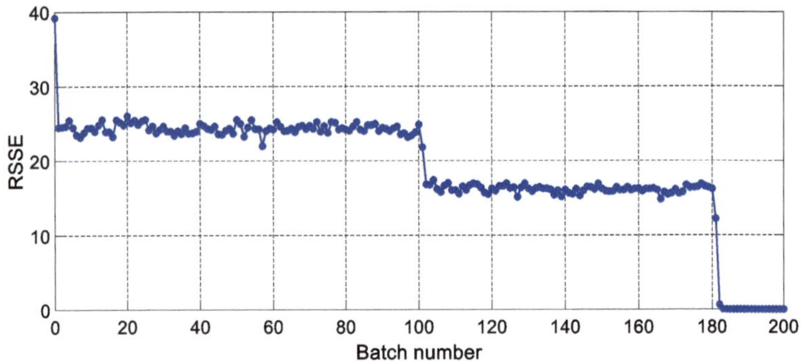

Fig. 4.4 Plot of RSSE under state feedback for Case 2

4.5 Summary

A state feedback based 2D ILC design (Liu and Wang 2012) has been presented in this Chapter for industrial nonlinear batch processes that may be practically divided into a series of piecewise affine operating regions. Based on an equivalent 2D system description of such a process, a novel 2D ILC scheme has been given for the use of state feedback. The proposed ILC scheme can guarantee robust convergence along both the time and batch directions, while maintaining no steady-state output tracking error from the first cycle, compared with the previous open- or closed-loop ILC methods developed in terms of a 1D system description that can only guarantee robust stability in the time or batch direction. By introducing a few control objectives for performance optimization or robust control, the corresponding robust stability theorems have been provided based on a polyhedral or norm-bounded description of process uncertainties. Unsteady initial (or switched) process conditions are therefore allowed for the application of the proposed 2D ILC scheme in practice. In fact, these stability theorems can directly be used in a closed-loop robust ILC design for linear batch processes with time-varying uncertainties and nonrepetitive load disturbance. It is a distinct merit that there are adjustable convergence indices for both the time and batch directions, and an adjustable robust control performance level of the closed-loop system, to solve the 2D ILC controller from the LMI conditions given in these stability theorems. Regarding the process input and state constraints in practice, the corresponding computation constraints are quantitatively formulated into the on- or off-line control algorithms presented herein. Therefore, these control algorithms can be adopted for practical applications with different requirements of perfect or robust tracking, on-line performance optimization, or constrained robust control. The applications to a piecewise affine CSTR with different disturbance cases from cycle to cycle have well demonstrated the effectiveness and merits of the proposed ILC method.

References

Ahn H-S, Chen YQ, Moore KL (2007) Iterative learning control: brief survey and categorization. IEEE Trans Syst, Man, Cybern Part C Appl Rev 37:1099–1121

Bonvin D, Srinivasan B, Hunkeler D (2006) Control and optimization of batch processes. IEEE Trans Control Syst Mag 26(1):34–45

Busch J, Cruse A, Marquardt W (2007) Run-to-run control of membrane filtration processes. AIChE J 53(9):2316–2326

Chin I, Qin SJ, Lee KS, Cho M (2004) A two-stage iterative learning control technique combined with real-time feedback for independent disturbance rejection. Automatica 40(11):1913–1922

Du C, Xie L, Zhang C (2001) Control and robust stabilization of two-dimensional systems in Roesser models. Automatica 37(2):205–211

Forni F, Galeani S (2010) Gain-scheduled, model-based anti-windup for LPV systems. Automatica 46:222–225

Galkowski K, Rogers E, Xu S, Lam J, Owens DH (2002) LMIs-a fundamental tool in analysis and controller design for discrete linear repetitive processes. IEEE Trans Circuits Syst I: Fundam Theory Appl 49(6):768–778

Hermanto MW, Braatz RD, Chiu M-S (2011) Integrated batch-to-batch and nonlinear model predictive control for polymorphic transformation in pharmaceutical crystallization. AIChE J 57(4):1008–1019

Li X, Chen Y, Sun H-J, Liu W (2024) Adaptive iterative learning control for high-order nonlinear systems with different types of uncertainties. Int J Robust Nonlinear Control 34(8):5399–5418

Li X, Huang D (2024) Adaptive iterative learning control for nonsquare nonlinear systems with various nonrepetitive uncertainties: a unified approach. IEEE Trans Autom Control 69(3):1736–1743

Liu T, Gao F (2010) Robust two-dimensional iterative learning control for batch processes with state delay and time-varying uncertainties. Chem Eng Sci 65(23):6134–6144

Liu T, Gao F (2012) Industrial process identification and control design: step-test and relay-experiment-based methods. Springer, London

Liu T, Wang Y (2012) A synthetic approach for robust constrained iterative learning control of piecewise affine batch processes. Automatica 48(11):2762–2775

Ljung L (1999) System identification: theory for the user, 2nd edn. Prentice Hall, Englewood Cliff, NJ

Lu Y, Arkun Y (2002) A scheduling quasi-min-max model predictive control for nonlinear systems. J Process Control 12:589–604

Nagy ZK, Chew JW, Fujiwara M, Braatz RD (2008) Comparative performance of concentration and temperature controlled batch crystallizations. J Process Control 18:399–407

Nagy ZK, Mahn B, Franke R, Allgöwer F (2007) Efficient output feedback nonlinear model predictive control for temperature control of industrial batch reactors. Control Eng Pract 15:839–850

Özkan L, Kothare MV, Georgakis C (2003) Control of a solution copolymerization reactor using multi-model predictive control. Chem Eng Sci 58:1207–1221

Pakshin P, Mandra S, Emelianova J, Rogers E, Erwinski K, Galkowski K (2024) Experimentally validated vector Lyapunov function-based iterative learning control design under input saturation. IEEE Trans Control Syst Technol 32(1):189–201

Park P, Jeong SC (2004) Constrained RHC for LPV systems with bounded rates of parameter variations. Automatica 40:865–872

Qu Z, Xu J-X (2002) Asymptotic learning control for a class of cascaded nonlinear uncertain systems. IEEE Trans Autom Control 47(8):1369–1376

Rogers E, Galkowski K, Owens DH (2007) Control systems theory and applications for linear repetitive processes. Springer, Berlin

Rugh WJ, Shamma JS (2000) Research on gain scheduling. Automatica 36:1401–1425

Shi J, Gao F, Wu T-J (2005) Robust design of integrated feedback and iterative learning control of a batch process based on a 2D Roesser system. J Process Control 15(8):907–924
Shi J, Gao F, Wu T-J (2006) Robust iterative learning control design for batch processes with uncertain perturbations and initialization. AIChE J 52(6):2171–2187
Spiegel I, Strijbosch N, de Rozario R, Oomen T, Barton K (2024) Stable inversion of piecewise affine systems with application to feedforward and iterative learning control. IEEE Trans Autom Control 69(10):6836–6851
Tayebi A, Chien C-J (2007) A unified adaptive iterative learning control framework for uncertain nonlinear systems. IEEE Trans Autom Control 52(10):1907–1913
Wang Y, Gao F, Doyle F (2009) Survey on iterative learning control, repetitive control, and run-to-run control. J Process Control 19(10):1589–1600
Wan Z, Kothare MV (2003) Efficient scheduled stabilizing model predictive control for constrained nonlinear systems. Int J Robust Nonlinear Control 13:331–346
Wan Z, Kothare MV (2008) A framework for design of scheduled output feedback model predictive control. J Process Control 18:391–398
Zhang J (2008) Batch-to-batch optimal control of a batch polymerisation process based on stacked neural network models. Chem Eng Sci 63(5):1273–1281

Chapter 5
Closed-Loop ILC Scheme with Output Feedback

5.1 Introduction

There usually exist time- and/or batch-varying factors in industrial batch processes arising from plant uncertainties and load disturbances. The developed ILC designs for completely repetitive processes cannot achieve satisfactory performance in the presence of these uncertainties or disturbance, even fail to guarantee robust stability of such a closed-loop control system for batch run. In the literature, convergence analysis for the typical open-loop feedforward ILC designs subject to nonrepetitive model uncertainties, external disturbances and initial conditions was studied in Meng and Moore (2017a, b) based on a so-called double-dynamic analysis approach. However, only error convergence along the batch direction was addressed, while the dynamics in the time domain was overlooked. Based on the lifting technique, an $\mathcal{H}_\infty$-norm condition was established in Meng (2019) to guarantee the asymptotic convergence of ILC along the batch direction under nonrepetitive plant uncertainties, such that the tracking performance could be gradually improved from batch to batch. Nevertheless, the solvability of the resulting LMI condition in Meng (2019) depends on the pole location of plant model. In contrast, the real-time feedback control strategy was incorporated into the ILC design to enhance system performance along the time direction in each batch. For example, Shi et al. (2005, 2006) proposed the state feedback (SF) and two-dimensional (2D) dynamic output feedback (DOF) based ILC methods, respectively, based on the 2D system theory. For the presence of nonrepetitive uncertainties and/or disturbances, robust 2D $\mathcal{H}_\infty$ based ILC designs were presented in Liu and Wang (2012), Hao et al. (2019), by assuming that the lumped "2D disturbance" consisting of nonrepetitive model uncertainties and external disturbances in two sequential batches belongs to $\mathcal{L}_2$ space. By combining the ILC design with an observer-based disturbance compensation, a robust SF-based ILC control design was presented in Mandra and Aschemann (2017) such that the initial tracking errors in the early learning stage could be apparently reduced. Lu et al. (2015) presented a two-stage design of 2D model predictive ILC based on

T. Liu et al., *Robust Iterative Learning Control of Industrial Batch Systems*, Intelligent Control and Learning Systems 22, https://doi.org/10.1007/978-981-96-9778-6_5

SF for disturbance attenuation. Subsequently, an indirect-type ILC design for batch processes with time- and batch-varying uncertainties was developed in Hao et al. (2020) by using a generalized extended state observer, such that the system tracking and disturbance rejection performance could be evidently improved. Besides, if the batch-varying dynamics could be modeled, internal model based ILC methods were proposed to deliver perfect or robust tracking (Yin et al. 2010; Zhu et al. 2015).

Note that frequency domain specifications are widely adopted to assess system performance in engineering practice, but they are difficult to be converted into tractable stability conditions for controller synthesis. The Kalman-Yakubovich-Popov (KYP) lemma (Rantzer 1996) established the relationship between frequency domain constraints and tractable stability conditions of LMI over the entire frequency range. However, the frequency domain specifications are usually related to a finite-frequency range in practical applications. For example, noise suppression is typically required in the high frequency range, while the batch-to-batch tracking error convergence is usually related to the low frequency range. Iwasaki and Hara (2005) extended the KYP lemma to a more generalized case, which facilitates the finite-frequency-range control synthesis. Based on the generalized KYP lemma, numerous results have been reported for control design with finite-frequency-range specifications. To enumerate a few, static output feedback (SOF) control designs with restricted frequency-domain specifications were explored for linear systems without uncertainties (Li and Gao 2014) or with polytopic-type uncertainties (Li and Gao 2015), and then were extended to the finite-frequency-range SOF design for time-delay systems in the absence of uncertainties (Hao and Duan 2015). In the ILC field, finite frequency specifications were introduced to the robust ILC designs for delay-free systems (Paszke et al. 2013) and state-delay systems (Tao et al. 2019). However, state measurement is required for these ILC designs, which may not be available in practice. To circumvent this deficiency, an output-feedback-based ILC method was proposed in Hladowski et al. (2012) without considering the finite frequency specifications, and the corresponding convergence conditions were established in terms of LMI and additional equality constraints.

In this Chapter, a closed-loop ILC scheme with DOF is presented for industrial batch processes subject to time-varying uncertainties and nonrepetitive disturbances. Based on a 2D system description for the batch operation, an LMI-based sufficient condition is derived to guarantee the robust convergence of the resulting ILC system along both the time and batch directions, which could be effectively solved to determine the learning controller gains. Concerning the finite-frequency nature of load disturbance in engineering practice, another SOF-based ILC design (Hao et al. 2021) with a finite-frequency-range 2D $\mathcal{H}_\infty$ performance specification is also presented for the convenience of implementation. Based on a 2D system description, sufficient conditions in terms of nonlinear matrix inequalities (NLMIs) are established to guarantee robust stability of the resulting ILC system along both the time and batch directions, by using the matrix dilatation technique. A two-stage heuristic approach is subsequently developed to derive the feasible ILC controller gains, based on predesigning the corresponding SF-based ILC gains for iterative computation. An illustrative example of the injection molding process is adopted to validate the effectiveness and advantages of the proposed ILC designs.

5.2 Dynamic Output Feedback-Based ILC Design

5.2.1 Dynamic Output Feedback Control Scheme

Consider a batch process with time-varying uncertainties subject to external disturbance described by

$$\begin{cases} x(t+1,k) = [A + \Delta A(t,k)]x(t,k) + [B + \Delta B(t,k)]u(t,k) + w(t,k), \\ y(t,k) = Cx(t,k), \quad 0 \le t \le N, \\ x(0,k) = x(0), \quad k = 0, 1, \ldots \end{cases} \tag{5.1}$$

where $x(t,k) \in \Re^{n_x}$, $y(t,k) \in \Re^{n_y}$, $u(t,k) \in \Re^{n_u}$, t and k denote the time and batch indices, respectively; $w(t,k) \in \Re^{n_x}$ denotes unmodeled process dynamics and load disturbance; N is the total number of batch time steps, and $x(0)$ denotes the initial resetting condition of each batch that may be reset to zero; $\Delta A(t,k)$ and $\Delta B(t,k)$ are bounded uncertainties specified as

$$\Delta A(t,k) = F\Delta(t,k)H_{\mathrm{A}}, \; \Delta B(t,k) = F\Delta(t,k)H_{\mathrm{B}}, \tag{5.2}$$

where F, H_{A}, and H_{B} are known matrices related to the estimated upper bounds of uncertainties in practice, and $\Delta(t,k)$ is a time-varying matrix satisfying $\Delta^{\mathrm{T}}(t,k)\Delta(t,k) \le \mathbf{I}$.

For a batch process as described above, when an ILC strategy is applied, the updating control law has a general form of

$$\begin{cases} u(t,k) = u(t,k-1) + \delta u(t,k), \\ u(0,0) = u(0), 0 \le t \le N, \quad k = 0, 1, \ldots \end{cases} \tag{5.3}$$

where $\delta u(t,k)$ is the control output to be determined in the current batch (k), and $u(0)$ denotes the initial value of iteration that is generally reset to zero for implementation.

To facilitate a closed-loop ILC design that allows for time-varying process uncertainties from batch to batch and nonrepetitive load disturbance, the output tracking error in the current batch is defined by

$$e(t,k) \triangleq y(t,k) - y_{\mathrm{d}}(t), \tag{5.4}$$

where $y_{\mathrm{d}}(t)$ denotes the set-point profile (or the desired output trajectory).

It follows from (5.1) and (5.3) along with the definition in (5.4) that

$$e(t,k) = e(t,k-1) + C\delta x(t,k), \tag{5.5}$$

$$\delta x(t+1,k) = A(t,k)\delta x(t,k) + B(t,k)\delta u(t,k) + \varpi(t,k), \tag{5.6}$$

$$\begin{aligned}\varpi(t,k) =& [A(t,k) - A(t,k-1)]x(t,k-1) \\ &+ [B(t,k) - B(t,k-1)]u(t,k-1) + \delta w(t,k),\end{aligned} \tag{5.7}$$

where $\delta x(t,k) = x(t,k) - x(t,k-1)$, $A(t,k) \triangleq A + \Delta A(t,k)$, and $B(t,k) \triangleq B + \Delta B(t,k)$. Obviously, there exists $\varpi(t,k) \neq 0$ in the presence of nonrepetitive process uncertainties and external disturbance. Hereafter, it will be used to denote various sources of load disturbances.

To meet with a practical requirement of no steady-state tracking error for each batch, the information of tracking error integral is introduced as an additional state variable for the ILC design, which is defined by

$$\Sigma e(t,k) \triangleq \sum_{i=0}^{t} e(i,k),\ 1 \leq t \leq N. \tag{5.8}$$

It follows that

$$\Sigma e(t,k) = \Sigma e(t-1,k) + e(t,k). \tag{5.9}$$

Therefore, an equivalent 2D system description of the above batch process shown in (5.1), together with two augmented state variables of the tracking error information, are expressed by

$$\begin{cases} \hat{x}(t+1,k) = (A_1 + \Delta A_1(t,k))\hat{x}(t,k) + A_2\hat{x}(t+1,k-1) \\ \qquad + \big(B_{\mathrm{g}} + \Delta B_{\mathrm{g}}(t,k)\big)\delta u(t,k) + D_{\mathrm{g}}\varpi(t,k), \\ y_{\mathrm{e}}(t,k) = \begin{bmatrix} e(t,k-1) \\ e(t,k) \\ \Sigma e(t,k) \end{bmatrix} = G\hat{x}(t,k),\ 1 \leq t \leq N, \\ z(t,k) = e(t,k) = L\hat{x}(t,k), \\ \delta\hat{x}(0,k) = \mathbf{0}, \quad k = 0, 1, \ldots, \end{cases} \tag{5.10}$$

where $L = [\mathbf{0}, \mathbf{I}, \mathbf{0}]$, and

$$A_1 = \begin{bmatrix} A & \mathbf{0} & \mathbf{0} \\ CA & \mathbf{0} & \mathbf{0} \\ \mathbf{0} & \mathbf{I} & \mathbf{I} \end{bmatrix}, A_2 = \begin{bmatrix} \mathbf{0} & \mathbf{0} & \mathbf{0} \\ \mathbf{0} & \mathbf{I} & \mathbf{0} \\ \mathbf{0} & \mathbf{0} & \mathbf{0} \end{bmatrix}, B_g = \begin{bmatrix} B \\ CB \\ \mathbf{0} \end{bmatrix}, D_{\mathrm{g}} = \begin{bmatrix} \mathbf{I} \\ C \\ \mathbf{0} \end{bmatrix}, G = \begin{bmatrix} -C & \mathbf{I} & \mathbf{0} \\ \mathbf{0} & \mathbf{I} & \mathbf{0} \\ \mathbf{0} & \mathbf{I} & \mathbf{I} \end{bmatrix},$$

$$\Delta A_1(t,k) = \overline{F}\Delta(t,k)\overline{H}_{\mathrm{A}}, \Delta B_{\mathrm{g}}(t,k) = \overline{F}\Delta(t,k)\overline{H}_{\mathrm{B}}, \overline{F} = \left[F^{\mathrm{T}}, (CF)^{\mathrm{T}}, 0\right]^{\mathrm{T}},$$

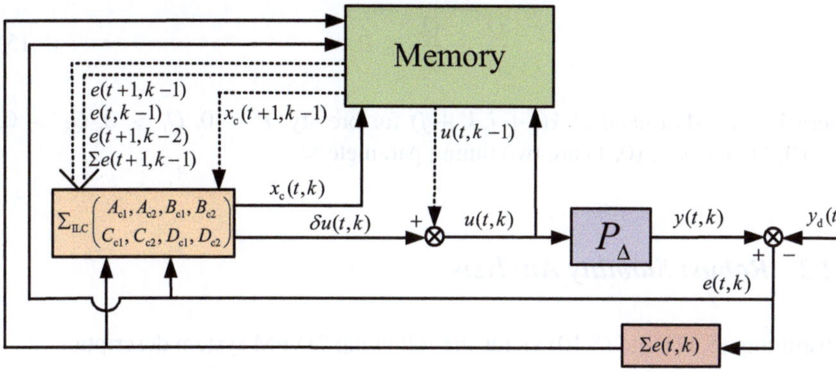

Fig. 5.1 Schematic of 2D dynamic output feedback-based ILC scheme

$$\overline{H}_{\mathrm{A}} = [H_{\mathrm{A}}, \mathbf{0}, \mathbf{0}], \overline{H}_{\mathrm{B}} = H_{\mathrm{B}}, \hat{x}(t,k) = \left[\delta x(t,k)^{\mathrm{T}}, e(t,k), \Sigma e(t-1,k)\right]^{\mathrm{T}}.$$

When only the process output measurement is available, an ILC scheme with DOF is proposed as shown in Fig. 5.1, where an output-error-based ILC updating law and the corresponding controller are constructed as

$$\begin{cases} x_{\mathrm{c}}(t+1,k) = A_{\mathrm{c}1}x_{\mathrm{c}}(t,k) + A_{\mathrm{c}2}x_{\mathrm{c}}(t+1,k-1) + B_{\mathrm{c}1}y_{\mathrm{e}}(t,k) + B_{\mathrm{c}2}y_{\mathrm{e}}(t+1,k-1), \\ \delta u(t,k) = C_{\mathrm{c}1}x_{\mathrm{c}}(t,k) + C_{\mathrm{c}2}x_{\mathrm{c}}(t+1,k-1) + D_{\mathrm{c}1}y_{\mathrm{e}}(t,k) + D_{\mathrm{c}2}y_{\mathrm{e}}(t+1,k-1), \end{cases} \tag{5.11}$$

where $A_{ci}, B_{ci}, C_{ci}, D_{ci}, i = 1, 2$, are the controller gains to be determined.

To deal with the process uncertainties and load disturbance, the robust $\mathcal{H}_\infty$ control performance is considered here

$$\|z(t,k)\|_2 \leq \gamma_{\mathrm{rob}} \|\varpi(t,k)\|_2, \tag{5.12}$$

where $\gamma_{\mathrm{rob}} > 0$ is a prescribed level of disturbance attenuation. This leads to the following robust batch control objective

$$J_{\mathrm{BP}} = \sum_{t=0}^{N_1=N} \sum_{k=0}^{N_2\to\infty} \left(\gamma_{\mathrm{rob}}^{-1} \|z(t,k)\|_2^2 - \gamma_{\mathrm{rob}} \|\varpi(t,k)\|_2^2\right) < 0. \tag{5.13}$$

Analogous to the state feedback based ILC in Chap. 4, the following Lyapunov–Krasovskii functions and inequalities of state energy are defined

$$V_P(\hat{x}_{i+1,j}) - V_{Q_t}(\hat{x}_{i,j}) - V_{Q_k}(\hat{x}_{i+1,j-1}) \leq -\left(\|z(t+i,k+j)\|_2^2 - \|\varpi(t+i,k+j)\|_2^2\right), \tag{5.14}$$

$$\frac{Q_t}{\rho_t} + \frac{Q_k}{\rho_k} \le P, \tag{5.15}$$

where $\hat{x}_{i,j}$ is a shorthand of $\hat{x}(t+i, k+j)$ for brevity; $P > 0$, $Q_t > 0$, $Q_k > 0$; $\rho_t \in (0, 1)$ and $\rho_k \in (0, 1)$ are two tuning parameters.

5.2.2 *Robust Stability Analysis*

Substituting (5.11) into (5.10) yields the following 2D FM system description

$$\begin{cases} \hat{x}_{\mathrm{g}}(t+1,k) = \left(\hat{A}_1 + \Delta\hat{A}_1\right)\hat{x}_{\mathrm{g}}(t,k) + \left(\hat{A}_2 + \Delta\hat{A}_2\right)\hat{x}_{\mathrm{g}}(t+1,k-1) + \hat{D}\varpi(t,k), \\ y_{\mathrm{e}}(t,k) = \hat{G}\hat{x}_{\mathrm{g}}(t,k), \\ z(t,k) = \hat{L}\hat{x}_{\mathrm{g}}(t,k), \end{cases} \tag{5.16}$$

where $\hat{G} = GH$, $\hat{L} = LH$, $\hat{D} = H^{\mathrm{T}}D_{\mathrm{g}}$, $H = [\,\mathbf{I}\ \mathbf{0}\,]$, and

$$\hat{A}_i = \begin{bmatrix} A_i + B_{\mathrm{g}}D_{\mathrm{c}i}G & B_{\mathrm{g}}C_{\mathrm{c}i} \\ B_{\mathrm{c}i}G & A_{\mathrm{c}i} \end{bmatrix}, \hat{x}_{\mathrm{g}}(t,k) = \begin{bmatrix} \hat{x}(t,k) \\ \hat{x}_c(t,k) \end{bmatrix} i = 1, 2,$$

$$\Delta\hat{A}_1 = \begin{bmatrix} \Delta A_1 + \Delta B_{\mathrm{g}}D_{\mathrm{c}1}G & \Delta B_{\mathrm{g}}C_{\mathrm{c}1} \\ \mathbf{0} & \mathbf{0} \end{bmatrix}, \Delta\hat{A}_2 = \begin{bmatrix} \Delta B_{\mathrm{g}}D_{\mathrm{c}2}G & \Delta B_{\mathrm{g}}C_{\mathrm{c}2} \\ \mathbf{0} & \mathbf{0} \end{bmatrix}.$$

Theorem 5.1 The closed-loop 2D system in (5.16) is guaranteed robustly stable with an $\mathcal{H}_\infty$ control performance level, γ_{rob}, and a TCI no larger than $\rho_t \in (0, 1)$ and a KCI no larger than $\rho_k \in (0, 1)$, if there exist $\widehat{P} > 0$, $\widehat{Q}_t > 0$, $\widehat{Q}_k > 0$, $\widehat{R} > 0$, $X > 0$, matrices $\widehat{A}_{ci}$, $\widehat{B}_{ci}$, $\widehat{C}_{ci}$, $\widehat{D}_{ci}$, $i = 1, 2$, and a positive scalar ε, such that the following LMI conditions hold

$$\begin{bmatrix} \Gamma_{11} & \Gamma_{12} & \Pi_{1,1} & \Pi_{1,2} & \Pi_{2,1} & \Pi_{2,2} & D_{\mathrm{g}} & \mathbf{0} & \mathbf{0} \\ * & \Gamma_{22} & \Pi_{1,3} & \Pi_{1,4} & \Pi_{2,3} & \Pi_{2,4} & D_{\mathrm{g}} & \mathbf{0} & \mathbf{0} \\ * & * & -\widehat{Q}_t(1,1) & -\widehat{Q}_t(1,2) & \mathbf{0} & \mathbf{0} & \mathbf{0} & L^{\mathrm{T}} & \overline{H}_{\mathrm{A}}^{\mathrm{T}} + G^{\mathrm{T}}\widehat{D}_{\mathrm{c}1}^{\mathrm{T}}\overline{H}_{\mathrm{B}}^{\mathrm{T}} \\ * & * & * & -\widehat{Q}_t(2,2) & \mathbf{0} & \mathbf{0} & \mathbf{0} & XL^{\mathrm{T}} & X\overline{H}_{\mathrm{A}}^{\mathrm{T}} + \widehat{C}_{\mathrm{c}1}^{\mathrm{T}}\overline{H}_{\mathrm{B}}^{\mathrm{T}} \\ * & * & * & * & -\widehat{Q}_k(1,1) & -\widehat{Q}_k(1,2) & \mathbf{0} & \mathbf{0} & G^{\mathrm{T}}\widehat{D}_{\mathrm{c}2}^{\mathrm{T}}\overline{H}_{\mathrm{B}}^{\mathrm{T}} \\ * & * & * & * & * & -\widehat{Q}_k(2,2) & \mathbf{0} & \mathbf{0} & \widehat{C}_{\mathrm{c}2}^{\mathrm{T}}\overline{H}_{\mathrm{B}}^{\mathrm{T}} \\ * & * & * & * & * & * & -\gamma_{\mathrm{rob}}\mathbf{I} & \mathbf{0} & 0 \\ * & * & * & * & * & * & * & -\gamma_{\mathrm{rob}}\mathbf{I} & 0 \\ * & * & * & * & * & * & * & * & -\varepsilon\mathbf{I} \end{bmatrix} < 0, \tag{5.17}$$

$$\frac{\widehat{Q}_t}{\rho_t} + \frac{\widehat{Q}_k}{\rho_k} \le 2\widehat{R} - \widehat{P}, \tag{5.18}$$

$$\widehat{P} = \begin{bmatrix} \widehat{P}_{11} & \widehat{P}_{12} \\ * & \widehat{P}_{22} \end{bmatrix}, \widehat{Q}_t = \begin{bmatrix} \widehat{Q}_t(1,1) & \widehat{Q}_t(1,2) \\ * & \widehat{Q}_t(2,2) \end{bmatrix}, \widehat{Q}_k = \begin{bmatrix} \widehat{Q}_k(1,1) & \widehat{Q}_k(1,2) \\ * & \widehat{Q}_k(2,2) \end{bmatrix}, \widehat{R} = \begin{bmatrix} \mathbf{I} & \mathbf{I} \\ * & X \end{bmatrix},$$

$$\Gamma_{11} = -\widehat{P}_{11} + \varepsilon\overline{F}\overline{F}^{\mathrm{T}}, \Gamma_{12} = -\widehat{P}_{12} + \varepsilon\overline{F}\overline{F}^{\mathrm{T}}, \Gamma_{22} = -\widehat{P}_{22} + \varepsilon\overline{F}\overline{F}^{\mathrm{T}},$$

$$\Pi_{i,1} = A_i + B_g\widehat{D}_{ci}G + \widehat{B}_{ci}G, \Pi_{i,2} = A_iX + B_g\widehat{C}_{ci} + \widehat{A}_{ci},$$

$$\Pi_{i,3} = A_i + B_g\widehat{D}_{ci}G, \Pi_{i,4} = A_iX + B_g\widehat{C}_{ci}, i = 1, 2.$$

If the above LMI conditions are feasible, the controller matrices can be retrieved by

$$\begin{cases} D_{ci} = \widehat{D}_{ci}, \\ B_{ci} = \Psi^{-1}\widehat{B}_{ci}, \\ C_{ci} = (\widehat{C}_{ci} - \widehat{D}_{ci}GX)\Phi^{-\mathrm{T}}, & i = 1, 2, \\ A_{ci} = \Psi^{-1}(\widehat{A}_{ci} - \widehat{B}_{ci}GX)\Phi^{-\mathrm{T}}, \end{cases} \tag{5.19}$$

where Φ and Ψ are any nonsingular matrices satisfying $\Phi\Psi^{\mathrm{T}} = \mathbf{I} - X$.

Proof Note that (5.13) can be rewritten as

$$\sum_{t=0}^{N_1=N}\sum_{k=0}^{N_2\to\infty}\left(\gamma_{\mathrm{rob}}^{-1}\|z(t,k)\| - \gamma_{\mathrm{rob}}\|\varpi(t,k)\| + \Delta V\right) - \sum_{t=0}^{N_1=N}\sum_{k=0}^{N_2\to\infty}\Delta V < 0, \tag{5.20}$$

where $\Delta V = V_P[\hat{x}_g(t+1,k)] - V_{Q_t}[\hat{x}_g(t,k)] - V_{Q_k}[\hat{x}_g(t+1,k-1)]$ is the 2D difference of a Lyapunov function as previously mentioned.

With the normalized initial condition shown in (5.10), there exist $\hat{x}_g(t,k) = 0$ for $t \le 0$ and $\forall k$, and $\hat{x}_g(t,k) = 0$ for $0 \le t \le N$, Therefore, one obtains

$$\begin{aligned} \sum_{t=0}^{N_1=N}\sum_{k=0}^{N_2\to\infty}\Delta V = {} & V_P[\hat{x}_g(N_1+1,N_2)] + \sum_{t=1}^{N_1}\sum_{k=1}^{N_2-1}V_{P-Q_t-Q_k}[\hat{x}_g(t,k)] \\ & + \sum_{k=1}^{N_2-1}V_{P-Q_k}[\hat{x}_g(N_1+1,k)], \end{aligned} \tag{5.21}$$

where $\hat{x}_g(N_1+1,k) = \hat{x}_g(N_1,k)$ with respect to any k.

It follows from (5.15) that $P > Q_t + Q_k$, and then

$$\sum_{t=0}^{N_1=N}\sum_{k=0}^{N_2\to\infty}\Delta V > 0. \tag{5.22}$$

Hence, a sufficient condition to guarantee (5.20) is that

$$\gamma_{\rm rob}^{-1}\|z(t,k)\| - \gamma_{\rm rob}\|\varpi(t,k)\| + \Delta V < 0. \tag{5.23}$$

Substituting (5.16) into (5.23) yields

$$\xi^{\rm T}(t,k)\Xi\xi(t,k) < 0, \tag{5.24}$$

where $\xi(t,k) = [\hat{x}_g^{\rm T}(t,k),\ \hat{x}_g^{\rm T}(t+1,k-1),\ \varpi^{\rm T}(t,k)]^{\rm T}$ and

$$\Xi = \begin{bmatrix} (\hat{A}_1+\Delta\hat{A}_1)^{\rm T} \\ (\hat{A}_2+\Delta\hat{A}_2)^{\rm T} \\ \hat{D}^{\rm T} \end{bmatrix} P\big[\hat{A}_1+\Delta\hat{A}_1\ \hat{A}_2+\Delta\hat{A}_2\ \hat{D}\big] - \begin{bmatrix} Q_t - \gamma_{\rm rob}\hat{L}\hat{L}^{\rm T} & \mathbf{0} & \mathbf{0} \\ * & Q_k & \mathbf{0} \\ * & * & \gamma_{\rm rob}\mathbf{I} \end{bmatrix}.$$

By the Schur complement, it can be seen that (5.24) is guaranteed by

$$\begin{bmatrix} -P^{-1} & \hat{A}_1+\Delta\hat{A}_1 & \hat{A}_2+\Delta\hat{A}_2 & \hat{D} & \mathbf{0} \\ * & -Q_t & \mathbf{0} & \mathbf{0} & \hat{L}^{\rm T} \\ * & * & -Q_k & \mathbf{0} & \mathbf{0} \\ * & * & * & -\gamma_{\rm rob}\mathbf{I} & \mathbf{0} \\ * & * & * & * & -\gamma_{\rm rob}\mathbf{I} \end{bmatrix} < 0. \tag{5.25}$$

By introducing a slack variable R into (5.25) by pre- and post-multiplying both sides with $\hat{R}^{\rm T} = \operatorname{diag}\big(\mathbf{I}, R^{\rm T}, R^{\rm T}, \mathbf{I}, \mathbf{I}\big)$ and $\hat{R}$, one obtains

$$\begin{bmatrix} -P^{-1} & (\hat{A}_1+\Delta\hat{A}_1)R & (\hat{A}_2+\Delta\hat{A}_2)R & \hat{D} & \mathbf{0} \\ * & -R^{\rm T}Q_tR & \mathbf{0} & \mathbf{0} & R^{\rm T}\hat{L}^{\rm T} \\ * & * & -R^{\rm T}Q_kR & \mathbf{0} & \mathbf{0} \\ * & * & * & -\gamma_{\rm rob}\mathbf{I} & \mathbf{0} \\ * & * & * & * & -\gamma_{\rm rob}\mathbf{I} \end{bmatrix} < 0. \tag{5.26}$$

Let $R = \Lambda_2\Lambda_1^{-1}$, where

$$\Lambda_1 = \begin{bmatrix} \mathbf{I} & \mathbf{I} \\ \Psi^{\rm T} & \mathbf{0} \end{bmatrix},\ \Lambda_2 = \begin{bmatrix} \mathbf{I} & X \\ \mathbf{0} & \Phi^{\rm T} \end{bmatrix},\ \Phi\Psi^{\rm T} = \mathbf{I} - X.$$

It can be easily verified from $\widehat{R} = \Lambda_1^{\rm T}R\Lambda_1$ and (5.18) that

$$\begin{bmatrix} \mathbf{I} & \mathbf{I} \\ * & X \end{bmatrix} > 0.$$

Therefore, $\mathbf{I} - X$ is guaranteed nonsingular by the Schur complement and so it is for Λ_1 and Λ_2.

By performing a congruence transformation to (5.26) by taking a pre-multiplication with $\hat{\Lambda}_1^{\mathrm{T}} = \mathrm{diag}(\Lambda_1^{\mathrm{T}}, \Lambda_1^{\mathrm{T}}, \Lambda_1^{\mathrm{T}}, \mathbf{I}, \mathbf{I})$ and a post-multiplication with $\hat{\Lambda}_1$, and using Lemma 2.1, one obtains the LMI in (5.17), where $\widehat{P} = \Lambda_1^{\mathrm{T}} P^{-1} \Lambda_1$, $\widehat{Q}_t = \Lambda_1^{\mathrm{T}} R^{\mathrm{T}} Q_t R \Lambda_1, \widehat{Q}_k = \Lambda_1^{\mathrm{T}} R^{\mathrm{T}} Q_k R \Lambda_1$. Meanwhile, the following matrix variables are defined for computation

$$\begin{cases} \widehat{D}_{ci} = D_{ci}, \\ \widehat{B}_{ci} = \Psi B_{ci}, \\ \widehat{C}_{ci} = \widehat{D}_{ci} G X + C_{ci} \Phi^{\mathrm{T}}, \qquad i = 1, 2, \\ \widehat{A}_{ci} = \widehat{B}_{ci} G X + \Psi A_{ci} \Phi^{\mathrm{T}}. \end{cases}$$

With the above definitions of $\widehat{P}, \widehat{Q}_t$, and $\widehat{Q}_k$, it follows from the fact $R + R^{\mathrm{T}} - \widehat{P} \leq R^{\mathrm{T}} \widehat{P}^{-1} R$ that (5.18) guarantees (5.15). This completes the proof. □

To reduce the computation effort or conservativeness for practical application, an off-line control algorithm can be similarly carried out by using the robust control objective in (5.12). For instance, an off-line control algorithm based on the output feedback can be established by solving the following optimization program

$$\begin{gathered} \underset{\rho_t \in (0,1), \rho_k \in (0,1)}{\text{Minimize}} \ \gamma_{\mathrm{rob}} \\ (\text{subject to } (5.17) \text{ and} (5.18)) \end{gathered} \tag{5.27}$$

Generally, a smaller value of γ_{rob} corresponds to a larger controller gain and vice versa. Note that the convergence indices of TCI and KCI are user specified in practice, which may be prescribed to determine the cycling periods and the on- or off-line control algorithms, in consideration of the process input and output constraints.

5.2.3 An Illustrative Example

Example 5.1 Consider the batch packing process of an injection molding machine as shown in Fig. 2.2, the cyclic V/P transfer process of injection molding studied in Shi et al. (2005) is used for illustration

$$P_\Delta:\begin{cases} x(t+1,k)=\left(\begin{bmatrix} 1.607 & 1 \\ -0.6086 & 0 \end{bmatrix}+\Delta A(t,k)\right)x(t,k) \\ \qquad +\left(\begin{bmatrix} 1.239 \\ -0.9282 \end{bmatrix}+\Delta B(t,k)\right)u(t,k)\ +w(t,k), \\ y(t,k)=\begin{bmatrix} 1 & 0 \end{bmatrix}x(t,k), \end{cases}$$

$$\Delta A(t,k)=\begin{bmatrix} 0.0804\delta(t,k) & 0 \\ -0.0304\delta(t,k) & 0 \end{bmatrix}=\begin{bmatrix} 1 & 0 \\ 0 & 1 \end{bmatrix}\begin{bmatrix} \delta(t,k) & 0 \\ 0 & \delta(t,k) \end{bmatrix}\begin{bmatrix} 0.0804 & 0 \\ -0.0304 & 0 \end{bmatrix},$$

$$\Delta B(t,k)=\begin{bmatrix} 0.062\delta(t,k) \\ -0.0464\delta(t,k) \end{bmatrix}=\begin{bmatrix} 1 & 0 \\ 0 & 1 \end{bmatrix}\begin{bmatrix} \delta(t,k) & 0 \\ 0 & \delta(t,k) \end{bmatrix}\begin{bmatrix} 0.062 \\ -0.0464 \end{bmatrix},$$

where $|\delta(t,k)|\leq 1$ represents the nonrepetitive process uncertainties.

The target profile (y_d) takes the following form as studied in the references (Shi et al. 2005; Liu et al. 2014)

$$y_d(t)=\begin{cases} 200, & 0\leq t\leq 100; \\ 200+5(t-100), & 100<t\leq 120; \\ 300, & 120<t\leq N=200. \end{cases}$$

Note that the initial part of $y_d(t)$ is smoothed by a user-specified prefilter $G_f(z)=(z^{-1}+z^{-2})/(3-z^{-1})$ for practical implementation.

By specifying the convergence indices for the time and batch directions as $\rho_t=0.98$ and $\rho_k=0.8$, respectively, and taking $\gamma_{rob}=100$ to comply with the operating constraints, the corresponding 2D dynamic output feedback-based ILC controller gains are computed as listed in Table 5.1, by solving the LMI conditions in (5.17) and (5.18).

For illustration, the following two cases are tested.

Case 1. Time-invariant uncertainties and repetitive load disturbance. Assume that the process uncertainties $\Delta A(t,k)$ and $\Delta B(t,k)$ are fixed as their upper bounds, and meanwhile, the disturbance is taken as $w(t,k)=\sin(t)$. The tracking results are shown in Fig. 5.2, and the corresponding ATE is plotted in Fig. 5.3. It is seen that perfect tracking could be realized after almost five batches. Note that due to the introduction of tracking error integral in the ILC law design, the tracking error in the initial batch is significantly reduced, compared to the above cited references.

Case 2. Time-varying uncertainties and nonrepetitive load disturbance. Assume that the uncertain parameter $\delta(t,k)$ is randomly varying within [−0.1,0.1], and the nonrepetitive disturbance is considered as $w(t,k)=\sin(t+\theta(k))$, where $\theta(k)$ is a random variable uniformly distributed in a range of $[0,2\pi]$. To evaluate the tracking performance and robustness of the proposed design under the stochastic disturbance, one hundred Monte Carlo tests are conducted. The averaged results of ATE are shown

Table 5.1 Dynamic output feedback-based ILC controller matrices

	$i = 1$	$i = 2$
A_{ci}	$\begin{bmatrix} 0.2103 & -0.1941 & 0.0061 & 0.1626 \\ -0.1827 & 0.1840 & 0.0123 & -0.1480 \\ 0.4577 & -0.4480 & -0.0116 & 0.3768 \\ 0.3090 & -0.3264 & -0.0355 & 0.2841 \end{bmatrix}$	$\begin{bmatrix} -0.0015 & -0.0012 & 0.0040 & 0.0013 \\ -0.0001 & 0.0005 & -0.0004 & -0.0002 \\ -0.0037 & -0.0005 & 0.0070 & 0.0020 \\ 0.0059 & 0.0027 & -0.0129 & -0.0039 \end{bmatrix}$
B_{ci}	$\begin{bmatrix} -0.0675 & 0.0666 & 0.1402 \\ 0.0611 & -0.0596 & 0.1599 \\ -0.1505 & 0.1488 & 0.0089 \\ -0.1059 & 0.1039 & 0.0510 \end{bmatrix}$	$\begin{bmatrix} 0.0001 & 0.1536 & 0.0001 \\ 0.0001 & -0.1208 & -0.0000 \\ 0.0007 & -0.0676 & -0.0000 \\ -0.0008 & 0.0330 & -0.0001 \end{bmatrix}$
C_{ci}	$\begin{bmatrix} -2.4456 & 1.5155 & -1.1744 & -0.4824 \end{bmatrix}$	$\begin{bmatrix} -0.0002 & -0.0080 & -0.0108 & 0.0112 \end{bmatrix}$
D_{ci}	$\begin{bmatrix} 1.3147 & -1.3039 & -0.1893 \end{bmatrix}$	$\begin{bmatrix} -0.0009 & -0.6447 & -0.0002 \end{bmatrix}$

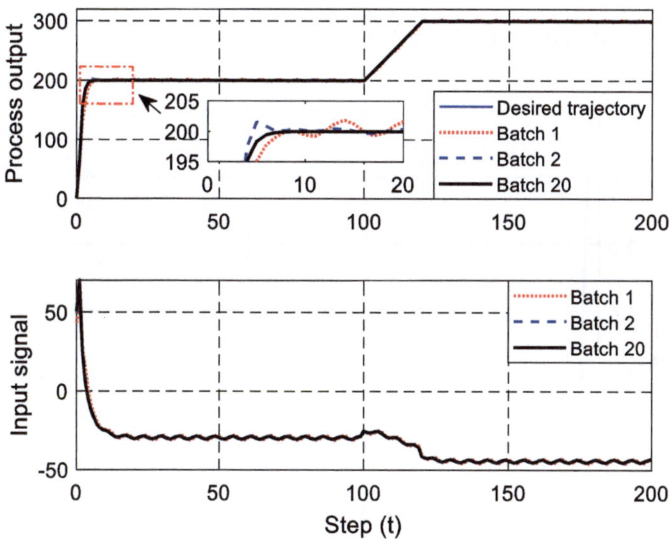

Fig. 5.2 Tracking results in Case 1 of Example 5.1

in Fig. 5.4. It is seen that the closed-loop ILC system well maintains robust stability along both the time and batch directions, and the process output could track the desired reference trajectory with a small error band.

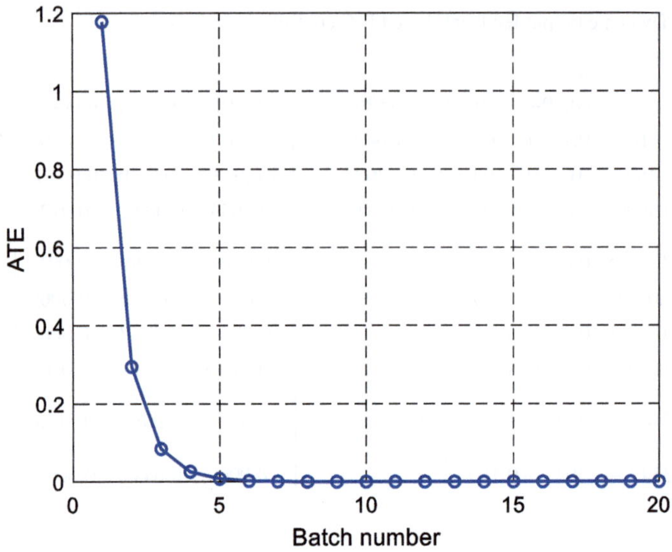

Fig. 5.3 Plot of ATE in Case 1 of Example 5.1

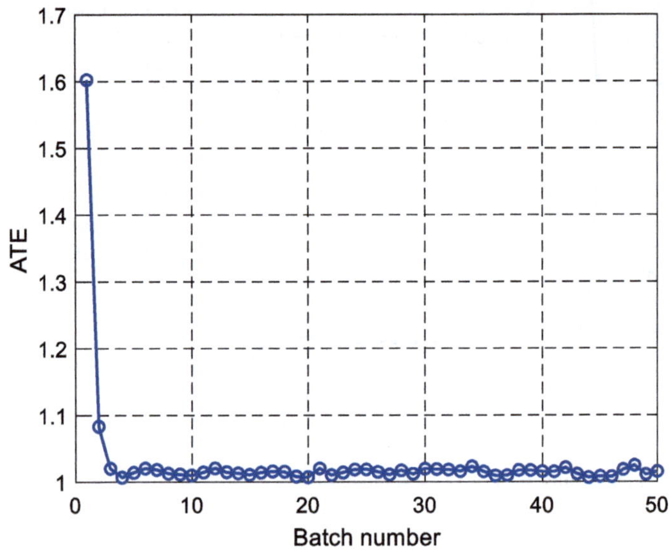

Fig. 5.4 Plot of ATE in Case 2 of Example 5.1

5.3 Static Output Feedback Based ILC Design

5.3.1 Control Synthesis with Finite-Frequency-Range Performance Specification

Consider the batch process in (5.1) with repetitive plant uncertainties (i.e., $\Delta(t,k) = \Delta(t)$) subject to nonrepetitive disturbance satisfying $\delta w(t,k) \in l_2[0,+\infty)$, an SOF-based ILC updating law is designed as

$$\delta u(t,k) = K_1\delta y(t,k) + K_2 e(t+1,k-1), \tag{5.28}$$

where $\delta y(t,k) = y(t,k) - y(t,k-1)$, $K_1 \in \Re^{n_u\times n_y}$ and $K_2 \in \Re^{n_u\times n_y}$ are the learning gains to be determined.

Applying the ILC law in (5.28) to the error dynamic system in (5.6) leads to the following equivalent 2D Roesser system description of the resulting ILC system

$$\begin{cases} \begin{bmatrix} \delta x(t+1,k) \\ e(t+1,k) \end{bmatrix} = \mathbb{A}\begin{bmatrix} \delta x(t,k) \\ e(t+1,k-1) \end{bmatrix} + \mathbb{B}\delta w(t,k), \\ z(t,k) = \mathbb{C}\begin{bmatrix} \delta x(t,k) \\ e(t+1,k-1) \end{bmatrix}, \end{cases} \tag{5.29}$$

where

$$\mathbb{A} = \begin{bmatrix} \mathcal{A} & \mathcal{B} \\ \mathcal{C} & \mathcal{D} \end{bmatrix}, \mathbb{B} = \begin{bmatrix} \mathbf{I} \\ C \end{bmatrix}, \mathbb{C} = \begin{bmatrix} 0 & \mathbf{I} \end{bmatrix},$$

$$\begin{aligned} \mathcal{A} &= A + \Delta A + (B+\Delta B)K_1 C,\ \mathcal{B} = (B+\Delta B)K_2, \\ \mathcal{C} &= C(A + \Delta A + (B+\Delta B)K_1 C),\ \mathcal{D} = \mathbf{I} + C(B+\Delta B)K_2. \end{aligned}$$

Accordingly, the time-varying uncertainties associated with $\Delta A(t)$ and $\Delta B(t)$ could be lumped into $\delta w(t,k)$ for the convenience of stability analysis.

The 2D transfer-function matrix (Li and Gao 2012) relating the disturbance $\delta w(t,k)$ to the controlled output $z(t,k)$ is derived as

$$\mathbb{G}_v(e^{j\omega_1}, e^{j\omega_2}) \triangleq \mathbb{C}(\mathrm{diag}\{e^{j\omega_1}\mathbf{I}, e^{j\omega_2}\mathbf{I}\} - \mathbb{A})^{-1}\mathbb{B},\ (\omega_1,\omega_2) \in [-\pi,\pi]\times[-\pi,\pi]. \tag{5.30}$$

For ease of analysis, $\mathbb{A}$ is reformulated as

$$\mathbb{A} = \hat{A}_\Delta + \hat{B}_\Delta K \hat{C}, \tag{5.31}$$

where $\hat{A}_\Delta = \hat{A} + \hat{F}\Delta\hat{H}_{\mathrm{A}}$, $\hat{B}_\Delta = \hat{B} + \hat{F}\Delta H_{\mathrm{B}}$, $K = \begin{bmatrix} K_1 & K_2 \end{bmatrix}$, $\hat{H}_{\mathrm{A}} = \begin{bmatrix} H_{\mathrm{A}} & 0 \end{bmatrix}$,

$$\hat{A} = \begin{bmatrix} A & 0 \\ CA & \mathbf{I} \end{bmatrix}, \hat{B} = \begin{bmatrix} B \\ CB \end{bmatrix}, \hat{F} = \begin{bmatrix} F \\ CF \end{bmatrix}, \hat{C} = \begin{bmatrix} C & 0 \\ 0 & \mathbf{I} \end{bmatrix}. \tag{5.32}$$

Here it aims at designing an ILC scheme based on only the output measurement, such that

(i) the system output can realize perfect tracking of the desired output trajectory in the absence of nonrepetitive disturbances;
(ii) a finite-frequency-range 2D $\mathcal{H}_\infty$ performance specification of disturbance attenuation satisfies

$$\left\| \mathbb{G}_v(e^{\mathrm{j}\omega_1}, e^{\mathrm{j}\omega_2}) \right\|_\infty \triangleq \sup_{(\omega_1,\omega_2)\in\Omega_1\times\Omega_2} \overline{\sigma}\left[\mathbb{G}_v(e^{\mathrm{j}\omega_1}, e^{\mathrm{j}\omega_2})\right] < \gamma, \tag{5.33}$$

where $\gamma > 0$ is a prescribed 2D $\mathcal{H}_\infty$ performance level, and

$$\begin{aligned} \Omega_1 &\triangleq \{\omega_1 | \omega_{11} \le \omega_1 \le \omega_{12}, \omega_{11}, \omega_{12} \in [-\pi, \pi]\}, \\ \Omega_2 &\triangleq \{\omega_2 | \omega_{21} \le \omega_2 \le \omega_{22}, \omega_{21}, \omega_{22} \in [-\pi, \pi]\} \end{aligned} \tag{5.34}$$

with $\omega_{11}, \omega_{12}, \omega_{21}, \omega_{22}$ being user-specified lower and upper bounds based on a priori knowledge of disturbance frequency that may be estimated from process operation or modeling; ω_1 and ω_2 represent the frequencies of disturbance variation along the time and batch directions, respectively.

To facilitate analysis, the following technical lemmas are presented.

Lemma 5.1 (Gahinet and Apkarian 1994) Given a symmetric matrix $\Upsilon \in \Re^{p\times p}$ and two matrices Λ and Σ of column dimension p, there exists a matrix W that satisfies

$$\Upsilon + \mathrm{sym}\{\Lambda^{\mathrm{T}} W \Sigma\} < 0,$$

if and only if the following two projection inequalities hold

$$\begin{aligned} \Lambda^{\perp^{\mathrm{T}}} \Upsilon \Lambda^{\perp} < 0, \text{ or } \Lambda^{\mathrm{T}} \Lambda > 0, \\ \Sigma^{\perp^{\mathrm{T}}} \Upsilon \Sigma^{\perp} < 0, \text{ or } \Sigma^{\mathrm{T}} \Sigma > 0, \end{aligned}$$

where $\Lambda^{\perp}$ and $\Sigma^{\perp}$ are arbitrary matrices whose columns form a basis of the null spaces of Λ and Σ, respectively.

Lemma 5.2 (Li and Gao 2012) The finite-frequency-range 2D $\mathcal{H}_\infty$ performance specification in (5.33) is satisfied if there exist matrices $\mathbb{P}_1 = \mathbb{P}_1^{\mathrm{T}}$, $\mathbb{P}_2 = \mathbb{P}_2^{\mathrm{T}}$, $\mathbb{Q}_1 > 0$ and $\mathbb{Q}_2 > 0$ such that

$$\begin{bmatrix} \mathbb{A} & \mathbb{B} \\ \mathbf{I} & 0 \end{bmatrix}^{\mathrm{T}} \begin{bmatrix} -\mathbb{P} & \tilde{\Lambda}\mathbb{Q} \\ \mathbb{Q}\tilde{\Lambda}^* & \mathbb{P} - \tilde{\Gamma}\mathbb{Q} \end{bmatrix} \begin{bmatrix} \mathbb{A} & \mathbb{B} \\ \mathbf{I} & 0 \end{bmatrix} + \begin{bmatrix} \mathbb{C} & 0 \\ 0 & \mathbf{I} \end{bmatrix}^{\mathrm{T}} \begin{bmatrix} \mathbf{I} & 0 \\ 0 & -\gamma^2\mathbf{I} \end{bmatrix} \begin{bmatrix} \mathbb{C} & 0 \\ 0 & \mathbf{I} \end{bmatrix} < 0,$$

where

$$\begin{aligned} \mathbb{P} &= \mathrm{diag}\{\mathbb{P}_1, \mathbb{P}_2\},\ \mathbb{Q} = \mathrm{diag}\{\mathbb{Q}_1, \mathbb{Q}_2\}, \\ \tilde{\Lambda} &= \mathrm{diag}\{e^{\mathrm{j}\omega_1^s\mathbf{I}}, e^{\mathrm{j}\omega_2^s\mathbf{I}}\},\ \tilde{\Gamma} = \mathrm{diag}\{2\cos(\omega_1^a)\mathbf{I},\ 2\cos(\omega_2^a)\mathbf{I}\}, \\ \omega_1^s &= (\omega_{12} + \omega_{11})/2,\ \omega_1^a = (\omega_{12} - \omega_{11})/2, \\ \omega_2^s &= (\omega_{22} + \omega_{21})/2,\ \omega_2^a = (\omega_{22} - \omega_{21})/2. \end{aligned}$$

5.3.2 Robust Stability Analysis

Based on the 2D system description in (5.29), robust stability of the resulting 2D Roesser system is analyzed herein, such that the convergence of tracking error could be clarified.

Theorem 5.2 The 2D Roesser system in (5.29) with $\delta w(t, k) = 0$ is robustly stable if there exist matrices $P_1 > 0$, $P_2 > 0$, W and scalars $\alpha, \beta \in (0, 1]$ such that

$$\Upsilon + \mathrm{sym}\{\Lambda^\top W \Sigma\} < 0, \tag{5.35}$$

where $\Lambda = \begin{bmatrix} \mathbb{A} & -\mathbf{I} \end{bmatrix}$ and

$$\Upsilon = \mathrm{diag}\{-\Pi_\alpha^\beta P, P\},\ P = \mathrm{diag}\{P_1, P_2\},\ \Pi_\alpha^\beta = \mathrm{diag}\{\alpha\mathbf{I}, \beta\mathbf{I}\},\ \Sigma = \begin{bmatrix} 0 & \mathbf{I} \end{bmatrix}.$$

Proof Suppose that the inequality in (5.35) holds. By Lemma 5.1, it follows that

$$\Lambda^{\perp^\top} \Upsilon \Lambda^\perp < 0,\ \Sigma^{\perp^\top} \Upsilon \Sigma^\perp < 0, \tag{5.36}$$

where $\Lambda^\perp$ and $\Sigma^\perp$ are the null spaces of Λ and Σ, respectively, which can be taken as

$$\Lambda^\perp = \begin{bmatrix} \mathbf{I} \\ \mathbb{A} \end{bmatrix},\ \Sigma^\perp = \begin{bmatrix} \mathbf{I} & 0 \end{bmatrix}^\top.$$

The second inequality in (5.36) is always satisfied for any $P_1 > 0$ and $P_2 > 0$, the first inequality in (5.36) is rewritten as

$$\mathbb{A}^\top \mathrm{P} \mathbb{A} - \Pi_\alpha^\beta \mathrm{P} < 0. \tag{5.37}$$

Based on Theorem 3.5 in Shi et al. (2005), it is concluded that the 2D Roesser system in (5.29) is robustly stable. The proof is complete. □

It is seen from the proof of Theorem 5.2 that the 2D Roesser system in (5.29) is robustly stable if the following condition is satisfied

$$E^{\top}\Upsilon E < 0, \tag{5.38}$$

where $E = \begin{bmatrix} \mathbf{I} & \mathbb{A}^{\top} \end{bmatrix}^{\top}$. Inspired by the previous work in Li and Gao (2015), it is found that the inequality in (5.38) is equivalent to the following dilated matrix inequality

$$\left[\frac{E}{K\hat{C}-L}\right]^{\top}\hat{\Upsilon}\left[\frac{E}{K\hat{C}-L}\right] < 0, \tag{5.39}$$

where $L = [L_1\ L_2]$, $L_1 \in \Re^{n_u \times n_x}$, $L_2 \in \Re^{n_u \times n_y}$ and

$$\hat{\Upsilon} \triangleq \begin{bmatrix} \Upsilon & 0 \\ 0 & 0 \end{bmatrix}. \tag{5.40}$$

Now it comes to the position to present the following sufficient condition guaranteeing robust stability of the resulting ILC system in (5.29).

Theorem 5.3 Given K_1, K_2, and $\alpha, \beta \in (0, 1]$, the 2D Roesser system described by (5.29) with $\delta w(t, k) = 0$ is robustly stable if there exist matrices $P_1 > 0$, $P_2 > 0$, M, R, L_1, and L_2, such that

$$\hat{\Upsilon} + \mathrm{sym}\{\hat{M}\hat{\Sigma}\} < 0, \tag{5.41}$$

where $\hat{\Upsilon}$ is defined in (5.40), and

$$\hat{\Sigma} = \begin{bmatrix} \hat{A}_{\Delta} + \hat{B}_{\Delta}L & -\mathbf{I} & \hat{B}_{\Delta} \\ K\hat{C} - L & 0 & -\mathbf{I} \end{bmatrix}, \hat{M} = \begin{bmatrix} M & 0 \\ 0 & R \end{bmatrix}.$$

Proof Note that the 2D Roesser system in (5.29) with $\delta w(t, k) = 0$ is robustly stable if the inequality in (5.39) is satisfied. Therefore, it suffices to prove that the condition in (5.41) is equivalent to that in (5.39).

(*Sufficiency*) Suppose that the inequality in (5.41) holds true. Multiplying $\hat{\Sigma}^{\perp^{\top}}$ and its transpose on both sides of inequality (5.41) results in

$$\hat{\Sigma}^{\perp^{\top}}\hat{\Upsilon}\hat{\Sigma}^{\perp} < 0. \tag{5.42}$$

Moreover, it is noted that the null space of $\hat{\Sigma}^{\perp}$ can be chosen as

$$\hat{\Sigma}^{\perp} = \left[\frac{E}{K\hat{C} - L}\right]. \tag{5.43}$$

Thus, it is seen that (5.42) is exactly the condition in (5.39).

(*Necessity*) Suppose that the inequality in (5.39) is satisfied. Let $\Sigma = \left[\mathbb{A} \ -\mathbf{I}\right]$, then it is easy to find $\Sigma^{\perp} = E$. Thus, the inequality in (5.38) can be rewritten as

$$\Sigma^{\perp^{\top}} \Upsilon \Sigma^{\perp} < 0. \tag{5.44}$$

By using Lemma 5.1 with $\Lambda = \mathbf{I}$, it follows that there exists a matrix M such that

$$\Upsilon + \mathrm{sym}\{M\Sigma\} < 0. \tag{5.45}$$

Therefore, there exists a sufficiently large $\varepsilon > 0$ such that $\Upsilon + \mathrm{sym}\{M\Sigma\} + \frac{1}{2\varepsilon} M\hat{B}_{\Delta}(M\hat{B}_{\Delta})^{\top} < 0$. Then by the Schur complement, it follows that

$$\begin{bmatrix} \Upsilon + \mathrm{sym}\{M\Sigma\} & M\hat{B}_{\Delta} \\ * & -2\varepsilon\mathbf{I} \end{bmatrix} < 0, \tag{5.46}$$

which is equivalent to

$$\hat{\Upsilon} + \mathrm{sym}\left\{\begin{bmatrix} M & 0 \\ 0 & \varepsilon\mathbf{I} \end{bmatrix}\begin{bmatrix} \Sigma & \hat{B}_{\Delta} \\ 0 & -\mathbf{I} \end{bmatrix}\right\} < 0. \tag{5.47}$$

By taking

$$R = \varepsilon\mathbf{I}, L_1 = K_1 C, L_2 = K_2, \tag{5.48}$$

it is concluded that (5.47) follows from the condition in (5.41). The proof is complete. □

Remark 5.1 It can be seen that the SOF-based ILC law in (5.28) reduces to the following state feedback (SF) based ILC law

$$\delta u(t,k) = L_1 \delta x(t,k) + L_2 e(t+1, k-1), \tag{5.49}$$

based on the use of (5.48). This fact will be used later to develop a two-stage heuristic algorithm for the SOF based ILC design.

The following result gives another characterization of the stability condition for the resulting 2D Roesser system. ◊

Theorem 5.4 Given $\alpha, \beta \in (0, 1]$, the 2D Roesser system described by (5.29) with $\delta w(t,k) = 0$ is robustly stable if there exist matrices $P_1 > 0$, $P_2 > 0$, M_1, M_2, L_1, L_2, $\hat{K}$, R and a scalar $\eta > 0$, such that the following matrix inequality is satisfied

$$\Psi \triangleq \begin{bmatrix} \phi_{11} & \phi_{12} & \phi_{13} & M_1\hat{F} & \eta(\hat{H}_A + H_B L)^T \\ * & \phi_{22} & \phi_{23} & M_2\hat{F} & 0 \\ * & * & \phi_{33} & 0 & \eta H_B^T \\ * & * & * & -\eta\mathbf{I} & 0 \\ * & * & * & * & -\eta\mathbf{I} \end{bmatrix} < 0, \tag{5.50}$$

where $\hat{A}$, $\hat{B}$, $\hat{F}$ and $\hat{H}_A$ are defined in (5.31), Π_α^β is defined in Theorem 5.2 and

$$\begin{aligned} \phi_{11} &= \text{sym}\{M_1(\hat{A} + \hat{B}L)\} - \Pi_\alpha^\beta P, \\ \phi_{12} &= -M_1 + (\hat{A} + \hat{B}L)^T M_2^T, \phi_{23} = M_2\hat{B}, \\ \phi_{13} &= M_1\hat{B} + (\hat{K}\hat{C} - RL)^T, \phi_{33} = -R - R^T, \\ \phi_{22} &= -M_2 - M_2^T + P, P = \text{diag}\{P_1, P_2\}. \end{aligned} \tag{5.51}$$

Moreover, if the above inequality in (5.50) is feasible, then the SOF based ILC gains can be retrieved by

$$\begin{bmatrix} K_1 & K_2 \end{bmatrix} = R^{-1}\hat{K}. \tag{5.52}$$

Proof Suppose that the inequality in (5.50) holds. Applying the Schur complement to (5.50) yields

$$\phi + \eta \begin{bmatrix} (\hat{H}_A + H_B L)^T \\ 0 \\ H_B^T \end{bmatrix} \begin{bmatrix} (\hat{H}_A + H_B L)^T \\ 0 \\ H_B^T \end{bmatrix}^T + \eta^{-1} \begin{bmatrix} M_1\hat{F} \\ M_2\hat{F} \\ 0 \end{bmatrix} \begin{bmatrix} M_1\hat{F} \\ M_2\hat{F} \\ 0 \end{bmatrix}^T < 0, \tag{5.53}$$

where

$$\phi = \begin{bmatrix} \phi_{11} & \phi_{12} & \phi_{13} \\ * & \phi_{22} & \phi_{23} \\ * & * & \phi_{33} \end{bmatrix}.$$

By Lemma 2.1, the above inequality in (5.53) implies that

$$\hat{\Upsilon} + \text{sym}\{\hat{\Gamma}\} < 0, \tag{5.54}$$

where $\hat{\Upsilon}$ is defined in (5.40) and

$$\hat{\Gamma} = \begin{bmatrix} M(\hat{A}_\Delta + \hat{B}_\Delta L) & -M & M\hat{B}_\Delta \\ \hat{K}\hat{C} - RL & 0 & -R \end{bmatrix}, M = \begin{bmatrix} M_1 \\ M_2 \end{bmatrix}. \tag{5.55}$$

By letting $\hat{K} = RK$, one obtains $\hat{M}\hat{\Sigma} = \hat{\Gamma}$. Moreover, it follows from (5.50) that $-R - R^{\mathrm{T}} < 0$, which indicates that R is invertible. According to Theorem 5.3, the SOF based ILC law with $K = R^{-1}\hat{K}$ ensures the robust stability of the resulting ILC system. The proof is complete. □

Next, a finite-frequency-range 2D $\mathcal{H}_\infty$ performance specification of disturbance attenuation is introduced to realize the above control objective (ii), i.e.

$$\left\| \mathbb{G}_v(e^{j\omega_1}, e^{j\omega_2}) \right\|_\infty < \gamma, \forall(\omega_1, \omega_2) \in \Omega_1 \times \Omega_2. \tag{5.56}$$

By Lemma 5.2, the inequality in (5.56) holds if there exist matrices $\mathbb{P}_1 = \mathbb{P}_1^{\mathrm{T}}$, $\mathbb{P}_2 = \mathbb{P}_2^{\mathrm{T}}$, $\mathbb{Q}_1 > 0$ and $\mathbb{Q}_2 > 0$ such that the following matrix inequality is satisfied

$$\begin{bmatrix} \mathbb{A} & \mathbb{B} \\ \mathbf{I} & 0 \end{bmatrix}^{\mathrm{T}} \Xi \begin{bmatrix} \mathbb{A} & \mathbb{B} \\ \mathbf{I} & 0 \end{bmatrix} + \begin{bmatrix} \mathbb{C} & 0 \\ 0 & \mathbf{I} \end{bmatrix}^{\mathrm{T}} \Pi^{\mathrm{Hinf}} \begin{bmatrix} \mathbb{C} & 0 \\ 0 & \mathbf{I} \end{bmatrix} < 0, \tag{5.57}$$

where $\Pi^{\mathrm{Hinf}} = \mathrm{diag}\{\mathbf{I}, -\gamma^2\mathbf{I}\}$, $\Xi = \left[\, \Xi_{11} \; \Xi_{12}; \; \Xi_{12}^* \; \Xi_{22} \,\right]$ with $\Xi_{11} = -\mathbb{P}$

$$\Xi_{12} = \tilde{\Lambda}\mathbb{Q}, \Xi_{22} = \mathbb{P} - \tilde{\Gamma}\mathbb{Q} \tag{5.58}$$

and $\mathbb{P}$, $\mathbb{Q}$, $\tilde{\Lambda}$, $\tilde{\Gamma}$ are defined in Lemma 5.2.

Following the same way as above, the inequality in (5.57) can be rewritten as

$$\mathbb{E}^{\mathrm{T}}\Upsilon^{\mathrm{Hinf}}\mathbb{E} < 0, \tag{5.59}$$

where

$$\mathbb{E} = \begin{bmatrix} \mathbb{A}^{\mathrm{T}} & \mathbf{I} & \mathbb{C}^{\mathrm{T}} & 0 \\ \mathbb{B}^{\mathrm{T}} & 0 & 0 & \mathbf{I} \end{bmatrix}^{\mathrm{T}}, \Upsilon^{\mathrm{Hinf}} = \begin{bmatrix} \Xi & 0 \\ 0 & \Pi^{\mathrm{Hinf}} \end{bmatrix}, \tag{5.60}$$

which can be further formulated as the following dilated matrix inequality

$$\begin{bmatrix} \mathbb{E} \\ \hline K\hat{C} - L \;\; 0 \end{bmatrix}^{\mathrm{T}} \hat{\Upsilon}^{\mathrm{Hinf}} \begin{bmatrix} \mathbb{E} \\ \hline K\hat{C} - L \;\; 0 \end{bmatrix} < 0, \tag{5.61}$$

where

$$\hat{\Upsilon}^{\mathrm{Hinf}} \triangleq \begin{bmatrix} \Upsilon^{\mathrm{Hinf}} & 0 \\ 0 & 0 \end{bmatrix}. \tag{5.62}$$

The following theorem gives a sufficient condition to guarantee the finite-frequency-range 2D $\mathcal{H}_\infty$ performance specification.

Theorem 5.5 Given K_1, K_2 and a state-space realization of $\mathbb{G}_v(e^{j\omega_1}, e^{j\omega_2})$, the 2D performance specification in (5.56) is satisfied over a specified frequency range $\Omega_1 \times \Omega_2$ if there exist matrices $\mathbb{P}_1^T = \mathbb{P}_1$, $\mathbb{P}_2^T = \mathbb{P}_2$, $\mathbb{Q}_1 > 0$, $\mathbb{Q}_2 > 0$, $N_i, i = 1, 2, 3, 4$,L_1,L_2, and R with appropriate dimension, such that

$$\hat{\Upsilon}^{\text{Hinf}} + \text{sym}\{\hat{N}^{\text{Hinf}}\} < 0, \tag{5.63}$$

where $\hat{\Upsilon}^{\text{Hinf}}$ is defined in (5.61) and

$$\hat{\Sigma}^{\text{Hinf}} = \begin{bmatrix} -\mathbf{I} & \hat{A}_\Delta + \hat{B}_\Delta L & 0 & \mathbb{B} & \hat{B}_\Delta \\ 0 & \mathbb{C} & -\mathbf{I} & 0 & 0 \\ 0 & K\hat{C} - L & 0 & 0 & -\mathbf{I} \end{bmatrix}, \hat{N} = \begin{bmatrix} N_1 & N_2 & 0 \\ N_3 & N_4 & 0 \\ 0 & 0 & R \end{bmatrix}. \tag{5.64}$$

Proof The proof of this theorem is similar to that of Theorem 5.3 and, therefore, is omitted here. □

Similarly, the following parametrization of the finite-frequency-range 2D $\mathcal{H}_\infty$ performance specification is presented.

Theorem 5.6 Given a state-space realization of $\mathbb{G}_v(e^{j\omega_1}, e^{j\omega_2})$, the 2D performance specification in (5.56) is satisfied over a specified frequency range $\Omega_1 \times \Omega_2$ if there exist matrices $\mathbb{P}_1^T = \mathbb{P}_1$, $\mathbb{P}_2^T = \mathbb{P}_2$, $\mathbb{Q}_1 > 0$, $\mathbb{Q}_2 > 0$, N_{11}, N_{12}, N_{21}, N_{22}, N_{31}, N_{32}, N_{41}, N_{42}, L_1, L_2, $\hat{K}$, R with appropriate dimension and a scalar $\lambda > 0$, such that the following matrix inequality is satisfied

$$\Psi^{\text{Hinf}} \triangleq \begin{bmatrix} \psi_{11} & \psi_{12} & \psi_{13} & \psi_{14} & \psi_{15} & N_{11}\hat{F} & 0 \\ * & \psi_{22} & \psi_{23} & \psi_{24} & \psi_{25} & N_{12}\hat{F} & \lambda(\hat{H}_A + H_B L)^T \\ * & * & \psi_{33} & \psi_{34} & \psi_{35} & N_{31}\hat{F} & 0 \\ * & * & * & \psi_{44} & \psi_{45} & N_{32}\hat{F} & 0 \\ * & * & * & * & \psi_{55} & 0 & \lambda H_B^T \\ * & * & * & * & * & -\lambda\mathbf{I} & 0 \\ * & * & * & * & * & * & -\lambda\mathbf{I} \end{bmatrix} < 0, \tag{5.65}$$

where Ξ is defined in (5.57), $\hat{A}$, $\hat{B}$, $\hat{F}$, $\hat{H}_A$, $\mathbb{B}$, and $\mathbb{C}$ are defined in (5.30) and

$$\begin{aligned}
&\psi_{11} = -\mathrm{sym}\{N_{11}\} + \Xi_{11},\ \psi_{12} = N_{11}(\hat{A} + \hat{B}L) + N_{21}\mathbb{C} - N_{12}^{\mathrm{T}} + \Xi_{12},\\
&\psi_{13} = -N_{21} - N_{31}^{\mathrm{T}},\ \psi_{14} = N_{11}\mathbb{B} - N_{32}^{\mathrm{T}},\ \psi_{15} = N_{11}\hat{B},\\
&\psi_{22} = \mathrm{sym}\{N_{12}(\hat{A} + \hat{B}L) + N_{22}\mathbb{C}\} + \Xi_{22},\\
&\psi_{23} = -N_{22} + (\hat{A} + \hat{B}L)^{\mathrm{T}}N_{31}^{\mathrm{T}} + \mathbb{C}^{\mathrm{T}}N_{41}^{\mathrm{T}},\\
&\psi_{24} = N_{12}\mathbb{B} + (\hat{A} + \hat{B}L)^{\mathrm{T}}N_{32}^{\mathrm{T}} + \mathbb{C}^{\mathrm{T}}N_{42}^{\mathrm{T}},\\
&\psi_{25} = N_{12}\hat{B} + (\hat{K}\hat{C} - RL)^{\mathrm{T}},\ \psi_{33} = -\mathrm{sym}\{N_{41}\} + \mathbf{I},\ \psi_{34} = N_{31}\mathbb{B} - N_{42}^{\mathrm{T}},\\
&\psi_{35} = N_{31}\hat{B},\ \psi_{44} = \mathrm{sym}\{N_{32}\mathbb{B}\} - \gamma^2\mathbf{I},\ \psi_{45} = N_{32}\hat{B},\ \psi_{55} = -R - R^{\mathrm{T}}.
\end{aligned} \tag{5.66}$$

Moreover, if the above inequality in (5.65) is feasible, then the SOF based ILC gains can be retrieved by the expression in (5.52).

Proof Suppose that the matrix inequality in (5.65) holds. Applying the Schur complement to (5.65) yields

$$\psi + \lambda \begin{bmatrix} 0 \\ (\hat{H}_{\mathrm{A}} + H_{\mathrm{B}}L)^{\mathrm{T}} \\ 0 \\ 0 \\ H_{\mathrm{B}}^{\mathrm{T}} \end{bmatrix} \begin{bmatrix} 0 \\ (\hat{H}_{\mathrm{A}} + H_{\mathrm{B}}L)^{\mathrm{T}} \\ 0 \\ 0 \\ H_{\mathrm{B}}^{\mathrm{T}} \end{bmatrix}^{\mathrm{T}} + \lambda^{-1} \begin{bmatrix} N_{11}\hat{F} \\ N_{12}\hat{F} \\ N_{31}\hat{F} \\ N_{32}\hat{F} \\ 0 \end{bmatrix} \begin{bmatrix} N_{11}\hat{F} \\ N_{12}\hat{F} \\ N_{31}\hat{F} \\ N_{32}\hat{F} \\ 0 \end{bmatrix}^{\mathrm{T}} < 0, \tag{5.67}$$

where

$$\psi = \begin{bmatrix} \psi_{11} & \psi_{12} & \psi_{13} & \psi_{14} & \psi_{15} \\ * & \psi_{22} & \psi_{23} & \psi_{24} & \psi_{25} \\ * & * & \psi_{33} & \psi_{34} & \psi_{35} \\ * & * & * & \psi_{44} & \psi_{45} \\ * & * & * & * & \psi_{55} \end{bmatrix}.$$

By Lemma 2.1, there follows

$$\hat{\Upsilon}^{\mathrm{Hinf}} + \mathrm{sym}\{\hat{\Gamma}^{\mathrm{Hinf}}\} < 0, \tag{5.68}$$

where $\hat{\Upsilon}^{\mathrm{Hinf}}$ is defined in (5.61) and

$$\hat{\Gamma}^{\text{Hinf}} = \begin{bmatrix} -N_1 & N_1(\hat{A}_\Delta + \hat{B}_\Delta L) + N_2\mathbb{C} & -N_2 & N_1\mathbb{B} & N_1\hat{B}_\Delta \\ -N_3 & N_3(\hat{A}_\Delta + \hat{B}_\Delta L) + N_4\mathbb{C} & -N_4 & N_3\mathbb{B} & N_3\hat{B}_\Delta \\ 0 & \hat{K}\hat{C} - RL & 0 & 0 & -R \end{bmatrix},$$
$$N_1 = \begin{bmatrix} N_{11} \\ N_{12} \end{bmatrix}, N_2 = \begin{bmatrix} N_{21} \\ N_{22} \end{bmatrix}, N_3 = \begin{bmatrix} N_{31} \\ N_{32} \end{bmatrix}, N_4 = \begin{bmatrix} N_{41} \\ N_{42} \end{bmatrix}. \tag{5.69}$$

Then by letting $\hat{K} = RK$, it follows that $\hat{N}\hat{\Sigma}^{\text{Hinf}} = \hat{\Gamma}^{\text{Hinf}}$. According to Theorem 5.5, the SOF based ILC law with $K = R^{-1}\hat{K}$ guarantees that the performance specification in (5.56) is satisfied over a specified frequency range $\Omega_1 \times \Omega_2$. The proof is complete. □

Remark 5.2 For the entire frequency range, i.e., $\mathbb{Q}_1 = 0$ and $\mathbb{Q}_2 = 0$, the 2D $\mathcal{H}_\infty$ performance specification can be regarded as a special case of the finite-frequency-range performance analysis presented as above. ◊

It is seen that both inequalities in (5.50) and (5.65) are nonlinear with respect to their arguments, but could become linear if L_1 and L_2 are specified in advance. To find suitable L_1 and L_2, that is, the related SF-based ILC gains, for implementing the proposed algorithm with iteration to determine these ILC gains, the 2D Roesser system in (5.29) is modified as the following form by applying the SF-based ILC law in (5.49),

$$\begin{cases} \begin{bmatrix} \delta x(t+1,k) \\ e(t+1,k) \end{bmatrix} = \begin{bmatrix} \overline{\mathcal{A}} & \overline{\mathcal{B}} \\ \overline{\mathcal{C}} & \overline{\mathcal{D}} \end{bmatrix} \begin{bmatrix} \delta x(t,k) \\ e(t+1,k-1) \end{bmatrix} + \mathbb{B}\delta w(t,k), \\ z(t,k) = \mathbb{C}\begin{bmatrix} \delta x(t,k) \\ e(t+1,k-1) \end{bmatrix}, \end{cases} \tag{5.70}$$

where

$$\overline{\mathcal{A}} = A + \Delta A + (B + \Delta B)L_1,\ \overline{\mathcal{B}} = (B + \Delta B)L_2,$$
$$\overline{\mathcal{C}} = C(A + \Delta A + (B + \Delta B)L_1),\ \overline{\mathcal{D}} = \mathbf{I} + C(B + \Delta B)L_2.$$

Correspondingly, the following theorem is presented for establishing stability conditions for the 2D Roesser system in (5.70) under an SF-based ILC law in (5.49), from which the feasible solutions of L_1 and L_2 could be derived.

Theorem 5.7 Consider the batch process described by (5.1) is robustly stable by a control law in (5.49) with a 2D $\mathcal{H}_\infty$ performance specification in (5.33) over a specified frequency range $\Omega_1 \times \Omega_2$, if there exist matrices $\hat{P}_1 > 0$, $\hat{P}_2 > 0$, $\widehat{\mathbb{P}}_1 > 0$, $\widehat{\mathbb{P}}_2 > 0$, $\widehat{\mathbb{Q}}_1 > 0$, $\widehat{\mathbb{Q}}_2 > 0$, $\hat{W}_1$, $\hat{W}_2$, $\hat{L}$ and scalars $\alpha, \beta \in (0, 1]$, $\gamma > 0$, $\varepsilon_1 > 0$, $\varepsilon_2 > 0$, such that the following LMIs are satisfied,

$$\begin{bmatrix} -\Pi_\alpha^\beta \hat{P} & (\hat{A}\hat{W}+\hat{B}\hat{L})^{\mathrm{T}} & (\hat{H}_{\mathrm{A}}\hat{W}+H_{\mathrm{B}}\hat{L})^{\mathrm{T}} \\ * & \hat{P}-\mathrm{sym}\{\hat{W}\}+\varepsilon_1\hat{F}\hat{F}^{\mathrm{T}} & 0 \\ * & * & -\varepsilon_1\mathbf{I} \end{bmatrix} < 0, \tag{5.71}$$

$$\begin{bmatrix} -\widehat{\mathbb{P}} & \hat{\Xi}_{12}-\hat{W}^{\mathrm{T}} & 0 & 0 & 0 \\ * & \hat{\Xi}_{22}+\mathrm{sym}\{\hat{A}\hat{W}+\hat{B}\hat{L}\}+\varepsilon_2\hat{F}\hat{F}^{\mathrm{T}} & \mathbb{B} & \hat{W}^{\mathrm{T}}\mathbb{C}^{\mathrm{T}} & (\hat{H}_{\mathrm{A}}\hat{W}+H_{\mathrm{B}}\hat{L})^{\mathrm{T}} \\ * & * & -\gamma^2\mathbf{I} & 0 & 0 \\ * & * & * & -\mathbf{I} & 0 \\ * & * & * & * & -\varepsilon_2\mathbf{I} \end{bmatrix} < 0, \tag{5.72}$$

where $\hat{A}$, $\hat{B}$, $\hat{F}$, $\hat{H}_{\mathrm{A}}$, H_{B}, $\mathbb{B}$, and $\mathbb{C}$ are defined in (5.2), (5.29) and (5.32), Π_α^β is defined in Theorem 5.2, $\tilde{\Lambda}$ and $\tilde{\Gamma}$ are defined in Lemma 5.2

$$\begin{aligned} &\hat{P}=\mathrm{diag}\{\hat{P}_1,\hat{P}_2\},\ \hat{W}=\mathrm{diag}\{\hat{W}_1,\hat{W}_2\}, \\ &\widehat{\mathbb{P}}=\mathrm{diag}\{\widehat{\mathbb{P}}_1,\widehat{\mathbb{P}}_2\},\ \widehat{\mathbb{Q}}=\mathrm{diag}\{\widehat{\mathbb{Q}}_1,\widehat{\mathbb{Q}}_2\}, \\ &\hat{\Xi}_{12}=\tilde{\Lambda}\widehat{\mathbb{Q}},\ \hat{\Xi}_{22}=\widehat{\mathbb{P}}-\tilde{\Gamma}\mathbb{Q}, \end{aligned} \tag{5.73}$$

and correspondingly, the SF-based ILC gains are retrieved by

$$\begin{bmatrix} L_1 & L_2 \end{bmatrix} = \hat{L}\hat{W}^{-1}. \tag{5.74}$$

Proof The proof of this theorem follows immediately from Lemmas 2.1, 5.2, and Theorem 5.2, and therefore is omitted for brevity.

Motivated by the heuristic approach studied in Li and Gao (2015); Mehdi et al. (2004), a two-stage heuristic algorithm is presented to iteratively solve the feasible SOF-based ILC gains in (5.28) as follows.

Algorithm I (Design of the SOF-based ILC law)

Step 1. Given $\alpha, \beta \in (0, 1]$ and $\gamma > 0$, solve the feasible LMI conditions in (5.71) and (5.72) to obtain L_1 and L_2;

Step 2. Give a specified tolerance $\kappa > 0$, set $L_1^1 = L_1$ and $L_2^1 = L_2$ and $k = 1$;

Step 3. Solve the following LMI problem to obtain ε_1^k,

$$\begin{aligned}
&\min \varepsilon_1^k = \varepsilon \\
&s.t.\ \Psi < \varepsilon I,\ \Psi^{\text{Hinf}} < \varepsilon I, \\
&\text{for } P_1 > 0, P_2 > 0, \mathbb{P}_1^\top = \mathbb{P}_1, \mathbb{P}_2^\top = \mathbb{P}_2, \mathbb{Q}_1 > 0, \mathbb{Q}_2 > 0, \\
&\quad M_j, N_{ij}, i = 1, 2, 3, 4, j = 1, 2, \hat{K}_1, \hat{K}_2, R, \eta > 0, \lambda > 0, \varepsilon, \\
&\text{with fixed } L_1 = L_1^k, L_2 = L_2^k.
\end{aligned} \tag{5.75}$$

If $\varepsilon_1^k \leq 0$, then $K_1 = R^{-1}\hat{K}_1$ and $K_2 = R^{-1}\hat{K}_2$ are the desired SOF based ILC gains, and exit; otherwise, denote $M_j^k = M_j, N_{ij}^k = N_{ij}, i = 1, 2, 3, 4, j = 1, 2, R^k = R, \eta^k = \eta$ and $\lambda^k = \lambda$, go to the next step;

Step 4. Solve the following LMI problem to obtain ε_2^k,

$$\begin{aligned}
&\min \varepsilon_2^k = \varepsilon \\
&\text{s.t.} \Psi < \varepsilon \mathbf{I},\ \Psi^{\text{Hinf}} < \varepsilon \mathbf{I}, \\
&\text{for } P_1 > 0, P_2 > 0, \mathbb{P}_1^\top = \mathbb{P}_1, \mathbb{P}_2^\top = \mathbb{P}_2, \mathbb{Q}_1 > 0, \mathbb{Q}_2 > 0, \mathrm{L}_1, \mathrm{L}_2, \hat{\mathrm{K}}_1, \hat{\mathrm{K}}_2, \varepsilon, \\
&\text{with fixed } M_j = M_j^k, N_{ij} = N_{ij}^k, i = 1, 2, 3, 4, j = 1, 2, R = R^k, \eta = \eta^k, \lambda = \lambda^k,
\end{aligned} \tag{5.76}$$

If $\varepsilon_2^k \leq 0$, then $K_1 = R^{-1}\hat{K}_1$ and $K_2 = R^{-1}\hat{K}_2$ are the desired SOF-based ILC gains, and exit; otherwise, set $L_1^{k+1} = L_1$ and $L_2^{k+1} = L_2$, go to the next step;

Step 5. If $|\varepsilon_1^k - \varepsilon_2^k|/\varepsilon_2^k < \kappa$, no feasible SOF-based ILC gains could be found, and exit; else, set $k \leftarrow k + 1$, and then go back to **Step 3**.

Remark 5.3 It is easy to see from Algorithm I that ε_1^k and ε_2^k satisfy $\varepsilon_1^k \geq \varepsilon_2^k \geq \varepsilon_1^{k+1}$ for all $k = 1, 2, \ldots$, indicating that both ε_1^k and ε_2^k are nonincreasing. Different from the developed control synthesis methods for designing the SOF controller (see e.g., Mehdi et al. 2004), the proposed algorithm computes the feasible solutions iteratively by introducing an additional parameter ε, which is nonincreasing during iteration. ◇

5.3.3 *An Illustrative Example*

Example 5.2 Consider again the above batch packing process of an injection molding machine as studied in Shi et al. (2006); Liu et al. (2014); Hao et al. (2019), described by

$$\begin{cases} x(t+1,k) = \left(\begin{bmatrix} 1.607 & 1 \\ -0.6086 & 0 \end{bmatrix} + \Delta A\right)x(t,k) \\ \qquad + \left(\begin{bmatrix} 1.239 \\ -0.9282 \end{bmatrix} + \Delta B\right)u(t,k) + \begin{bmatrix} 1 \\ 0 \end{bmatrix} w(t,k), \\ y(t,k) = \begin{bmatrix} 1 & 0 \end{bmatrix} x(t,k), \quad x_0 = \begin{bmatrix} 0 & 0 \end{bmatrix}. \end{cases} \tag{5.77}$$

where the model uncertainties are in the form of (5.2) with

$$F = \begin{bmatrix} 0.5 & 0 \\ 0 & 0.5 \end{bmatrix}, H_{\mathrm{A}} = \begin{bmatrix} 0.0804 & 0 \\ -0.0304 & 0 \end{bmatrix}, H_{\mathrm{B}} = \begin{bmatrix} 0.062 \\ -0.0464 \end{bmatrix}, \Delta(t) = \begin{bmatrix} \delta_1(t) & 0 \\ 0 & \delta_2(t) \end{bmatrix},$$

and $\delta_i(t), i = 1, 2$ are time-varying factors within the interval [0, 1]. For illustration, the disturbance is assumed as $w(t,k) = 5\sin(0.1\delta_3 t + 0.2\delta_4 k)$, $k \le 20$, where δ_3 and δ_4 are randomly varying in the interval [0, 1]. Moreover, the desired reference trajectory is taken as

$$y_{\mathrm{d}}(t) = \begin{cases} 200, & 0 \le t \le 100, \\ 200 + 5(t - 100), & 100 < t \le 120, \\ 300, & 120 < t \le 200. \end{cases} \tag{5.78}$$

It should be mentioned that the initial part of $y_{\mathrm{d}}(t)$ is smoothed by a user-specified prefilter $G_{\mathrm{f}}(z) = (z^{-1} + z^{-2})/(3 - z^{-1})$ for practical implementation.

Based on the upper bounds of disturbance frequency along both the time and batch directions, the frequency ranges of interest are taken as $[-0.1, 0.1] \times [-0.2, 0.2]$. The SF based ILC controller gains are computed by solving the LMI conditions in (5.71) and (5.72) in Theorem 5.7 with $\alpha = 1$, $\beta = 1$ and $\gamma = 10$, obtaining

$$L_1 = \begin{bmatrix} -1.2759 & -0.8227 \end{bmatrix}, L_2 = -1.2175. \tag{5.79}$$

Then by using the proposed heuristic algorithm with the obtained L_1 and L_2 as initial values, feasible SOF based ILC controller gains are computed as

$$K_1 = -0.8343, K_2 = -0.8743. \tag{5.80}$$

Note that for the entire frequency range, that is, $\mathbb{Q}_1 = 0$ and $\mathbb{Q}_2 = 0$, no feasible SOF-based ILC controller gains could be found by Algorithm I. Figure 5.5 shows the averaged maximum singular value of $\mathbb{G}_v(e^{\mathrm{j}\omega_1}, e^{\mathrm{j}\omega_2})$ with respect to 100 random cases of the above model uncertainties, indicating that the control objective in (5.33) is satisfied. For comparison, the conventional open-loop ILC (Meng and Moore, 2017a, 2017b), output feedback-based ILC (Hladowski et al. 2012) and the 2D DOF-based ILC (Shi et al. 2006) are also applied to the above process with uncertainties and disturbances, where the ILC controller structure and learning gains are listed in Table 5.2 by solving the corresponding convergence conditions given therein.

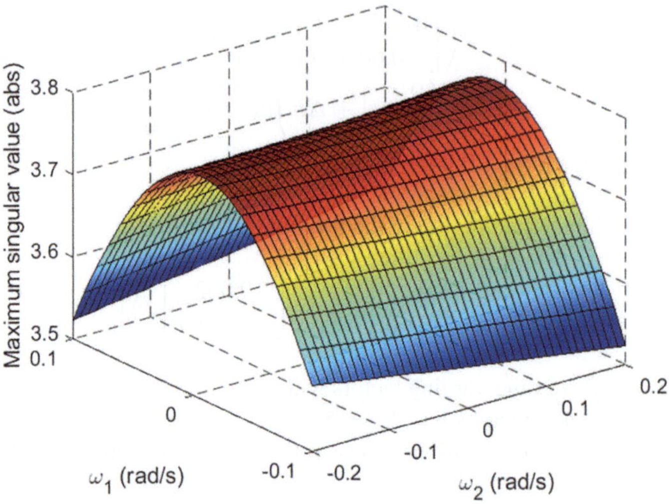

Fig. 5.5 Averaged maximum singular value of $\mathbb{G}_v(e^{j\omega_1}, e^{j\omega_2})$ with respect to 100 uncertain cases in Example 5.2

For the nominal case with no uncertainties and disturbances, the control results are shown in Fig. 5.6, along with the output tracking error in terms of the averaged tracking error index ($\text{ATE}(k) = \sum_{t=1}^{N} |e_k(t)|/N$) in Fig. 5.7. It is seen that perfect tracking can be achieved after three batches by the proposed method similar to that by the 2D DOF based ILC in Shi et al. (2006), whereas almost six batches are needed to reach the same tracking performance by the output feedback based ILC design in Hladowski et al. (2012). Note that almost 160 batches are needed to realize perfect tracking by the conventional open-loop ILC method (Meng and Moore 2017a, b) due to the existence of a slow pole in the process model.

For the presence of time-varying uncertainties and nonrepetitive disturbances, the tracking results are shown in Fig. 5.8, while the corresponding ATE indices are plotted in Fig. 5.9. It is shown that the closed-loop system well maintains robust stability by the proposed method, and the process output can robustly track the desired trajectory with a small tracking error. Compared with the open-loop ILC in Meng and Moore (2017a, b) and output feedback-based ILC in Hladowski et al. (2012), the tracking performance is significantly improved by the proposed method. It should be noted that the ILC method in Shi et al. (2006) slightly outperforms the proposed method in the initial batches due to the use of 2D DOF for ILC implementation. However, the complexity of such a control structure of 2D DOF based ILC (Shi et al. 2006), as shown in Table 5.2, is much higher than the proposed method based on only SOF for practical implementation.

Table 5.2 Parameters setting in the cited ILC methods

References	ILC updating law	Learning gains
Meng and Moore (2017a, b)	$\delta u(t,k) = \Gamma e(t+1,k-1)$	$\Gamma = -0.5$
Hladowski et al. (2012)	$\delta u(t,k) = K_1 \delta y(t,k) + K_2 \delta y(t-1,k) + K_3 e(t+1,k-1)$	$K_1 = -0.9434$, $K_2 = 7.7872 \times 10^{-6}$, $K_3 = -0.5794$
Shi et al. (2006)	$\begin{cases} x^{c}(t+1,k) = A_{c1} x^{c}(t,k) + A_{c2} x^{c}(t+1,k-1) \\ \quad + B_{c1} \hat{y}(t,k) + B_{c2} \hat{y}(t+1,k-1) \\ \delta u(t,k) = C_{c1} x^{c}(t,k) + C_{c2} x^{c}(t+1,k-1) \\ \quad + D_{c1} \hat{y}(t,k) + D_{c2} \hat{y}(t+1,k-1) \\ \hat{y}(t,k) = \left[e^{\mathrm{T}}(t,k-1) \; e^{\mathrm{T}}(t,k) \right]^{\mathrm{T}} \end{cases}$	$A_{c1} = \begin{bmatrix} 0.0145 & -0.0204 & 0.0003 \\ -0.5264 & 0.7540 & 0.0002 \\ -0.0013 & -0.0003 & 0.0117 \end{bmatrix}$, $D_{c1} = \begin{bmatrix} 1.3089 \\ -1.3133 \end{bmatrix}^{\mathrm{T}}$, $D_{c2} = \begin{bmatrix} 0.0000 \\ -0.7686 \end{bmatrix}^{\mathrm{T}}$ $A_{c2} = \begin{bmatrix} 0.0002 & 0.0000 & 0.0000 \\ 0.0002 & 0.0000 & 0.0000 \\ -0.0010 & -0.0000 & 0.0006 \end{bmatrix}$, $B_{c1} = \begin{bmatrix} -0.0031 & 0.0030 \\ 0.1103 & -0.1111 \\ 0.0005 & 0.0002 \end{bmatrix}$ $B_{c2} = \begin{bmatrix} -0.0001 & 0.4550 \\ -0.0001 & 0.1863 \\ 0.0005 & -0.0002 \end{bmatrix}$, $C_{c1} = \begin{bmatrix} -3.1147 \\ 4.4602 \\ 0.0000 \end{bmatrix}^{\mathrm{T}}$, $C_{c2} = 10^{-5} \cdot \begin{bmatrix} -0.3639 \\ -0.0051 \\ 0.2753 \end{bmatrix}^{\mathrm{T}}$

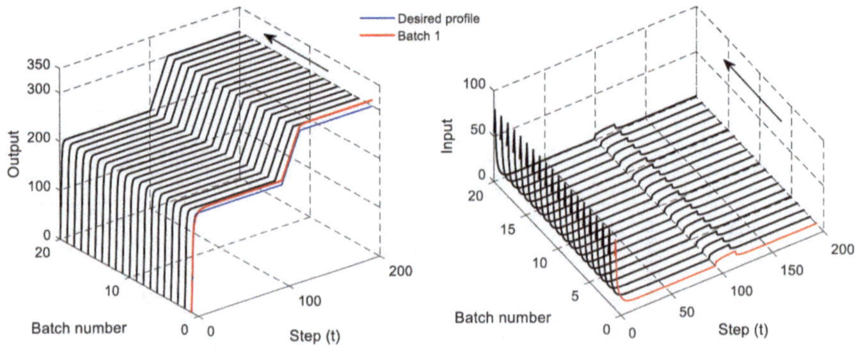

Fig. 5.6 Tracking results for the nominal case in Example 5.2

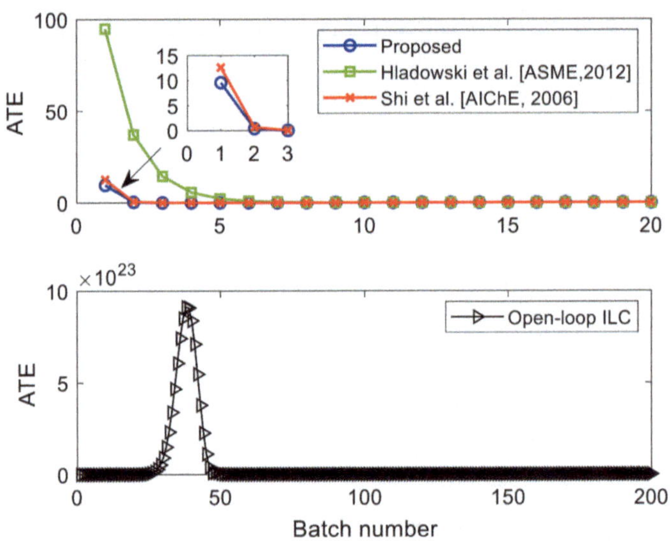

Fig. 5.7 ATE performance index for the nominal case

5.4 Summary

For batch processes with time-varying uncertainties subject to load disturbance, a DOF-based ILC scheme has been presented in this chapter, based on a 2D FM system description of the batch process operation (Liu and Wang 2012). By using the 2D Lyapunov stability theory, a sufficient condition in terms of LMI has been established, such that the robust convergence along both the time and batch directions can be guaranteed, which could be effectively solved to determine the controller gain matrices.

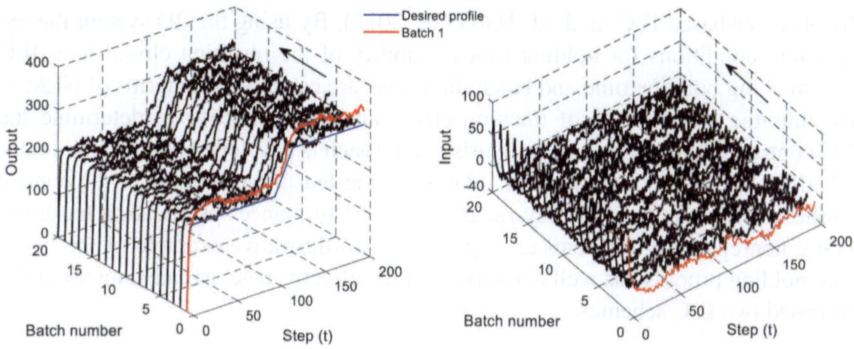

Fig. 5.8 Tracking results under time-varying uncertainties and nonrepetitive disturbances in Example 5.2

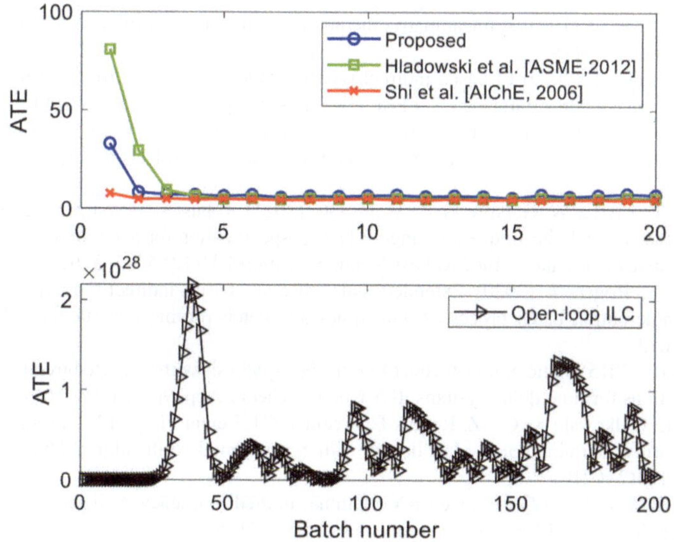

Fig. 5.9 Plots of ATE performance index under time-varying uncertainties and nonrepetitive disturbances in Example 5.2

Concerning the finite frequency nature of external disturbance in practice, a robust SOF-based ILC design with a finite-frequency-range 2D $\mathcal{H}_\infty$performance specification has also been presented for batch processes with time-varying uncertainties subject to nonrepetitive disturbances (Hao et al. 2021). Compared with the previous SF-based ILC designs with finite-frequency-range specifications (Paszke et al. 2013; Tao et al. 2019), only output measurement is required in the proposed control scheme, thus facilitating practical applications. Moreover, the proposed control scheme is much simpler for implementation, with no need to store the historical controller states as required by the previous 2D DOF-based ILC method (Shi et al. 2006) and

2D observer-based ILC method (Hao et al. 2016). By using the 2D system theory, sufficient conditions for holding robust stability of the resulting closed-loop ILC system along both the time and batch directions are established in terms of NLMIs, and thus the convergence of tracking error can be guaranteed. To determine the ILC controller gains, a two-stage heuristic approach is developed to solve the established NLMI conditions, which facilitates finding feasible solutions for the control synthesis with a finite-frequency-range 2D $\mathcal{H}_\infty$performance specification for attenuating nonrepetitive disturbances in practice. An illustrative example of the injection molding process has well demonstrated the effectiveness and advantages of the proposed two ILC schemes.

References

Gahinet P, Apkarian P (1994) A linear matrix inequality approach to control. Int J Robust Nonlinear Control 4(4):421–448

Hao S, Liu T, Gao F (2019) PI based indirect-type iterative learning control for batch processes with time-varying uncertainties: a 2D FM model based approach. J Process Control 78:57–67

Hao S, Liu T, Paszke W, Galkowski K (2016) Robust iterative learning control for batch processes with input delay subject to time-varying uncertainties. IET Control Theory Appl 10(15):1904–1915

Hao S, Liu T, Paszke W, Galkowski K, Wang QG (2021) Robust static output feedback based ILC design with a finite-frequency-range 2-D $\mathcal{H}_\infty$ specification for batch processes subject to nonrepetitive disturbances. Int J Robust Nonlinear Control 31(12):5745–5761

Hao S, Liu T, Rogers E (2020) Extended state observer based indirect-type ILC for single-input single-output batch processes with time- and batch-varying uncertainties. Automatica 112:108673

Hao Y, Duan Z (2015) Static output-feedback controller synthesis with restricted frequency domain specifications for time-delay systems. IET Control Theory Appl 9(10):1608–1614

Hladowski L, Galkowski K, Cai Z, Rogers E, Freeman CT, Lewin PL (2012) Output information based iterative learning control law design with experimental verification. J Dyn Syst Measur Control 134(2):021012

Iwasaki T, Hara S (2005) Generalized KYP lemma: unified frequency domain inequalities with design applications. IEEE Trans Autom Control 50(1):41–59

Li X, Gao H (2012) Robust finite frequency filtering for uncertain 2-D Roesser systems. Automatica 48(6):1163–1170

Li X, Gao H (2014) A heuristic approach to static output-feedback controller synthesis with restricted frequency-domain specifications. IEEE Trans Autom Control 59(4):1008–1014

Li X, Gao H (2015) Robust frequency-domain constrained feedback design via a two-stage heuristic approach. IEEE Trans Cybern 45(10):2065–2075

Liu T, Wang X-Z, Chen J (2014) Robust PID based indirect-type iterative learning control for batch processes with time-varying uncertainties. J Process Control 24(12):95–106

Liu T, Wang Y (2012) A synthesis approach for robust constrained iterative learning control of piecewise affine batch processes. Automatica 48(11):2762–2775

Lu J, Wang Z, Gao F (2015) A two-stage design of two-dimensional model predictive iterative learning control for nonrepetitive disturbance attenuation. Ind Eng Chem Res 54(21):5683–5689

Mandra S, Aschemann H (2017) Robust guaranteed cost ILC with dynamic feedforward and disturbance compensation for accurate PMSM position control. Control Eng Pract 65:36–47

Mehdi D, Boukas EK, Bachelier O (2004) Static output feedback design for uncertain linear discrete time systems. IMA J Math Control Inf 21:1–13

Meng D, Moore KL (2017a) Convergence of iterative learning control for SISO nonrepetitive systems subject to iteration-dependent uncertainties. Automatica 79:167–177

Meng D, Moore KL (2017b) Robust iterative learning control for nonrepetitive uncertain systems. IEEE Trans Autom Control 62(2):907–913

Meng D (2019) Convergence conditions for solving robust iterative learning control problems under nonrepetitive model uncertainties. IEEE Trans Neural Netw Learn Syst 30(6):1908–1919

Paszke W, Rogers E, Galkowski K, Cai Z (2013) Robust finite frequency range iterative learning control design and experimental verification. Control Eng Pract 21(10):1310–1320

Rantzer A (1986) On the Kalman-Yakubovich-Popov lemma. Syst Control Lett 28(1):7–10

Shi J, Gao F, Wu T-J (2005) Robust design of integrated feedback and iterative learning control of a batch process based on a 2D Roesser system. J Process Control 15(8):907–924

Shi J, Gao F, Wu T-J (2006) Robust iterative learning control design for batch processes with uncertain perturbations and initialization. AIChE J 52(6):2171–2187

Tao H, Paszke W, Yang H, Galkowski K (2019) Finite frequency range robust iterative learning control of linear discrete system with multiple time-delays. J Franklin Inst 356(5):2690–2708

Yin C, Xu J-X, Hou Z (2010) A high-order internal model based iterative learning control scheme for nonlinear systems with time-iteration-varying parameters. IEEE Trans Autom Control 55(11):2665–2670

Zhu Q, Xu JX, Huang D, Hu GD (2015) Iterative learning control design for linear discrete-time systems with multiple high-order internal models. Automatica 62:65–76

Chapter 6
Extended State Observer (ESO) Based ILC Design Under Process Uncertainties and Disturbance

6.1 Introduction

In many industrial batch processes, batch-varying dynamics arise from various sources, e.g., model uncertainties, uncertain initial states, and the presence of external disturbances. These issues could provoke non-repetitive system behavior or unstable performance, see, e.g., Altın et al. (2017); Meng (2019); Meng and Moore (2017a, b); Liu et al. (2023); Zhang and Meng (2023). Robust designs in this area, e.g., Paszke et al. (2016); Sebastian et al. (2019) and Shi et al. (2006) adopted the feedback control to maintain closed-loop system stability and a feedforward control based on previous batch information for reference tracking, formulating a 2D system setting to improve the control performance along both the time and batch directions. In cases where the batch-varying dynamics can be modeled, internal model-based ILC methods were developed to realize perfect or robust tracking (Zhu et al. 2015). Convergence conditions for the classical feedforward ILC designs under non-repetitive system uncertainties were analyzed based on the contraction mapping principle in the existing work (Meng 2019; Meng and Moore 2017a, b), which, however, only addressed error convergence along the batch direction while the system dynamics in the time direction were not considered. This may result in very poor or unacceptable performance in the initial batch run, especially for open-loop unstable processes (Sebastian et al. 2019).

In contrast to the above direct-type ILC methods, indirect-type ILC has received much attention since the feedback and learning control parts can be separately designed. Using the internal model control (IMC) structure, an indirect-type ILC scheme was developed in the frequency domain for the application to batch processes with uncertain time delay (Liu et al. 2010). Based on a model predictive control (MPC) scheme, a set-point tuning strategy was developed to form an indirect-type ILC design (Shi et al. 2014). For time-varying linear systems, a quadratic criterion was introduced to analyze the convergence of an MPC-based indirect-type ILC design (Lee et al. 2000). In Chi et al. (2023), an indirect-type data-driven set-point learning

T. Liu et al., *Robust Iterative Learning Control of Industrial Batch Systems*, Intelligent Control and Learning Systems 22, https://doi.org/10.1007/978-981-96-9778-6_6

control scheme was proposed for nonlinear repetitive systems with a priori P-type feedback control loop. By formulating the overall closed-loop system dynamics as a 2D Roesser state–space model, the robust stability condition of a learning-type set-point design for a PI control loop was analyzed in Wang et al. (2012), and then an enhanced ILC-based PI controller was designed by considering the state variations in the ILC design (Wang et al. 2013). Subsequently, Liu et al. (2014) developed a robust PID-based indirect-type ILC design by adding feedforward controllers to adjust the process input. A critical point emerging from a review of the literature is that a one-degree-of-freedom (1DOF) feedback controller was mostly adopted in the existing indirect-type ILC designs to support the learning controller design, which leads to a water-bed effect between the set-point tracking and disturbance rejection in the inner loop. This chapter develops a two-degree-of-freedom (2DOF) structure for the set-point learning controller design to address this issue.

A generalized extended state observer (GESO) based indirect-type ILC is presented herein to improve the tracking performance against time- and batch-varying uncertainties arising in the process dynamics and unsteady initial states together with the presence of external disturbances. A GESO-based 2DOF control is designed in the inner loop to ensure robust stability of the controlled system in the time direction, and then a simple P-type ILC law is used to solely update the set-point command in the inner loop for batch run optimization. A sufficient condition for holding the system robust stability together with bounded output tracking error and system input is established in terms of a tractable linear matrix inequality (LMI), for the convenience of practical applications. An illustrative example based on an injection molding process is given to demonstrate the achievable performance by the new design.

6.2 ESO-Based Feedback Control Design

Consider a single-input-single-output (SISO) batch process with time- and batch-varying uncertainties described by the state-space model

$$\begin{cases} x(t+1,k) = A(t,k)x(t,k) + B(t,k)[u(t,k) + \omega(t,k)], \\ y(t,k) = Cx(t,k), t \in \mathbb{Z}_N, k \in \mathbb{Z}_+, \end{cases} \tag{6.1}$$

where t and k, respectively, denote the time and batch indices. $x(t,k) \in \Re^n$, $u(t,k) \in \Re$ and $y(t,k) \in \Re$ are, respectively, the state vector, input and output of the current batch process. $\omega(t,k) \in \Re$ is the external load disturbance. $A(t,k) = A + E\Delta(t,k)H_\mathrm{A}$ and $B(t,k) = B + E\Delta(t,k)H_\mathrm{B}$ denote time- and batch-varying system matrices, where E, H_A and H_B are known matrices and $\Delta(t,k)$ denotes time- and batch-varying uncertainty satisfying $\Delta^\mathrm{T}(t,k)\Delta(t,k) \leq \mathbf{I}$. It is generally assumed that the system relative degree is unity, i.e., $CB(t,k) \neq 0$ (see Meng and Moore 2017a). If this assumption does not hold, then specific changes to the ILC design are required, see e.g., Paszke et al. (2016).

Let $y_{\rm d}(t)$, $t \in \mathbb{Z}_N$ denote the reference trajectory satisfying $|y_{\rm d}(t)| \leq \beta_r < \infty$. Then perfect tracking under an ILC law applied to (6.1) requires that

$$\lim_{k\to\infty} y(t,k) = y_{\rm d}(t), \quad \forall t \in \mathbb{Z}_N. \tag{6.2}$$

The above property cannot be acquired in the presence of batch-varying uncertainties. Instead, it is practically required that the tracking error $e(t,k) = y_{\rm d}(t) - y(t,k)$ is bounded for $k \in \mathbb{Z}_+$ and $t \in \mathbb{Z}_N$, i.e.

$$\sup_{k\geq 0}|e(t,k)| \leq \beta_e, \quad \lim_{k\to\infty}\sup|e(t,k)| \leq \beta_{e_{\rm sup}}, \tag{6.3}$$

where $\beta_e > \beta_{e_{\rm sup}} \geq 0$ are finite bounds. Meanwhile, the input and state in (6.1) are also required to be bounded, i.e.

$$\begin{aligned}
&\sup_{k\geq 0}\max_{k\leq t\leq N-1}|u(t,k)| \leq \beta_u < \infty, \\
&\sup_{k\geq 0}\max_{0\leq t\leq N}\|x(t,k)\| \leq \beta_x < \infty,
\end{aligned} \tag{6.4}$$

where $\beta_u \geq 0$ and $\beta_x \geq 0$ are finite bounds.

Assumption 6.1 (Meng and Moore 2017a). It is assumed that $x(0,k), \omega(t,k), A(t,k)$ and $B(t,k)$ are bounded for $t \in \mathbb{Z}_N$ and $k \in \mathbb{Z}_+$ i.e.

$$\begin{aligned}
&\sup_{k\geq 0}\|x(0,k)\| \leq \beta_{x_0}, \quad \sup_{k\geq 0}\max_{0\leq t\leq N}|\omega(t,k)| \leq \beta_\omega, \\
&\sup_{k\geq 0}\max_{0\leq t\leq N}\|A(t,k)\| \leq \beta_{\rm A}, \quad \sup_{k\geq 0}\max_{0\leq t\leq N}\|B(t,k)\| \leq \beta_{\rm B},
\end{aligned}$$

where $\beta_{x0} \geq 0, \ \beta_\omega \geq 0, \ \beta_{\rm A} \geq 0, \ \beta_{\rm B} \geq 0$, and $x(0,k)$ is the initial state vector in (6.1).

In this section, a generalized extended state observer (GESO) based control scheme applicable to open-loop stable or unstable processes is designed to guarantee robust stability of the controlled system in the time direction and also attenuate the effect of external disturbance, thus facilitating a set-point learning ILC design. Since the batch index k is not relevant to the GESO design, it is omitted in this section. The z-domain transfer-function of the nominal system is described by

$$G(z) = \frac{b_1 z^{n-1} + \cdots + b_{n-1}z + b_n}{z^n + a_1 z^{n-1} + \cdots + a_{n-1}z + a_n}. \tag{6.5}$$

The uncertain system dynamics with regard to (6.5) can be written as

$$Y(z) = G(z)[1 + \Delta(z)][U(z) + W(z)], \tag{6.6}$$

where $\Delta(z)$ is the plant uncertainty and $W(z)$ is the load disturbance, or alternatively,

$$Y(z) = G(z)[U(z) + W_{\text{sum}}(z)], \tag{6.7}$$

where $W_{\text{sum}}(z) = \Delta(z)U(z)+\Delta(z)W(z)+W(z)$ denotes the total disturbance formed from the unmodeled dynamics and external disturbance. By such an arrangement, the multiplicative uncertainty $\Delta(z)$ is used to express the total disturbance effect on the system in the same channel as the control input.

To facilitate the observer design, the system in (6.7) is reformulated in the observable canonical form

$$\begin{cases} x(t+1) = A_0 x(t) + B_0[u(t) + \omega_{\text{sum}}(t)], \\ y(t) = C_0 x(t), \end{cases} \tag{6.8}$$

where $C_o = [\,1\ 0 \ldots 0\ 0\,]$, and

$$A_0 = \begin{bmatrix} -a_1 & 1\ 0 \cdots 0 \\ \vdots & \vdots\ \vdots\ \ddots\ \vdots \\ -a_{n-1} & 0\ 0 \cdots 1 \\ -a_n & 0\ 0 \cdots 0 \end{bmatrix}, \quad B_0 = \begin{bmatrix} b_1 \\ \vdots \\ b_{n-1} \\ b_n \end{bmatrix}. \tag{6.9}$$

These last two matrices represent the nominal parts, respectively, of $A(t,k)$ and $B(t,k)$ in (6.1), and $\omega_{\text{sum}}(t)$ is the inverse z-transform of $W_{\text{sum}}(z)$. To estimate the total disturbance $\omega_{\text{sum}}(t)$, an augmented system is written below by treating $\omega_{\text{sum}}(t)$ as an extended state vector

$$\begin{cases} X(t+1) = \hat{A}X(t) + \hat{B}u(t) + \hat{E}\Delta\omega_{\text{sum}}(t+1), \\ y(t) = \hat{C}X(t), \end{cases} \tag{6.10}$$

where $X(t) = \left[x^{\mathrm{T}}(t)\ \omega_{\text{sum}}^{\mathrm{T}}(t)\right]^{\mathrm{T}}$, $\Delta\omega_{\text{sum}}(t+1) = \omega_{\text{sum}}(t+1) - \omega_{\text{sum}}(t)$, and $\hat{A} = \begin{bmatrix} A_0 & B_0 \\ \mathbf{0} & 1 \end{bmatrix}$, $\hat{B} = \begin{bmatrix} B_0 \\ 0 \end{bmatrix}$, $\hat{C} = \begin{bmatrix} C_0^{\mathrm{T}} \\ 0 \end{bmatrix}^{\mathrm{T}}$, $\hat{E} = \begin{bmatrix} \mathbf{0} \\ 1 \end{bmatrix}$.

Using (6.10), the following observer is designed to estimate both the system state and the total disturbance

$$\begin{cases} \hat{X}(t+1) = \hat{A}\hat{X}(t) + \hat{B}u(t) + L[y(t) - \hat{y}(t)], \\ \hat{y}(t) = \hat{C}\hat{X}(t), \end{cases} \tag{6.11}$$

where $\hat{X}(t) = \left[\hat{x}^{\mathrm{T}}(t)\ \hat{\omega}_{\text{sum}}^{\mathrm{T}}(t)\right]^{\mathrm{T}}$, L is the observer gain that can be tuned by placing the eigenvalues of $\hat{A} - L\hat{C}$ at α_0, where $\alpha_0 \in (0, 1)$ is an adjustable parameter, $\hat{x}(t)$ and $\hat{\omega}_{\text{sum}}(t)$ are the estimates of $x(t)$ and $\omega_{\text{sum}}(t)$, respectively. The control law in the inner loop is

$$u(t) = Fy^s(t) - K\hat{X}(t), \tag{6.12}$$

where $K = [K_o \; 1]$ is the feedback controller with K_o tuned by placing the eigenvalues of $A_o - B_o K_o$ at α_c, where $\alpha_c \in (0, 1)$ is another adjustable parameter, and F is a set-point gain to eliminate the steady-state tracking error, which is given by $F = 1/[C_o(\mathbf{I} - A_o + B_o K_o)^{-1} B_o]$.

Note that there is only one tuning parameter associated with L or K_o and hence it is convenient to tune each of them to obtain a desired tradeoff between the control performance and robustness of the inner loop. Besides, the GESO-based control scheme is equivalent to a 2DOF control structure as discussed in Liu et al. (2019). Hence, such a 2DOF control scheme can be adopted to acquire better performance compared with the traditional 1DOF control schemes, e.g., the PID control loop used in the recent indirect-type ILC designs (Liu et al. 2014; Wang et al. 2012, 2013).

Correspondingly, the inner loop is written by

$$\begin{cases} \mathcal{X}(t+1, k) = \mathcal{A}(t, k)\mathcal{X}(t, k) + \mathcal{B}(t, k)y_s(t, k) + \mathcal{D}(t, k)\omega(t, k), \\ y(t, k) = \mathcal{C}\mathcal{X}(t, k), \end{cases} \tag{6.13}$$

where $\mathcal{X}(t, k) = \left[x^{\mathrm{T}}(t, k) \; \hat{X}^{\mathrm{T}}(t, k)\right]^{\mathrm{T}}$, and

$$\mathcal{A}(t, k) = \begin{bmatrix} A(t, k) & -B(t, k)K \\ LC & A - \hat{B}K - L\hat{C} \end{bmatrix}, \mathcal{B}(t, k) = \begin{bmatrix} B(t, k)F \\ \hat{B}F \end{bmatrix},$$
$$\mathcal{C} = \left[C \; \mathbf{0}\right], \mathcal{D}(t, k) = \left[B^{\mathrm{T}}(t, k) \; \mathbf{0}\right]^{\mathrm{T}},$$

where $y_s(t, k)$ is regarded as the set-point of the closed-loop system (6.13), which will be updated by the indirect-type ILC designed in the next section.

6.3 P-Type Set-Point Learning Control Scheme

Figure 6.1 shows a simple P-type indirect ILC scheme designed for the convenience of implementation, where only one constant-type learning controller Γ is used to adjust the set-point command, i.e.

$$y_s(t, k) = y_s(t, k-1) + \Gamma e(t+1, k-1), \tag{6.14}$$

where $y_s(t, k-1)$ and $e(t+1, k-1)$ denote, respectively, the set-point command and one-step ahead output error in the previous batch.

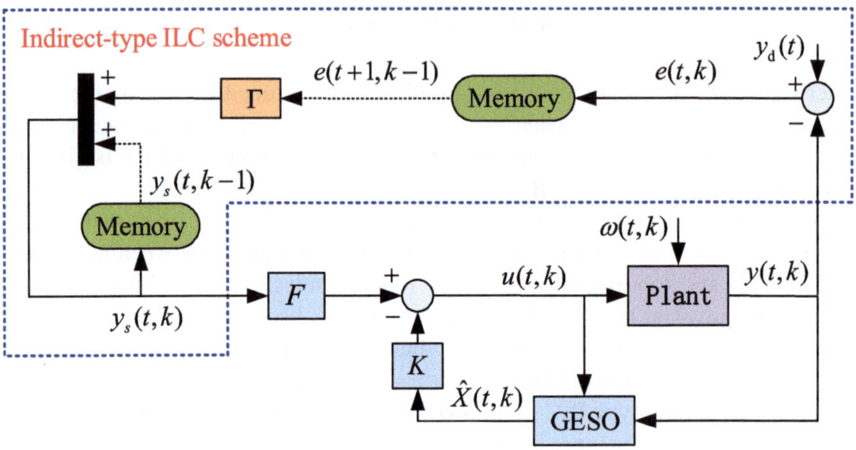

Fig. 6.1 Diagram of ESO-based indirect-type ILC scheme

Denote by $\delta f(t,k) \triangleq f(t,k) - f(t,k-1)$ an increment function along the batch direction, it follows from (6.13) that

$$\begin{aligned}\delta\mathcal{X}(t+1,k) &= \mathcal{A}(t,k)\delta\mathcal{X}(t,k) + \mathcal{B}(t,k)\delta y_s(t,k) + \varpi(t,k) \\ &= \mathcal{A}(t,k)\delta\mathcal{X}(t,k) + \mathcal{B}(t,k)\Gamma e(t+1,k-1) + \varpi(t,k),\end{aligned} \tag{6.15}$$

where

$$\begin{aligned}\varpi(t,k) = \delta\mathcal{A}(t,k)\mathcal{X}(t,k-1) + \delta\mathcal{B}(t,k)y_s(t,k-1) \\ + \mathcal{D}(t,k)\omega(t,k) - \mathcal{D}(t,k-1)\omega(t,k-1).\end{aligned} \tag{6.16}$$

Moreover,

$$\begin{aligned}e(t+1,k) &= e(t+1,k-1) - \mathcal{C}\delta\mathcal{X}(t+1,k) \\ &= [1 - \mathcal{C}\mathcal{B}(t,k)\Gamma]e(t+1,k-1) - \mathcal{C}\mathcal{A}(t,k)\delta\mathcal{X}(t,k) - \mathcal{C}\varpi(t,k).\end{aligned} \tag{6.17}$$

Hence, the ILC system described by (6.15) and (6.17) is equivalently expressed as a 2D Roesser system

$$\begin{bmatrix}\delta\mathcal{X}(t+1,k) \\ \hline e(t+1,k)\end{bmatrix} = \overline{\mathscr{A}}\begin{bmatrix}\delta\mathcal{X}(t,k) \\ \hline e(t+1,k-1)\end{bmatrix} + \mathscr{D}\varpi(t,k), \tag{6.18}$$

where $\mathscr{D} = \begin{bmatrix}\mathbf{I} & -\mathcal{C}^{\mathrm{T}}\end{bmatrix}^{\mathrm{T}}$ and

$$\overline{\mathscr{A}} = \left[\begin{array}{c|c} \mathcal{A}(t,k) & \mathcal{B}(t,k)\Gamma \\ \hline -\mathcal{C}\mathcal{A}(t,k) & 1-\mathcal{C}\mathcal{B}(t,k)\Gamma \end{array}\right] \triangleq \left[\begin{array}{c|c} \overline{\mathscr{A}}_{11} & \overline{\mathscr{A}}_{12} \\ \hline \overline{\mathscr{A}}_{21} & \overline{\mathscr{A}}_{22} \end{array}\right]. \tag{6.19}$$

The following result gives a tractable sufficient condition for convergence of the ILC system described by (6.13) and (6.14) along both the time and batch directions.

Theorem 6.1 Consider the system described by (6.13) under Assumption 6.1, where the set-point is updated by (6.14). Then the bounded tracking objective in (6.3) and the boundedness objective in (6.4) are guaranteed together with robust stability of (6.13), if there exist a matrix $P > 0$, scalars Γ, $\varepsilon > 0$ and $\lambda \in (0,\ 1)$ such that the following LMI is feasible

$$\begin{bmatrix} -P+\varepsilon\mathscr{E}_1\mathscr{E}_1^\top & \mathscr{A}_{11}P & \mathscr{A}_{12} & \varepsilon\mathscr{E}_1\mathscr{E}_2^\top & 0 \\ * & -P & 0 & P\mathscr{A}_{21}^\top & P\mathscr{F}_1^\top \\ * & * & -\lambda & \mathscr{A}_{22}^\top & \mathscr{F}_2^\top \\ * & * & * & -\lambda+\varepsilon\mathscr{E}_2\mathscr{E}_2^\top & 0 \\ * & * & * & * & -\varepsilon\mathbf{I} \end{bmatrix} < \mathbf{0}, \tag{6.20}$$

where $\mathscr{E}_1 = \begin{bmatrix} E^\top & \mathbf{0} \end{bmatrix}^\top$, $\mathscr{E}_2 = -CE$, $\mathscr{F}_1 = \begin{bmatrix} H_A & -H_B K \end{bmatrix}$, $\mathscr{F}_2 = H_B F\Gamma$, and

$$\mathscr{A}_{11} = \begin{bmatrix} A & -BK \\ LC & \hat{A}-\hat{B}K-L\hat{C} \end{bmatrix}, \mathscr{A}_{12} = \begin{bmatrix} BF\Gamma \\ \hat{B}F\Gamma \end{bmatrix},$$
$$\mathscr{A}_{21} = \begin{bmatrix} -CA & CBK \end{bmatrix}, \mathscr{A}_{22} = 1 - CBF\Gamma.$$

Proof Applying the Schur complement to (6.20) gives

$$\begin{bmatrix} -P & \mathscr{A}_{11}P & \mathscr{A}_{12} & 0 \\ * & -P & 0 & P\mathscr{A}_{21}^\top \\ * & * & -\lambda & \mathscr{A}_{22}^\top \\ * & * & * & -\lambda \end{bmatrix} + \varepsilon\xi_1^\top\xi_1 + \varepsilon^{-1}\xi_2^\top\xi_2 < 0,$$

where $\xi_1 = \begin{bmatrix} \mathscr{E}_1^\top & 0 & 0 & \mathscr{E}_2^\top \end{bmatrix}$ and $\xi_2 = \begin{bmatrix} 0 & \mathscr{F}_1 P & \mathscr{F}_2 & 0 \end{bmatrix}$.

Then by applying Lemma 2.1, it follows that

$$\begin{bmatrix} -P & \bar{\mathscr{A}}_{11}P & \bar{\mathscr{A}}_{12} & 0 \\ * & -P & 0 & P\bar{\mathscr{A}}_{21}^T \\ * & * & -\lambda & \bar{\mathscr{A}}_{22} \\ * & * & * & -\lambda \end{bmatrix} < 0. \tag{6.21}$$

Hence, there are $\rho(\overline{\mathscr{A}}_{11}) < 1$ and $\rho(\overline{\mathscr{A}}_{22}) < \lambda < 1$. Accordingly, it is routine to verify that $\left\|\mathcal{G}(e^{j\theta})\right\|_\infty < \lambda < 1$ for any $\theta \in [-\pi, \pi]$ by using the standard results in robust $\mathcal{H}_\infty$ control theory, where $\mathscr{G}(e^{j\theta}) = \overline{\mathscr{A}}_{21}\left(e^{j\theta}\mathbf{I} - \overline{\mathscr{A}}_{11}\right)^{-1}\overline{\mathscr{A}}_{12} + \overline{\mathscr{A}}_{22}$.

The boundedness objective in (6.4) for any $t \in \mathbb{Z}_N$ and $k \in \mathbb{Z}_+$ $t \in \mathbb{Z}_N$ and $k \in \mathbb{Z}_+$ can also be verified by mathematical induction on $t \in \mathbb{Z}_N$ $t \in \mathbb{Z}_N$. Firstly, the above (6.14) is reformulated as

$$\begin{aligned} y_s(t,k) &= y_s(t,k-1) + \Gamma\left[y_{\mathrm{d}}(t+1) - \mathcal{C}\mathcal{X}(t+1,k-1)\right] \\ &= [1 - \Gamma\mathcal{C}\mathcal{B}(t,k-1)]y_s(t,k-1) + \phi_s(t,k) - \Gamma\mathcal{C}\mathcal{A}(t,k-1)\mathcal{X}(t,k-1), \end{aligned} \tag{6.22}$$

where $\phi_s(t,k) \triangleq \Gamma y_{\mathrm{d}}(t+1) - \Gamma\mathcal{C}\mathcal{D}(t,k-1)\omega(t,k-1)$. Then it is immediate that $|\phi_s(t,k)| \le |\Gamma|(\beta_r + \|C\|\beta_{\mathrm{B}}\beta_\omega)$ for any $t \in \mathbb{Z}_N$ and $k \in \mathbb{Z}_+$. Moreover, both $A(t,k)$ and $B(t,k)$ are bounded for any $t \in \mathbb{Z}_N$ and $k \in \mathbb{Z}_+$ based on Assumption 6.1 plus the finite observer and controller gains denoted, respectively, by β_{A} and β_{B}. The remainder of the proof is presented in two steps.

Step (I): Let $t = 0$, there stands $|-\Gamma\mathcal{C}\mathcal{A}(0,k-1)\mathcal{X}(0,k-1)| \le |\Gamma|\|C\|\beta_{\mathcal{A}}\beta_{\mathcal{X}}(0)$ for any $k \in \mathbb{Z}_+$, where $\beta_{\mathcal{X}}(0) < \infty$ denotes a finite bound of the initial state vector in (6.13). Moreover, it follows from Lemma 9 in Meng and Moore (2017a) that $y_s(0,k)$ is bounded for all $k \in \mathbb{Z}_+$ and satisfies $\sup_{k\in\mathbb{Z}_+}|y_s(0,k)| \le \beta_s(0)$ for a finite bound $\beta_s(0) < \infty$ if $\rho\left(\overline{A}_{22}\right) < 1$, since $1 - \Gamma\mathcal{C}\mathcal{B}(t,k-1) = 1 - \mathcal{C}\mathcal{B}(t,k-1)\Gamma$ for an SISO system.

Step (II): For any given time instant $t \in \mathbb{Z}_{N-1}$, assume that $\sup_{k\in\mathbb{Z}_+}\|\mathcal{X}(t,k)\| \le \beta_{\mathcal{X}}(t) < \infty$ and $\sup_{k\in\mathbb{Z}_+}|y_s(t,k)| \le \beta_s(t) < \infty$, respectively, if $\rho(\overline{\mathscr{A}}_{22}) < 1$ holds, it follows from (6.13) that

$$\begin{aligned} \|\mathcal{X}(t+1,k)\| &\le \|\mathcal{A}(t,k)\|\|\mathcal{X}(t,k)\| + \|\mathcal{B}(t,k)\||y_s(t,k)| + \|\mathcal{D}(t,k)\||\omega(t,k)| \\ &\le \beta_{\mathcal{X}}(t)\beta_{\mathcal{A}} + \beta_s(t)\beta_{\mathcal{B}} + \beta_{\mathcal{B}}\beta_\omega \\ &\triangleq \beta_{\mathcal{X}}(t+1), \quad \forall k \in \mathbb{Z}_+, \end{aligned} \tag{6.23}$$

which results in

$$|-\Gamma\mathcal{C}\mathcal{A}(t+1,k-1)\mathcal{X}(t+1,k-1)| \le |\Gamma|\|C\|\beta_{\mathcal{A}}\beta_{\mathcal{X}}(t+1). \tag{6.24}$$

Applying Lemma 9 in Meng and Moore (2017a) to (6.22) for the time instant $t+1$ $t+1$ verifies $\sup_{k\in\mathbb{Z}_+}|y_s(t+1,k)| \le \beta_s(t+1)$ for a finite bound $\beta_s(t+1) < \infty$ based on $\rho(\overline{\mathscr{A}}_{22}) < 1$. Moreover, the boundedness of $X(N,k)$ $X(N,k)$ is ensured by considering (6.23) again.

By the mathematical induction, it follows that

$$\sup_{k\in\mathbb{Z}_+}\max_{t\in\mathbb{Z}_N-1}|y_s(t,k)| < \beta_s, \ \sup_{k\in\mathbb{Z}_+}\max_{t\in\mathbb{Z}_N}\|\mathcal{X}(t,k)\| \le \beta_{\mathcal{X}},$$

with $\beta_s = \max_{t\in\mathbb{Z}_{N-1}}\beta_s(t)$ and $\beta_{\mathcal{X}} = \max_{t\in\mathbb{Z}_N}\beta_{\mathcal{X}}(t)$.

Using the relationships between $y_s(t,k)$ and $u(t,k)$ given in (6.12), between $x(t,k)$ and $\mathcal{X}(t,k)$ in (6.13), the boundedness objective in (6.4) holds true when $\rho(\overline{\mathscr{A}}_{22}) < 1$.

It follows from (6.18) that

$$e(t+1,k) = \mathcal{G}(e^{j\theta})e(t+1,k-1) + \mathcal{G}_{\omega}(e^{j\theta})\varpi(t,k), \tag{6.25}$$

where $\mathcal{G}_{\omega}(e^{j\theta}) = \overline{\mathscr{A}}_{21}(e^{j\theta}\mathbf{I} - \overline{\mathscr{A}}_{11})^{-1} - \mathcal{C}$.

Thus, $\mathcal{G}_{\omega}(e^{j\theta})$ is bounded for any $\theta \in [-\pi,\pi]$, $t \in \mathbb{Z}_N$ $t \in \mathbb{Z}_N$ and $k \in \mathbb{Z}_+$ $k \in \mathbb{Z}_+$ based on Assumption 6.1 together with the finite observer and controller gains. With the bounded $\mathcal{X}(t,k)$ and $y_s(t,k)$ $y_s(t,k)$, the boundedness of $\varpi(t,k)$ can be easily verified, i.e., $\|\varpi(t,k)\| \le 2\beta_A\beta_{\mathcal{X}} + 2\beta_{\mathcal{B}}\beta_s + 2\beta_B\beta_{\omega}$. The bounded tracking objective in (6.3) is then guaranteed by applying Theorem 3 in Meng (2019) based on $\left\|\mathcal{G}(e^{j\theta})\right\|_{\infty} < 1$ for any $\theta \in [-\pi,\pi]$. Finally, with the bounded $y_s(t,k)$ and $\omega(t,k)$ for any $t \in \mathbb{Z}_N$ and $k \in \mathbb{Z}_+$, robust stability of the closed-loop system in (6.13) along the time direction is guaranteed since $\rho(\overline{\mathscr{A}}_{11}) < 1$. This completes the proof. □

In contrast to the developed robust feedforward ILC methods, e.g., Meng (2019); Meng and Moore (2017a, b), where only error convergence along the batch direction was considered, the above ILC scheme simultaneously ensures error convergence along the batch direction and robust stability of the closed-loop system in the time direction. Moreover, although error convergence in the batch direction could be guaranteed for an open-loop unstable process by the existing ILC methods (e.g., Sebastian et al. 2019), the tracking performance in the initial batches may be very poor or unacceptable in the time direction for implementation in practice.

To optimize the convergence rate with respect to specified uncertainty bounds on $A(t,k)$ and $B(t,k)$, the following optimization program can be implemented

$$\begin{aligned} \lambda^* = & \min_{P \succ 0, \varepsilon > 0; \lambda \in (0,1), \Gamma} \lambda \\ & \text{s.t. } (6.20), \end{aligned} \tag{6.26}$$

where λ^* is the optimal convergence rate.

If batch-varying uncertainties could 'settle down' along the batch direction, i.e.,

$$\begin{cases} \lim\limits_{k\to\infty} x(0,k) = x_0, & \lim\limits_{k\to\infty} \omega(t,k) = \omega(t), \\ \lim\limits_{k\to\infty} A(t,k) = A(t), & \lim\limits_{k\to\infty} B(t,k) = B(t), \end{cases} \tag{6.27}$$

where x_0, $\omega(t)$, $A(t)$, and $B(t)$ are batch-invariant, perfect tracking can be realized as stated below.

Corollary 6.1 Consider the system described by (6.13) under Assumption 6.1 and (6.27), and let the set-point command be updated by the indirect-type ILC law in (6.14). Then perfect tracking in the sense of (6.2), the boundedness objective in (6.4) and robust stability of the closed-loop system in (6.13) along the time direction hold if there exist a matrix $P > 0$ and scalars $\Gamma, \varepsilon > 0$ and $\lambda \in (0, 1)$ such that the LMI in (6.20) is feasible.

Proof If the LMI in (6.20) of Theorem 6.1 holds, then $\left\|\mathcal{G}(e^{j\theta})\right\|_\infty < \lambda < 1$. It follows from Assumption 6.1 and (6.27) that $\varpi(t,k) \to 0$, as $k \to \infty, \forall t \in \mathbb{Z}_N$, and from (6.25) that

$$\begin{aligned}\|e(t,k)\| &= \left\|\mathcal{G}(e^{j\theta})\right\|_\infty \|e(t,k-1)\| + \left\|\mathcal{G}_\omega(e^{j\theta})\right\|_\infty \|\varpi(t-1,k)\| \\ &\le \lambda\|e(t,k-1)\| + \beta_\delta \\ &\le \lambda^k\|e(t,0)\| + \frac{1-\lambda^k}{1-\lambda}\beta_\delta,\end{aligned} \tag{6.28}$$

where $\beta_\delta = \sup_{k\in\mathbb{Z}_+}\left\{\left\|\mathcal{G}_\omega(e^{j\theta})\right\|_\infty \|\varpi(t-1,k)\|\right\}$, which tends to zero as $k \to \infty$. Since $e(t,0)$ is bounded, it is concluded that $\lim\limits_{k\to\infty} e(t,k) = 0$ for any $t \in \mathbb{Z}_N$. This completes the proof. □

A necessary and sufficient (NAS) condition has been developed in Meng and Moore (2017a) for the use of a time- and batch-varying learning gain. The following theorem presents an NAS condition guaranteeing convergence of the above ILC system along the batch direction.

Theorem 6.2 Consider the system described by (6.13) under Assumption 6.1, where the set-point command is updated by (6.14). Then the bounded tracking objective in (6.3) and the boundedness objective in (6.4) hold if and only if there exists a time- and batch-varying learning gain $\Gamma(t,k)$ for any $t \in \mathbb{Z}_{N-1}$, such that there exists an iteration sequence $\{\xi(t,s) : s \in \mathbb{Z}_+\}$ where $\xi(t,0) = 0$ and $0 < \xi(t,s+1) - \xi(t,s) \le \sigma(t)$ for $\sigma(t) \in \mathbb{Z}$ satisfying for $\forall s \in \mathbb{Z}_+$

$$\left|\prod_{j=\xi(t,s)}^{\xi(t,s+1)-1}\left[1 - C\mathcal{B}(t,j)\Gamma(t,j)\right]\right| < 1. \tag{6.29}$$

Proof The proof follows immediately from that for Lemma 9 in Meng and Moore (2017a).

6.4 An Illustrative Example

Consider the batch packing process of an injection molding machine studied in the previous Sect. 2.2.3. For illustration, it is assumed herein that

$$\omega(t,k) = 10\sin(0.1t) + \delta_1(t,k),$$

$$\Delta A(t,k) = \begin{bmatrix} 0.08\delta_2(t,k) & 0 \\ 0.08\delta_3(t,k) & 0 \end{bmatrix}, \ \Delta B(t,k) = \begin{bmatrix} 0.1\delta_4(t,k) \\ 0.14\delta_5(t,k) \end{bmatrix},$$

$$x_0 = \begin{bmatrix} 0 & 0 \end{bmatrix}^{\mathrm{T}}, \ \lambda(t,k) = \begin{bmatrix} \delta_6(t,k) & \delta_7(t,k) \end{bmatrix}^{\mathrm{T}},$$

and $\delta_i(t,k), i = 1, 2, \ldots, 7$, are non-repetitive uncertain parameters varying arbitrarily within $[-0.1, 0.1]$ as assumed in Meng and Moore (2017a). The reference trajectory is taken as

$$y_{\mathrm{d}}(t) = \begin{cases} 200, & 0 \le t \le 100, \\ 200 + 5(t - 100), & 100 < t \le 120, \\ 300, & 120 < t \le N = 200, \end{cases}$$

which is representative of the set-point trajectories conducted in the injection molding machine. The initial part of $y_{\mathrm{d}}(t)$ is smoothed by a user-specified prefilter $G_f(z) = \left(z^{-1} + z^{-2}\right)/\left(3 - z^{-1}\right)$ for practical implementation.

By taking $\alpha_o = 0.45$ and $\alpha_c = 0.5$, the corresponding observer gain and feedback controller are computed as $L = [1.2570 \; -1.0144 \; 0.5353]^{\mathrm{T}}$ and $K = [2.1966 \; 2.2782 \; 1]$. As a result, the set-point gain is $F = 0.8044$. By solving the optimization program in (6.26) for a specified uncertainty bound of the process, the optimal learning gain is computed as $\Gamma^* = 0.8865$, corresponding to the optimal convergence rate $\lambda^* = 0.33$. For comparison, another feasible and smaller learning gain, e.g., $\Gamma = 0.5$, is also tested, and both are compared with the recently developed indirect-type ILC design in Liu et al. (2014). The tracking results are shown in Fig. 6.2 for the nominal case and the resulting output tracking errors in terms of $\mathrm{ATE}(k) = \sum_{t=1}^{N} |e(t,k)|/N$ are shown in Fig. 6.3. It is seen that perfect tracking is realized after 5 cycles by the new indirect-type ILC design, whereas the design in Liu et al. (2014) requires approximately 20 cycles to reach the same objective. Moreover, a faster convergence speed is obtained by the optimal learning gain given in the new design. Note that no feasible solution could be found using the LMI condition of Theorem 4 in Meng (2019) due to the existence of a slow pole in the process model.

To make a further comparison, consider the presence of time-and batch-varying uncertainties and disturbance, as assumed in Meng and Moore (2017a). One hundred Monte Carlo tests were performed, and the averaged results for the ATE index are shown in Fig. 6.4 together with those by Liu et al. (2014). It is seen that the controlled system maintains robust stability, and the tracking performance is significantly improved compared to that of Liu et al. (2014). Note that a relatively larger steady-state tracking error arises from the larger learning gain of $\Gamma^* = 0.8865$, which implies that a trade-off between the convergence speed and steady-state tracking error should be made for practical application.

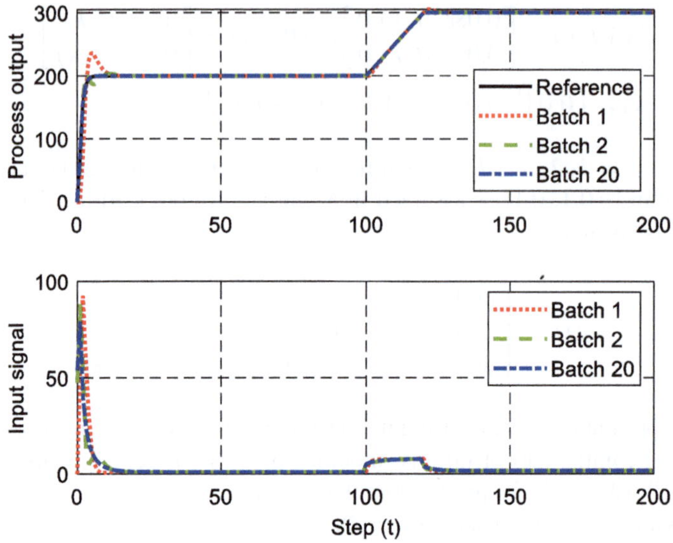

Fig. 6.2 Tracking performance for the nominal system in the absence of uncertainties and disturbance ($\Gamma^* = 0.8865$)

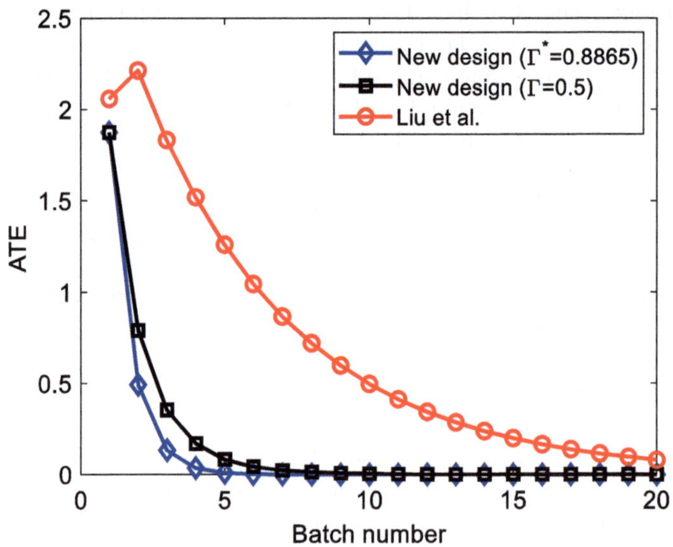

Fig. 6.3 ATE performance indices in the absence of system uncertainties and disturbance

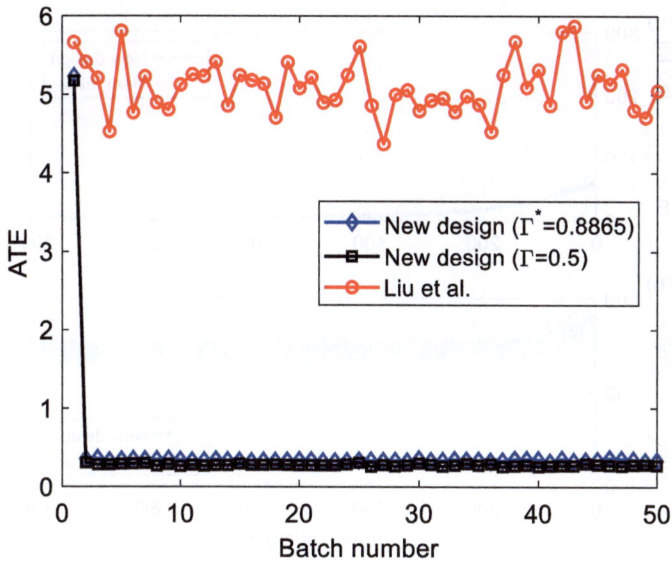

Fig. 6.4 ATE performance indices under time- and batch-varying uncertainties and disturbance

To further illustrate the performance in terms of the tracking objectives in (6.2) and (6.3), another reference trajectory $y_d(t) = 200\sin(t\pi/10)$ over $t \in Z_{20}$ as studied in Meng and Moore (2017a) is considered. The learning gain $\Gamma(t, k)$ satisfying (6.29) is chosen as

$$\Gamma(t, k) = \begin{cases} 0.0249, & \forall \text{ even } k, \forall t, \\ -0.0124, & \forall \text{ odd } k, \forall t, \end{cases}$$

which results in

$$1 - \mathcal{C}\mathcal{B}(t, k)\Gamma(t, k) = \begin{cases} [0.9750, 0.9754], \ \forall \text{ even } k, \forall t, \\ [1.0119, 1.0121], \ \forall \text{ odd } k, \forall t. \end{cases}$$

Note that the choice of $\Gamma(t, k)$ does not satisfy the conventional contraction mapping-based ILC condition (Ahn et al. 2007a, b). However, it can be easily verified that there exists an iteration sequence $\{\xi(t, s) : s \in Z_+\}$ with $\xi(t, s) = 2s$ for all t t satisfying that the left-hand side of (6.29) is no larger than 0.9880, similar to that of the robust feedforward ILC design in Meng and Moore (2017a) for fair comparison. The control results with respect to the evolution of tracking error evaluated by $\max_{1<t<N} |e(t, k)|$ are shown in Fig. 6.5. It is seen that the tracking performance in terms of speed and steady-state tracking error is significantly improved by the new indirect-type ILC design, especially for the initial cycles.

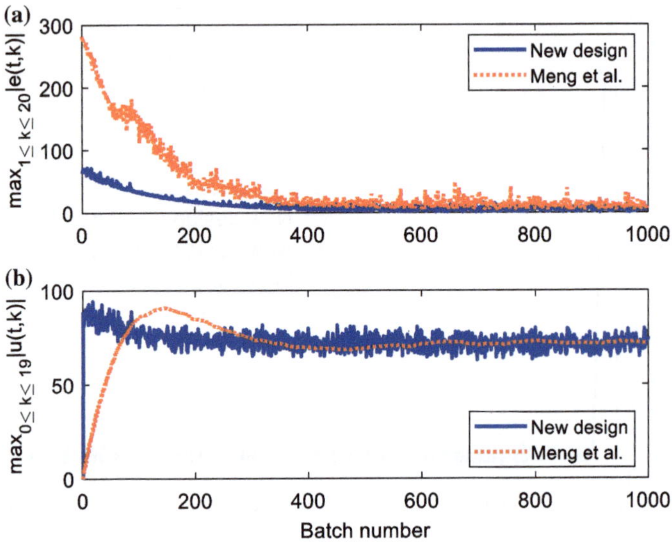

Fig. 6.5 Evolution of tracking error and control input evaluated by **a** $\max_{1\le t\le 20}|e(t,k)|$ and **b** $\max_{0\le t\le 19}|u(t,k)|$

6.5 Summary

An indirect-type ILC scheme (Hao et al. 2020) is presented in this chapter for open-loop stable or unstable batch processes with time- and batch-varying uncertainties. Differing from the previous PI or PID-based indirect-type ILC designs, where only the traditional unity feedback control is used, a GESO-based 2DOF design is proposed for the inner loop to enable a set-point learning controller design, which can significantly improve robust tracking performance along both the time and batch directions. A simple P-type ILC is then used for set-point learning, which can be quantitatively tuned to realize the optimal convergence rate, or monotonically tuned to make a trade-off between the convergence rate and robustness against batch-varying uncertainties. Moreover, a tractable LMI-based sufficient condition is established to guarantee boundedness of the output tracking error and system input in the presence of time- and batch-varying process uncertainties. Case studies on batch injection molding processes subject to time- and batch-varying uncertainties and disturbance have well demonstrated the effectiveness and merits of the new design in comparison with the previously developed robust feedforward ILC designs (Meng 2019; Meng and Moore 2017a, b) and indirect-type ILC designs (Hao et al. 2019; Liu et al. 2014; Wang et al. 2012, 2013).

References

Ahn H-S, Chen Y-Q, Moore KL (2007a) Iterative learning control: brief survey and categorization. IEEE Trans Syst Man Cybern Part C Appl Rev 37(6):1099–1121

Ahn HS, Moore KL, Chen YQ (2007b) Iterative learning control: robustness and monotonic convergence for interval systems. Springer-Verlag

Altın B, Willems J, Oomen T, Barton K (2017) Iterative learning control of iteration-varying systems via robust update laws with experimental implementation. Control Eng Pract 62:36–45

Chi R, Li H, Shen D, Hou Z, Huang B (2023) Enhanced P-type control: Indirect adaptive learning from set-point updates. IEEE Trans Autom Control 68(3):1600–1613

Hao S, Liu T, Gao F (2019) PI-based indirect-type iterative learning control for batch processes with time-varying uncertainties: a 2D FM model based approach. J Process Control 78:57–67

Hao S, Liu T, Rogers E (2020) Extended state observer based indirect-type ILC for single-input single-output batch processes with time- and batch-varying uncertainties. Automatica 112(108673):1–7

Lee J, Lee K, Kim W (2000) Model-based iterative learning control with a quadratic criterion for time-varying linear systems. Automatica 36:641–657

Liu J, Zhou Z, Hong W, Shi J (2023) Two-dimensional iterative learning control with deep reinforcement learning compensation for the non-repetitive uncertain batch processes. J Process Control 131(103106):1–11

Liu T, Gao F, Wang Y (2010) IMC-based iterative learning control for batch processes with uncertain time delay. J Process Control 20(2):173–180

Liu T, Hao S, Li D, Chen W-H, Wang Q-G (2019) Predictor-based disturbance rejection control for sampled systems with input delay. IEEE Trans Control Syst Technol 27(2):772–780

Liu T, Wang X, Chen J (2014) Robust PID based indirect-type iterative learning control for batch processes with time-varying uncertainties. J Process Control 24(12):95–106

Meng D (2019) Convergence conditions for solving robust iterative learning control problems under nonrepetitive model uncertainties. IEEE Trans Neural Netw Learn Syst 30(6):1908–1919

Meng D, Moore K (2017a) Convergence of iterative learning control for SISO nonrepetitive systems subject to iteration-dependent uncertainties. Automatica 79:167–177

Meng D, Moore K (2017b) Robust iterative learning control for nonrepetitive uncertain systems. IEEE Trans Autom Control 62(2):907–913

Paszke W, Roger E, Gałkowski K (2016) Experimentally verified generalized KYP based iterative learning control design. Control Eng Pract 53:57–67

Sebastian G, Tan Y, Oetomo D (2019) Convergence analysis of feedback based iterative learning control with input saturation. Automatica 101:44–52

Shi J, Gao F, Wu T-J (2006) Robust iterative learning control design for batch processes with uncertain perturbations and initialization. AIChE J 52(6):2171–2187

Shi J, Zhou H, Cao Z, Jiang Q (2014) A design method for indirect iterative learning control based on two-dimensional generalized predictive control algorithm. J Process Control 24(10):1527–1537

Wang Y, Liu T, Zhao Z (2012) Advanced PI control with simple learning set-point design: application on batch processes and robust stability analysis. Chem Eng Sci 71(26):153–165

Wang Y, Yang Y, Zhao Z (2013) Robust stability analysis for an enhanced ILC-based PI controller. J Process Control 23(2):201–214

Zhang J, Meng D (2023) Improving tracking accuracy for repetitive learning systems by high-order extended state observer. IEEE Trans Neural Netw Learn Syst 34(12):10398–10407

Zhu Q, Xu J-X, Huang D, Hu G-D (2015) Iterative learning control design for linear discrete-time systems with multiple high-order internal models. Automatica 62:65–76

Chapter 7
Robust ILC Design Under Process Input Constraints

7.1 Introduction

For the batch run of industrial processes, there often exist process input constraints on control implementation, which deserve special treatment to ensure system stability and eliminate the corresponding output tracking error. Since the model predictive control (MPC) methodology can explicitly deal with the input constraints and optimize the control performance (Qin and Badgwell 2003; Lee et al. 1999) developed a synthetic scheme to integrate ILC with MPC for the application to constrained batch processes, which could not only comply with the input constraints but also improve the tracking performance. Following this idea, Xu et al. (2013) designed an ILC-MPC-based run-to-run scheme to address time-varying uncertainties, where the optimal input sequence is determined during the interval between two sequential batches. Concerning the classical ILC design for a run-to-run scheme, the control within a batch run is indeed an open-loop manner, which implies that the disturbance involved with one batch cannot be rejected efficiently. To overcome the deficiency, real-time feedback ILC-MPC designs were studied by Shi et al. (2007) and Wang et al. (2008). A 2D system model was adopted in these references to describe the batch process dynamics along both the time and batch directions.

Concerning the convergence of ILC-MPC, the receding horizon constrained optimization and real-time feedback control within one batch bring difficulty for such analysis, especially for those batch systems not satisfying the "perfect tracking condition" or the desired output trajectory not reachable by the input constraints. To address this problem, Li et al. (2016) proposed a two-mode framework and an invariant updating strategy for implementing ILC-MPC. Besides, the mismatch of the system model from the real plant must be considered in practice. The unmodelled dynamics may deteriorate the control performance of ILC-MPC, and make the tracking error slowly convergent along the batch direction, or even divergent. According to the robust control theory, alternative ILC-MPC designs (Li et al. 2024a, b; Zhang et al. 2024) were developed to guarantee the convergence of batch run, but in exchange for tracking performance degradation.

T. Liu et al., *Robust Iterative Learning Control of Industrial Batch Systems*, Intelligent Control and Learning Systems 22, https://doi.org/10.1007/978-981-96-9778-6_7

In the past decade, there has been a trend to utilize the machine learning approach for ILC design (Li et al. 2024a, b). Lu et al. (2019) reviewed a variety of learning mechanisms for enhancing the ILC performance. Liu et al. (2022) introduced a notion of learnability for establishing an ILC system and presented a data-based learning scheme for batch optimization. Li et al. (2019) developed a data-driven ILC-MPC method where an input mapping scheme was provided to make use of the historical batch data. This scheme was further extended to deal with batch uncertainty in the reference (He et al. 2022). In order to efficiently use the historical data for batch optimization, Zhou et al. (2022) proposed a data preprocessing technique to select distinct data and incorporate them into the conic ILC design. Thereby, a good trade-off between the learning capability from historical data and computational efficiency in solving the complex optimization problem can be acquired.

7.2 Robust Model Predictive Control (MPC) Based ILC Scheme

Consider a constrained batch process described by

$$x(t+1,k) = Ax(t,k) + Bu(t,k) + f(u(t,k)), \tag{7.1}$$

$$y(t,k) = Cx(t,k), 0 \le t \le L, \tag{7.2}$$

where $x(t,k) \in R^n$ is the measurable system state, $u(t,k) \in R^m$ is the control input, $y(t,k) \in R^l$ $(l \le m)$ is the system output, and $f(u(t,k))$ is the unknown nonlinear dynamics related to the control input. The cycle time of each batch is denoted by L. The process input constraints are expressed below:

$$\underline{u}_i \le u_i(t,k) \le \overline{u}_i, \overline{u}_i > 0, \underline{u}_i \le 0, \tag{7.3}$$

$$\underline{\delta u}_i \le u_i(t,k) - u_i(t-1,k) \le \overline{\delta u}_i, \underline{\delta u}_i < 0, \overline{\delta u}_i > 0, \tag{7.4}$$

$$\underline{du}_i \le u_i(t,k) - u_i(t,k-1) \le \overline{du}_i, \underline{du}_i < 0, \overline{du}_i > 0, \tag{7.5}$$

where $i = 1, 2, \ldots, m$. The constraints in (7.3) and (7.4) are the limitations on the amplitudes and variation amplitudes of the control inputs, respectively, which are commonly concerned in engineering applications. The constraints in (7.5) are the limitations on variation amplitudes of control inputs between two neighbor batches for the purpose of smooth operation.

Regarding the nonlinear dynamics denoted by $f(u(t,k))$, the following assumption is made.

Assumption 7.1 The nonlinear function $f(\cdot)$ is globally Lipschitz, satisfying

$$\|f(u_1) - f(u_2)\| \le \mu \|u_1 - u_2\|, \tag{7.6}$$

where $\mu < \infty$ is a positive Lipschitz constant.

The control goal of the above batch process is to steer the system output to track the reference signal $y_{\mathrm{d}}(t)$ of which the time sequence is denoted by $\overrightarrow{y_{\mathrm{d}}}$. In addition, without loss of generality, the initial condition of each batch is reset to zero, i.e., $x(0,k) = 0$ and $u(i,k) = 0$ $(i < 0)$.

For the convenience of analysis, the above process description in (7.1) and (7.2) can be transferred into a 2D model to include the dynamics of both the horizontal (time direction) and vertical (batch direction) information. By taking the following notations

$$d_x(t,k) = x(t,k) - x(t-1,k),$$

$$d_u(t,k) = u(t,k) - u(t-1,k),$$

$$e(t,k) = y_{\mathrm{d}}(t) - y(t,k),$$

$$\delta x(t,k) = d_x(t,k) - d_x(t,k-1),$$

$$\delta u(t,k) = d_u(t,k) - d_u(t,k-1),$$

$$\delta e(t,k) = e(t,k) - e(t,k-1),$$

the above (7.1) can be rewritten as

$$d_x(t+1,k) = Ad_x(t,k) + Bd_u(t,k) + f(u(t,k)) - f(u(t-1,k)). \tag{7.7}$$

Then, a 2D model of the above batch process is written as

$$z(t+1,k) = \tilde{A}z(t,k) + \tilde{B}\delta u(t,k) + \tilde{D}w(t,k), \tag{7.8}$$

$$e(t,k) = e(t,k-1) + \tilde{C}z(t,k), \tag{7.9}$$

where $z(t,k) = \left[\delta x(t,k)^{\mathrm{T}}, \delta e(t,k)^{\mathrm{T}}\right]^{\mathrm{T}}$, and

$$w(t,k) = f(u(t,k)) - f(u(t-1,k)) - f(u(t-1,k) + f(u(t-1,k-1))a,$$

$$\tilde{A} = \begin{bmatrix} A & 0 \\ -CA & I \end{bmatrix}, \tilde{B} = \begin{bmatrix} B \\ -CB \end{bmatrix}, \tilde{C} = [0, I], \tilde{D} = \begin{bmatrix} I \\ -C \end{bmatrix}.$$

According to the 2D model in (7.8) and (7.9), the system state and control input can be written as

$$x(t,k) = x(t,k-1) + x(t-1,k) - x(t-1,k-1) + \delta x(t,k),$$

$$u(t,k) = u(t,k-1) + u(t-1,k) - u(t-1,k-1) + \delta u(t,k),$$

Corresponding to the updating control law $d_u(t,k) = d_u(t,k-1) + \delta u(t,k)$.

7.2.1 *Two-Mode Control Framework of Real-Time-Feedback ILC–MPC*

Concerning the above process described in (7.1) and (7.2), if the nonlinear part is not considered, a real-time feedback ILC-MPC can be designed to solve the following optimization problem to determine the updating law $d_u(t,k)$ at time t of batch k.

$$\min_{\delta u(t,k),\ldots,\delta u(t+h-1,k)} J(t,k)\ s.t.(7.8) - (7.9) \text{ without input nonlinearity}, \quad (7.10)$$

where h is the control horizon,

$$J(t,k) = \sum_{i=0}^{h-1} ||e(t+i+1,k|t,k)||_Q^2 + ||\delta u(t+i,k|t,k)||_{\mathrm{P}}^2,$$

Q and P are the weighting matrices, respectively. The above design is for the unconstrained case with $t+h \leq L$. If $t+h > L$, the problem will be correspondingly modified due to the zero initial states. Since the analytical solution of problem (7.10) can be given, its closed-loop convergence can be analyzed as studied in the reference (Shi et al. 2007). For the constrained case, the above optimization problem becomes variable due to the limited batch length and initial state resetting, causing difficulty to guarantee the tracking error convergence.

To overcome the above difficulty, a dual-mode control strategy is proposed, which is a common approach to design an MPC scheme with guaranteed closed-loop stability (Mayne et al. 2000). The key idea of dual-mode control is to extend the finite control horizon of MPC to the infinite horizon. To conduct online computation for implementing the dual-mode control, the optimization problem of MPC at each time instant constructs the future control inputs as some free control actions followed by a fixed (or designed online) terminal feedback control law. Thus, the system dynamics at infinite time can be formulated. Hence, if the terminal feedback control law satisfies some conditions (Mayne et al. 2000), the closed-loop system can be guaranteed stable.

Similarly, for designing the real-time feedback ILC-MPC, the following two-mode control framework is proposed

$$u(\cdot) = \begin{cases} u(i,k); & t \leq i \leq L-1, \\ u(i,k+s) + g(i,k+s); & 0 \leq i \leq L-1, s \geq 1, \end{cases} \tag{7.11}$$

where t is the current time and $g(i, k+s)$ is the updating strategy after batch k. The future control actions after time instant t of batch k are divided into two parts: the future control inputs within the current batch k (named the first mode) and an updating strategy after batch k (named the second mode).

Compared with the previous real-time feedback-based ILC-MPC designs, which can be seen as only having the first mode of the two-mode framework, the two-mode control design can take into account both the dynamics of the current batch and the future evolution along the batch direction. Therefore, the dynamics of the current batch can be improved by the iterative learning mechanism together with the convergence.

By the two-mode control framework in (7.11), the corresponding ILC-MPC should determine the $u(i, k|t, k)$ and $g(k+s)$ at each time instant of each batch. While the first mode has a similar optimization problem to that in (7.10), the resultant ILC-MPC should also solve the optimization problem for the second mode. This may lead to a problem with infinite multivariates, which could become impossible for an online solution. Hence, it is suggested to take the following run-to-run strategy for $g(k+s)$ to simplify the online computation

$$\Delta U(k+i+1) = \rho^i \Delta U(k+1), i > 0, \tag{7.12}$$

where $0 < \rho < 1$ is a specified scalar.

For the updating strategy in (7.12), $0 < \rho < 1$ means that $\Delta U(k+i+1)$ will be zeros when $i \to \infty$. Meanwhile, by using the process model in (7.1) and (7.2), the following yields

$$Y(k+1) = Sx(0,k) + GU(k+1) + \overline{G}F(U(k+1)),$$

where $Y(k+1) = \left[y^T(1,k+1), y^T(2,k+1), \ldots, y^T(L,k+1)\right]^T$,

$$F(U(k+1)) = \left[f^T(u(0,k+1)), f^T(u(1,k+1)), \ldots, f^T(u(L-1,k+1))\right]^T,$$

$$S=\begin{bmatrix} CA \\ CA^2 \\ \vdots \\ CA^L \end{bmatrix}, G=\begin{bmatrix} CB & 0 & \cdots & 0 \\ CAB & CB & & 0 \\ \vdots & & \ddots & \vdots \\ CA^{L-1}B & CA^{L-2}B & \cdots & CB \end{bmatrix}, \overline{G}=\begin{bmatrix} C & 0 & \cdots & 0 \\ CA & C & & 0 \\ \vdots & & \ddots & \vdots \\ CA^{L-1} & CA^{L-2} & \cdots & C \end{bmatrix},$$

and correspondingly,

$$\vec{e}(k+1)=\vec{e}(k)-G\Delta U(k+1)-\overline{G}\{F(U(k+1))-F(U(k))\}.$$

Then, the difference between the system output and the reference at batch ∞ will be

$$\vec{e}(k+\infty)=\vec{e}(k)-\sum_{i=0}^{\infty}G\rho^i\Delta U(k+1)-\sum_{i=0}^{\infty}\overline{G}\{F(U(k+i+1))-F(U(k+i))\}.$$

According to (7.12), one obtains

$$\vec{e}(k+\infty)=\vec{e}(k)-\frac{G\Delta U(k+1)}{1-\rho}-\sum_{i=0}^{\infty}\overline{G}\{F(U(k+i+1))-F(U(k+i))\}. \tag{7.13}$$

Concerning the first mode in (7.11), from the current state $z(t,k)$ and 2D system model in (7.8) and (7.9), one can get

$$\begin{aligned}\begin{bmatrix} e(t+1,k|t,k) \\ e(t+2,k|t,k) \\ \vdots \\ e(t+L,k|t,k) \end{bmatrix} &= \begin{bmatrix} e(t+1,k-1) \\ e(t+2,k-1) \\ \vdots \\ e(t+L,k-1) \end{bmatrix} + \tilde{S}_t z(t,k) \\ &\quad + \tilde{G}_t \begin{bmatrix} \delta u(t,k) \\ \delta u(t+1,k) \\ \vdots \\ \delta u(L-1,k) \end{bmatrix} + \tilde{E}_t \begin{bmatrix} w(t,k) \\ w(t+1,k) \\ \vdots \\ w(L-1,k) \end{bmatrix} \\ &= \begin{bmatrix} e^o(t+1,k|t,k) \\ e^o(t+2,k|t,k) \\ \vdots \\ e^o(t+L,k|t,k) \end{bmatrix} + \tilde{E}_t \begin{bmatrix} w(t,k) \\ w(t+1,k) \\ \vdots \\ w(L-1,k) \end{bmatrix}, \end{aligned} \tag{7.14}$$

where

$$\tilde{S}_t = \begin{bmatrix} \tilde{C}\tilde{A} \\ \tilde{C}\tilde{A}^2 \\ \vdots \\ \tilde{C}\tilde{A}^{L-t} \end{bmatrix}, \tilde{G}_t = \begin{bmatrix} \tilde{C}\tilde{B} & 0 & \cdots & 0 \\ \tilde{C}\tilde{A}\tilde{B} & \tilde{C}\tilde{B} & & 0 \\ \vdots & & \ddots & \vdots \\ \tilde{C}\tilde{A}^{L-t-1}\tilde{B} & \tilde{C}\tilde{A}^{L-t-2}\tilde{B} & \cdots & \tilde{C}\tilde{B} \end{bmatrix},$$

$$\tilde{E}_t = \begin{bmatrix} \tilde{C}\tilde{D} & 0 & \cdots & 0 \\ \tilde{C}\tilde{A}\tilde{D} & \tilde{C}\tilde{D} & & 0 \\ \vdots & & \ddots & \vdots \\ \tilde{C}\tilde{A}^{L-t-1}\tilde{D} & \tilde{C}\tilde{A}^{L-t-2}\tilde{D} & \cdots & \tilde{C}\tilde{D} \end{bmatrix}.$$

Since the states before instant t at batch k are known, one can further rewrite (7.13) as

$$\begin{aligned}
\vec{e}(k+\infty) &= \vec{e}(k) - \frac{G\Delta U(k+1)}{1-\rho} - \overline{G}\{F(U(k+\infty)) - F(U(k))\} \\
&= \begin{bmatrix} e(1,k) \\ \vdots \\ e(t,k) \\ e(t+1,k|t,k) \\ \vdots \\ e(t+L,k|t,k) \end{bmatrix} - \frac{G\Delta U(k+1)}{1-\rho} - \overline{G}\{F(U(k+\infty)) - F(U(k))\} \\
&= \begin{bmatrix} e(1,k) \\ \vdots \\ e(t,k) \\ e(t+1,k|t,k) \\ \vdots \\ e(t+L,k|t,k) \end{bmatrix} + \begin{bmatrix} 0 \\ \tilde{S}_t z(t,k) + \tilde{G}_t \begin{bmatrix} \delta u(t,k) \\ \delta u(t+1,k) \\ \vdots \\ \delta u(L-1,k) \end{bmatrix} \end{bmatrix} \\
&\quad + \begin{bmatrix} 0 \\ \tilde{E}_t \begin{bmatrix} w(t,k) \\ w(t+1,k) \\ \vdots \\ w(L-1,k) \end{bmatrix} \end{bmatrix} \\
&\quad - \frac{G\Delta U(k+1)}{1-\rho} - \overline{G}\{F(U(k+\infty)) - F(U(k))\}
\end{aligned}$$

$$= \vec{e}^{o}(k+\infty) + \left[\tilde{E}_t \begin{bmatrix} 0 \\ \begin{bmatrix} w(t,k) \\ w(t+1,k) \\ \vdots \\ w(L-1,k) \end{bmatrix} \end{bmatrix}\right] - \overline{G}\{F(U(k+\infty)) - F(U(k))\}. \tag{7.15}$$

Note that in the second mode of (7.11), an amplitude-decaying strategy expressed by (7.12) is introduced. The amplitude-decaying strategy satisfies the following lemma.

Lemma 7.1 Consider a scalar a, if there exists ρ $(1 > \rho \geq 0)$ satisfying $\underline{a} \leq a + \frac{\alpha}{1-\rho} \leq \overline{a}$ and $\underline{a} \leq a \leq \overline{a}$, then $\underline{a} \leq a + \sum_{i=0}^{s} \alpha\rho^i \leq \overline{a} (s = 0, 1, 2, \ldots, \infty)$.

It is obvious that the $a + \sum_{i=0}^{s} \alpha\rho^i$ in Lemma 7.1 increases (decreases) monotonically when $\alpha > 0$ $(\alpha < 0)$. Hence, its proof can be easily clarified and thus is omitted.

Proposition 7.1 Consider a batch process described by (7.1) and (7.2) subject to the constraints (7.3–7.5), these constraints can be implemented from batch k, if there exists an updating strategy in (7.12) such that the following conditions are satisfied

$$\underline{u_i} \leq u_i(k,t) \leq \overline{u}_i, \underline{u_i} \leq u_i(k,t) + \frac{\Delta u_i(t,k+1)}{1-\rho} \leq \overline{u}_i, \tag{7.16}$$

$$\underline{\delta u_i} \leq u_i(t,k) - u_i(t-1,k) \leq \overline{\delta u_i},$$

$$\underline{\delta u_i} \leq u_i(t,k) - u_i(t-1,k) + \frac{\Delta u_i(t,k+1) - \Delta u_i(t-1,k+1)}{1-\rho} \leq \overline{\delta u_i}, \tag{7.17}$$

$$\underline{du_i} \leq \Delta u_i(t,k+1) \leq \overline{du_i}, \tag{7.18}$$

where $i = 1, 2, \ldots, m$, $t = 1, 2, \ldots, L-1$, and $Y(k)$ and $U(k)$ are the output sequence and control input sequence at batch k, respectively. Correspondingly, the output error at batch ∞ will satisfy

$$\left\| \vec{e}(k+\infty) - \vec{e}(k) + \frac{G\Delta U(k+1)}{1-\rho} \right\| \leq \frac{\mu \left\|\overline{G}\right\| \|\Delta U(k+1)\|}{1-\rho}. \tag{7.19}$$

Proof For the condition in (7.18), since $0 \leq \rho < 1$ and $\underline{du_i} < 0$, $\underline{du_i} \leq \Delta u_i(t,k+1) \leq \overline{du_i}$ means

$$\underline{du_i} \leq \rho^j \underline{du_i} \leq \rho^j \Delta u_i(t,k+1) \leq \rho^j \overline{du_i} \leq \overline{du_i} \; (j > 0).$$

That is, the condition in (7.18) can ensure the updating strategy in (7.12) satisfies the constraint in (7.5) from batch k.

For the condition in (7.16), one gets

$$u_i(t, k+j) = u_i(t, k) + \sum_{s=0}^{j-1} \rho^s \Delta u_i(t, k+1)(j > 0).$$

Since $0 < \rho < 1$, from Lemma 7.1, if the conditions in (7.16) hold, the constraints in (7.3) can be ensured from batch k. Similarly, one can use Lemma 7.1 to prove that the condition in (7.17) ensures the constraints in (7.4) from batch k, respectively.

Concerning the conclusion in (7.19), one can obtain from Eq. (7.13) that

$$\vec{e}(k+\infty) = \vec{e}(k) - \frac{G\Delta U(k+1)}{1-\rho} - \overline{G}\{F(U(k+\infty)) - F(U(k))\}.$$

That is,

$$\vec{e}(k+\infty) - \vec{e}(k) + \frac{G\Delta U(k+1)}{1-\rho} = -\overline{G}\{F(U(k+\infty)) - F(U(k))\}.$$

Meanwhile, it follows from the strategy in (7.12) that

$$U(k+\infty) = U(k) + \sum_{i=0}^{\infty} \rho^i \Delta U(k+1) = U(k) + \frac{\Delta U(k+1)}{1-\rho}.$$

Hence, based on Assumption 7.1, the conclusion in (7.19) holds true. □

Furthermore, for the evolution procedure under the updating strategy in (7.12), the following proposition is presented.

Proposition 7.2 *For the batch process described by* (7.1) and (7.2) *controlled with the strategy in* (7.12), *if* $\rho^i \Delta U(1)$ *leads to* $\vec{e}^T(0)Q\vec{e}(0) > \vec{e}^T(\infty)Q\vec{e}(\infty)$, *then* $\vec{e}^T(0)Q\vec{e}(0) > \vec{e}^T(1)Q\vec{e}(1)$.

Proof Consider $i = 0$. For the nominal process, since $\vec{e}(\infty) = \vec{e}(0) - \frac{G\Delta U(1)}{1-\rho}$ and $\vec{e}^T(0)Q\vec{e}(0) > \vec{e}^T(\infty)Q\vec{e}(\infty)$, one obtains

$$\|\vec{e}(0)\|_Q^2 > \|\vec{e}(0)\|_Q^2 + \left\|\frac{G\Delta U(1)}{1-\rho}\right\|_Q^2 - 2\vec{e}^T(0)Q\frac{G\Delta U(1)}{1-\rho}.$$

That is, $0 > \frac{\|G\Delta U(1)\|_Q^2}{(1-\rho)^2} - 2\vec{e}^T(0)Q\frac{G\Delta U(1)}{1-\rho}$.

From $1 - \rho > 0$, one has

$$0 > \frac{\|G\Delta U(1)\|_Q^2}{1-\rho} - 2\vec{e}^T(0)Q\frac{G\Delta U(1)}{1-\rho}.$$

Meanwhile, $\vec{e}(1) = \vec{e}(0) - G\Delta U(1)$ yields

$$\|\vec{e}(1)\|_Q^2 = \|\vec{e}(0)\|_Q^2 + \|G\Delta U(1)\|_Q^2 - 2\vec{e}^T(0)QG\Delta U(1).$$

Thus, from $\frac{1}{1-\rho} > 1$, one concludes

$$0 > \frac{G\Delta U(1)_Q^2}{1-\rho} - 2\vec{e}^T(0)Q\frac{G\Delta U(1)}{1-\rho} > G\Delta U(1)_Q^2 - 2\vec{e}^T(0)QG\Delta U(1).$$

This yields $\vec{e}^T(0)Q\vec{e}(0) > \vec{e}^T(1)Q\vec{e}(1)$. □

Proposition 7.1 reflects that if the conditions in (7.16)–(7.18) hold, by the updating strategy in (7.12), the output sequence of the batch process can be steered to a neighbor region of $\vec{e}(k) - \frac{G\Delta U(k+1)}{1-\rho}$ at batch ∞ and the input constraints can always be ensured. In contrast, Proposition 7.2 means that the strategy in (7.12) can ensure the tracking error of the nominal process monotonically decreasing along the batch direction, which is the reason for introducing the amplitude-decaying strategy.

7.2.2 Real-Time Feedback ILC–MPC Algorithm

From the two-mode control framework in 7.2.1, the following optimization problem can be formulated

$$\min_{\delta u(t,k),\ldots,\delta u(t+L-1,k),\Delta U(k+1)} J(t,k) s.t. \delta u(t+i,k) \in \Omega_u, (7.16) - (7.18), \tag{7.20}$$

where Ω_u is the admissible set on $u(t+i,k)(i = 0, \ldots, L-1)$ determined by the constraints (7.3–7.5)

$$J(t,k) = q^o\left|\left|\vec{e}^o(k|t,k)\right|\right|^2 + q||\vec{e}(\infty|t,k)||^2 + \sum_{i=0}^{L-t-1} \lambda||\delta u(t+i,k)||^2 + \frac{\lambda\|\Delta U(k+1)\|}{1-\rho^2},$$

and q^o, q, and λ are the weighting parameters and $q^o \leq q$.

The objective function of the optimization problem in (7.20) includes not only the nominal tracking errors of the current batch, which is similar to the common design in (7.10), but also the tracking error of the terminal batch (i.e., batch ∞), which is taken to address the convergence of ILC. The terminal tracking error in the objection function will be presented in the convergence analysis of ILC-MPC.

Although the constraints of the optimization problem in (7.20) are linear, which can be tackled by Proposition 7.1, its objective function is not a quadratic function due to the unknown input linearity in the system model. To address this issue, the upper bound of this function is regarded as the objective function of ILC-MPC. By using Assumption 7.1, it can be derived, similar to (7.19) that

$$\begin{aligned} J(k,t) &\le q^o\left|\left|\vec{e}^o(k|t,k)\right|\right|^2 + q\left|\left|\vec{e}^o(\infty|t,k)\right|\right|^2 \\ &+ \sum_{i=0}^{L-t-1} \lambda||\delta u(t+i,k)||^2 + \frac{\lambda\|\Delta U(k+1)\|^2}{1-\rho^2} \\ &+ q\left|\left|\bar{G}\{F(U(k+\infty)) - F(U(k))\}\right|\right|^2 + q\left|\left|\left[\tilde{E}_t\begin{bmatrix}0\\ \begin{bmatrix} w(t,k)\\ w(t+1,k)\\ \vdots \\ w(L-1,k)\end{bmatrix}\end{bmatrix}\right]\right|\right|^2 \qquad (7.21) \\ &\le q^o\left|\left|\vec{e}^o(k|t,k)\right|\right|^2 + q\left|\left|\vec{e}^o(\infty|t,k)\right|\right|^2 \\ &+ \sum_{i=0}^{L-t-1} \lambda||\delta u(t+i,k)||^2 + \frac{\lambda\|\Delta U(k+1)\|^2}{1-\rho^2} \\ &+ q\tilde{E}_t^2\left|\left|\begin{bmatrix} w(t,k)\\ w(t+1,k)\\ \vdots \\ w(L-1,k)\end{bmatrix}\right|\right|^2 + \frac{q\mu^2\left\|\bar{G}\right\|^2\|\Delta U(k+1)\|^2}{(1-\rho)^2}. \end{aligned}$$

Based on $w(t,k) = f(u(t,k)) - f(u(t-1,k)) - (f(u(t,k-1)) - f(u(t-1,k-1)))$ and Assumption 7.1, it follows that $||w(k,t)|| \le \mu||\delta u(t,k)|| + \mu||\delta u(t,k-1)||$. Hence

$$
\begin{aligned}
J(k,t) &\le q^o\left\|\vec{e}^o(k|t,k)\right\|^2 + q\left\|\vec{e}^o(\infty|t,k)\right\|^2 \\
&+ \sum_{i=0}^{L-t-1} \lambda||\delta u(t+i,k)||^2 + \frac{\lambda\|\Delta U(k+1)\|^2}{1-\rho^2} \\
&+ q\mu^2\left\|\tilde{E}_t\right\|^2\left(\left\|\begin{bmatrix}\delta u(t,k)\\ \delta u(t+1,k)\\ \vdots \\ \delta u(L-1,k)\end{bmatrix}\right\|^2 + \left\|\begin{bmatrix}\delta u(t-1,k-1)\\ \delta u(t,k-1)\\ \vdots \\ \delta u(L-2,k-1)\end{bmatrix}\right\|^2\right) \\
&+ \frac{q\mu^2\left\|\overline{G}\right\|^2\|\Delta U(k+1)\|^2}{(1-\rho)^2} := \hat{J}(t,k).
\end{aligned}
$$

Since the input information of batch $k-1$ is known at batch k, which will not influence the solution of the above optimization problem, the above objective function $\hat{J}(t,k)$ can be rewritten as

$$
\begin{aligned}
\hat{J}(t,\ k) &= q^o\left\|\vec{e}^o(k|t,k)\right\|^2 \\
&\quad + q\left\|\vec{e}^o(\infty|t,k)\right\|^2 + \sum_{i=0}^{L-t-1} \lambda||\delta u(t+i,k)||^2 \\
&\quad + \frac{\lambda\|\Delta U(k+1)\|^2}{1-\rho^2} + q\mu^2\left\|\tilde{E}_t\right\|^2\left\|\begin{bmatrix}\delta u(t,k)\\ \delta u(t+1,k)\\ \vdots \\ \delta u(L-1,k)\end{bmatrix}\right\|^2 \\
&\quad + \frac{q\mu^2\left\|\bar{G}\right\|^2\|\Delta U(k+1)\|^2}{(1-\rho)^2}.
\end{aligned}
$$

Then, the optimization problem of ILC-MPC is written as

$$
\min_{\delta u(t,k),\ldots,\delta u(t+L-1,k),\,\Delta U(k+1)} \hat{J}(t,k) s.t. \delta u(t+i,k) \in \Omega_u, (7.16)-(7.18) \tag{7.22}
$$

which is a standard quadratic programming (QP) problem and can be efficiently solved by many commercial QP toolboxes.

For clarity, the robust real-time-feedback ILC-MPC algorithm is summarized below:

Algorithm I (Robust real-time feedback ILC-MPC)

Step 1. Choose the adjustable parameters $\rho(0 < \rho < 1)$, q^o, q, and λ;
Step 2. At time t of batch k, solve the optimization problem (7.22) to acquire $\delta u(t, k)$;
Step 3. Calculate $u(t, k)$ according to
$u(t, k) = u(t, k-1) + u(t-1, k) - u(t-1, k-1) + \delta u(t, k)$, for implementation in the controlled system;
Step 4. Record the control input $u(t, k)$, $y(t+1, k)$, and $x(t+1, k)$;
Step 5. Set $t = t + 1$, return to Step 2

The online computational burden should be mentioned for the real implementation of the robust real-time feedback ILC-MPC. For the above Algorithm I, the computational burden of the related QP optimization problem primarily depends on the number of online optimization variables. It can be easily verified that the maximum and minimum numbers of the online optimization variables of Algorithm I are $2mL$ and $mL + m$, respectively, which means that the computational burden of Algorithm I is similar to that of the previously developed ILC-MPC methods, such as (Shi et al. 2007).

7.2.3 Convergence Analysis

The following theorem is presented for analyzing the convergence of Algorithm I.

Theorem 7.1 Consider a batch process described by (7.1) and (7.2) subjected to the input constraints (7.3)–(7.5), the ILC-MPC in Algorithm I is feasible and guarantees the convergence if Assumption I holds and the optimization problem in (7.22) is feasible at time t of batch k.

Proof Since the optimization problem in (7.22) in Algorithm I is feasible for the above batch process at time t of batch k, the optimal solution is assumed to be

$$\Sigma^*(t) = \left\{\delta u^*(t, k), \ldots, \delta u^*(t + L - 1, k), \Delta U^*(k + 1)\right\},$$

which satisfies all constraints and corresponds to $\hat{J}^*(t, k)$.

If $L - 1 > t \geq 0$ holds true at time $t + 1$, a solution is obtained as

$$\Sigma(t + 1) = \left\{\delta u^*(t + 1, k), \ldots, \delta u^*(t + L - 1, k), \Delta U^*(k + 1)\right\},$$

by removing $\delta u^*(t, k)$ from the above solution at time t. Owing to the feasibility of $\Sigma^*(t)$ at time t, $\Sigma(t + 1)$ is also feasible at time $t + 1$.

Concerning $\hat{J}(t + 1, k)$, since the uncertainty in the item

$$q\left|\left|\vec{e}^o(\infty|t,k)\right|\right|^2 + q\mu^2\|\tilde{E}_t\|^2 \left\|\begin{bmatrix} \delta u(t,k) \\ \delta u(t+1,k) \\ \vdots \\ \delta u(L-1,k) \end{bmatrix}\right\|^2$$

does not exist at time t, it is easy to get

$$q\left|\left|\vec{e}^o(\infty|t,k)\right|\right|^2 + q\mu^2\|\tilde{E}_t\|^2 \begin{bmatrix} \delta u(t,k) \\ \delta u(t+1,k) \\ \vdots \\ \delta u(L-1,k) \end{bmatrix}^2$$

$$\geq q\left|\left|\vec{e}^o(\infty|t+1,k)\right|\right|^2 + q\mu^2\|\tilde{E}_{t+1}\|^2 \begin{bmatrix} \delta u(t+1,k) \\ \delta u(t+2,k) \\ \vdots \\ \delta u(L-1,k) \end{bmatrix}^2.$$

Meanwhile, the item of $\delta u^*(t,k)$ in $\hat{J}^*(t,k)$ is erased in $\hat{J}(t+1,k)$ and $q^o||\vec{e}^o(k|t,k)||^2$ is unchanged. Thus, $\Sigma(t+1)$ leads to $\hat{J}^*(t,k) \geq \hat{J}(t+1,k)$. Accordingly, it can be concluded that the optimal $\hat{J}(t+1,k)$ certainly satisfies $\hat{J}^*(t,k) \geq \hat{J}(t+1,k) \geq \hat{J}^*(t+1,k)$ when $L-1 > t \geq 0$.

If $t = L-1$, batch $k+1$ will begin at the next time instant. For the new batch, a solution can be obtained as $\Sigma(t+1) = \{\delta u^*(0,k+1),\ldots,\delta u^*(L-1,k+1),\rho\Delta U^*(k+1)\}$, where $\delta u^*(i,k+1)$ in $\Sigma(t+1)$ is the corresponding item of $\Delta U^*(k+1)$. According to the optimization theory, it can be easily verified that

$$q\left|\left|\vec{e}^o(k|0,k)\right|\right|^2 > q\left|\left|\vec{e}^o(\infty|0,k)\right|\right|^2.$$

Otherwise, the $\Sigma(t+1)$ will be zero. Therefore, it follows from Proposition 7.2 that

$$q\left|\left|\vec{e}^o(k|L-1,k)\right|\right|^2 > q\left|\left|\vec{e}^o(k+1|0,k+1)\right|\right|^2.$$

Note that $\Sigma(t+1)$ is exactly the second mode of ILC-MPC at batch k, which yields

$$\begin{aligned} &\left\|\left[\delta u^{*T}(0,k+1),\ldots,\delta u^{*T}(L-1,k+1)\right]^T\right\|^2 \\ &+\frac{\rho^2}{1-\rho^2}\left\|\left[\delta u^{*T}(0,k+1),\ldots,\delta u^{*T}(L-1,k+1)\right]^T\right\|^2 \\ &=\frac{1}{1-\rho^2}\left\|\left[\delta u^{*T}(0,k+1),\ldots,\delta u^{*T}(L-1,k+1)\right]^T\right\|^2. \end{aligned}$$

Meanwhile, $\tilde{E}_0 = \overline{G}\Gamma$, $\Gamma = \begin{bmatrix} I \cdots 0 \\ \vdots \ddots \vdots \\ I \cdots I \end{bmatrix}$, and $\Delta U^*(\mathrm{k}+1) = \Gamma\left[\delta u^{*T}(0,k+1),\ldots,\delta u^{*T}(L-1,k+1)\right]^T$. That is, $\left\|\tilde{E}_0\right\| \geq \left\|\overline{G}\right\|\|\Gamma\|$. Therefore, one obtains

$$\begin{aligned}
&\left\|\tilde{E}_0\right\|^2\left\|\left[\delta u^{*T}(0,k+1),\ldots,\delta u^{*T}(L-1,k+1)\right]^T\right\|^2 \\
&+\frac{\left\|\overline{G}\right\|^2\|\Gamma\|^2\|\rho\|^2}{1-\rho^2}\left\|\left[\delta u^{*T}(0,k+1),\ldots,\delta u^{*T}(L-1,k+1)\right]^T\right\|^2 \\
&\leq \frac{\left\|\overline{G}\right\|^2}{1-\rho^2}\left\|\Delta U^*(k+1)\right\|^2.
\end{aligned}$$

Since the uncertainty in the item

$$q\left\|\vec{e}^{o}(\infty|L-1,k)\right\|^2+q\mu^2\left\|\tilde{E}_0\right\|^2\left\|\left[\delta u^{*T}(0,k+1),\ldots,\delta u^{*T}(L-1,k+1)\right]^T\right\|^2$$

does not exist at time $L-1$ of batch k, the item $\delta u^*(L-1,k|L-1,k)$ in $\hat{J}^*(t,k)$ is erased in $\hat{J}(0,k)$. Hence, it is concluded that $\hat{J}^*(L-1,k) \geq \hat{J}(0,k) \geq \hat{J}^*(0,k)$.

The above analysis elucidates that the cost function in (7.22) will decrease along both the time and batch directions until the convergence, while the optimization problem remains feasible if it can be solved at time t of batch k. This completes the proof. □

7.2.4 An Illustrative Example

An injection molding machine is used for batch run tests, as shown in Fig. 7.1. For the molding process, the injection velocity is a key variable in the filling stage, which is controlled by manipulating the opening of a hydraulic valve. According to the reference (Shi et al. 2006), the injection velocity response to the proportional valve away from the operation point of resin polymer molding can be modeled as

$$p(z)=\frac{1.69z+1.419}{z^2-1.582z+0.5916},$$

which may be expressed in a state-space form of

$$x(t+1)=\begin{bmatrix}1.582 & -0.5916\\ 1 & 0\end{bmatrix}x(t)+\begin{bmatrix}1\\0\end{bmatrix}u(t),$$

$$y(t)=\begin{bmatrix}1.69 & 1.419\end{bmatrix}x(t).$$

Fig. 7.1 External view of an injection molding machine

Case 1: the nominal case without external disturbance

Consider the process input constraints,

$$|u(t)| \leq 2,$$

$$-0.2 \leq u(t,k) - u(t,k-1) \leq 0.2.$$

Since the nominal model is considered, there is $\mu = 0$, so the items related to μ in Algorithm I is erased. In order to illustrate the capability of Algorithm I, the above constraints on the input increments between two neighbour batches are considered here.

By taking the adjustable parameters $\rho = 0.9$, $q^o = q = 1$, $\lambda = 0.01$ for implementation, the control results are shown in Fig. 7.2 along with the corresponding inputs, where the representative batches 1, 2, and 14 are plotted. The evolution of tracking error is shown in Fig. 7.3, which illustrates that the tracking error is convergent after batch 12 and the final error sequence becomes small, close to zero. By investigating the output sequence in Fig. 7.2, especially the circled parts, it can be found that the output errors are erased or significantly reduced by iterative learning. Note that the tracking error in the circled part of batch 14 is caused by the input constraints, which is the reason why the final error is not zero. Meanwhile, it can also be verified from the inputs of batches 1 and 2 that the constraint on the input increments between two neighbor batches are satisfied.

Case 2: unknown nonlinear input

Consider the presence of input nonlinearity described by

$$f(u(t)) = 2\sin(u(t))$$

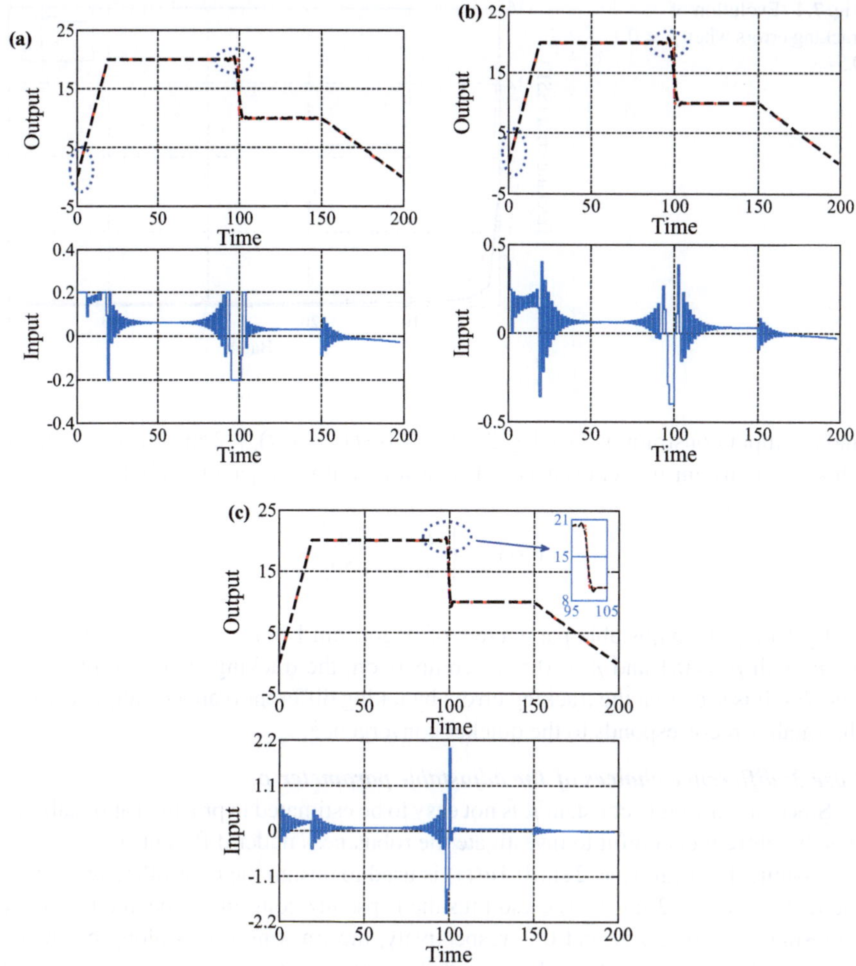

Fig. 7.2 The output and input sequences of (a) Batch 1; (b) Batch 2; (c) Batch 14 (the reference is denoted by a dash-dot line and the process outputs/inputs by solid lines)

Fig. 7.3 Evolution of tracking errors

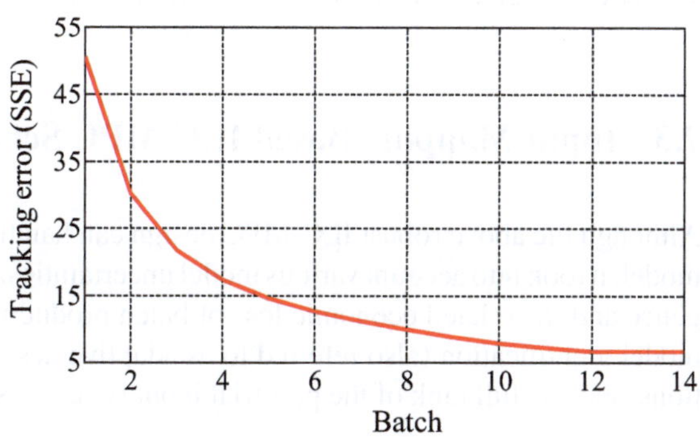

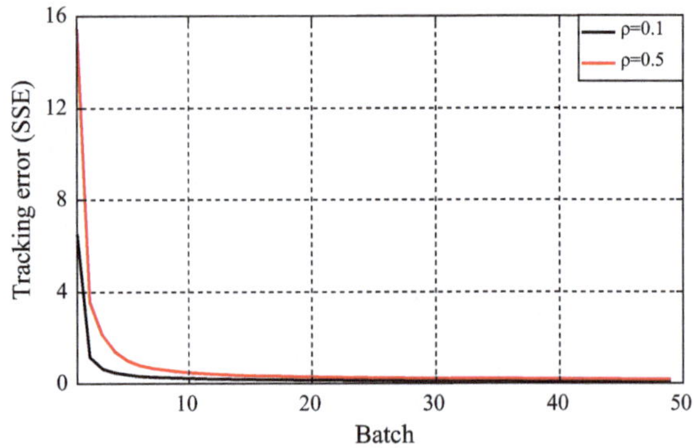

Fig. 7.4 Evolution of tracking errors when $\rho = 0.1, 0.5$

and the input constraint is $|u(t)| \leq 2$. Since $f(u(t)) = u(t) + 2\sin(u(t)) - u(t)$, the Lipschitz constant μ is chosen as 3. The desired output trajectory is in the form of

$$y_{\mathrm{d}}(t) = \begin{cases} t, \ t \leq 20, \\ 20, \ t > 20. \end{cases}$$

By taking the adjustable parameters of Algorithm I as $q^o = 1, q = 100, \lambda = 1$, along with $\rho = 0.1$ and $\rho = 0.5$ for comparison, the tracking errors are plotted in Fig. 7.4. It is seen that the tracking errors by taking different ρ are convergent, while the smaller ρ corresponds to the quicker convergence.

Case 3: difference choices of the adjustable parameter ρ.

Since the Lipschitz constant μ is not easy to be estimated in practical applications, it is therefore meaningful to investigate the robustness under different ρ.

Assume that $f(u(t)) = 2\sin(0.1u(t))$ is used to design the controller, away from the real $f(u(t)) = 2\sin(2u(t))$, and thus the Lipschitz constant is specified as 6. By choosing $\rho = 0.1, 0.5$, and 0.9, respectively, the tracking errors along batch run are plotted in Fig. 7.5. It can be seen that when $\rho = 0.1$, the convergence is not guaranteed. In fact, after batch 50, the tracking error becomes diverged. In contrast, the increase of ρ leads to improved robustness. That is, although a larger ρ may cause slower convergence, it can improve the robustness.

7.3 Input Mapping Based ILC-MPC Scheme

Although the above robust ILC-MPC design can handle the mismatch of the process model, it took into account various model uncertainties. This may lead to slow convergence and the related economic loss of batch production. Besides, the use of online model identification (also referred to as adaptive design) requires very strict conditions, such as full rank of the past trial input sequences (Janssens et al. 2013). These

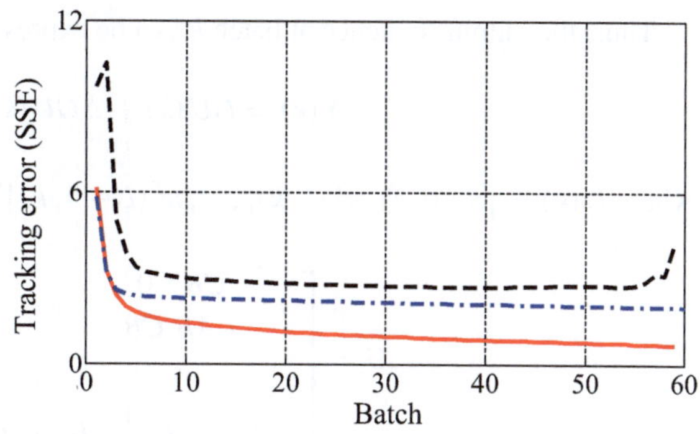

Fig. 7.5 Evolution of tracking errors by taking different ρ ($\rho = 0.1$ with dash line, $\rho = 0.5$ with blue line, and $\rho = 0.9$ with black line)

requirements bring difficulty to the practical application of the above robust ILC-MPC design. To circumvent the problem, an online data-driven approach, namely an input mapping scheme, is presented in this section to design run-to-run ILC-MPC.

Consider a discrete-time batch process described by

$$x(t+1,k) = (A + \Delta A(t))x(t,k) + (B + \Delta B(t))u(t,k), \tag{7.23}$$

$$y(t,k) = Cx(t,k), 0 \le t \le L. \tag{7.24}$$

The model (7.23) and (7.24) is similar to that in (7.1) and (7.2) except for the uncertain items $\Delta A(t)$ and $\Delta B(t)$, which reflect unknown process dynamics. The cycle time of each batch is also denoted by L,. while the input constraints in (7.3)–(7.5) are also considered here. Note that $\Delta A(t)$ and $\Delta B(t)$ are also periodic with regard to the same cycle time of a batch process.

The nominal model of the above batch process can be written as

$$x^o(t+1,k) = Ax^o(t,k) + Bu(t,k), \tag{7.25}$$

$$y^o(t,k) = Cx^o(t,k). \tag{7.26}$$

According to the model Eqs. (7.23) and (7.24) with $x(0,k) = 0$, the process state at batch k can be written as

$$x(1,k) = (B + \Delta B(0))u(0,k),$$

$$x(2,k) = (A + \Delta A(1))(B + \Delta B(0))u(0,k) + (B + \Delta B(1))u(1,k),$$

$$\vdots$$

Thus, the output sequence at batch k can be expressed by

$$Y(k) = GU(k) + \Delta GU(k), \tag{7.27}$$

where $U(k) = \left[u^T(0,k), u^T(1,k), \ldots, u^T(L-1,k)\right]^T$,

$$G = \begin{bmatrix} CB & 0 & \cdots & 0 \\ CAB & CB & & 0 \\ \vdots & & \ddots & \vdots \\ CA^{L-1}B & CA^{L-2}B & \cdots & CB \end{bmatrix},$$

and ΔG is a matrix including the unknown dynamics $\Delta A(t)$ and $\Delta B(t)$. Although $\Delta A(t)$ and $\Delta B(t)$ are time-varying within a batch, they are assumed to be fixed along the batch direction, and so is for ΔG.

For the convenience of analysis, the following assumption is made.

Assumption 7.2 The norm of ΔG in (7.27) is bounded by $\overline{\Delta G}$, while the actuator is assumed to be ideal without dead time.

From (7.27), it follows that

$$\vec{e}(k) = \vec{e}(k-1) - (G + \Delta G)\Delta U(k). \tag{7.28}$$

According to (7.28), a run-to-run ILC-MPC scheme can determine $\Delta U(k)$ at the interval between batch $k-1$ and batch k.

7.3.1 Data-Driven Control Structure with Input Mapping Scheme

Based on the above (7.28), the following relation can be established

$$\Delta e^o(\Delta U(k)) \triangleq \Delta G \Delta U(k) = \vec{e}(k-1) - \vec{e}(k) - G\Delta U(k). \tag{7.29}$$

After batch k, the vectors $\vec{e}(k-1)$, $\vec{e}(k)$, and $\Delta U(k)$ in the above relation becomes known. Thus, $\Delta e^o(\Delta U(k))$ can be exactly obtained, which reflects the information from the inputs to outputs caused by ΔG, although ΔG is unknown beforehand.

The updating control sequence $\Delta U(k)$ in (7.29), belongs to a linear space with dimension of mL and can be decomposed into a linear combination of mL linearly independent vectors such as $[1, 0, 0, \ldots, 0]^T$, $[0, 1, 0, 0, \ldots, 0]^T$ and so on, therefore called as the base vectors of R^n denoted by $\Delta \hat{U}_{\mathrm{i}}^{\mathrm{o}}$ $(i = 1, 2, \ldots, mL)$. For a base vector $\Delta \hat{U}_{\mathrm{i}}^{\mathrm{o}}$, if the corresponding $\Delta e^{\mathrm{o}}\left(\Delta \hat{U}_{\mathrm{i}}^{\mathrm{o}}\right)$ can be obtained from (7.29) based on the past data, it is easily obtained that $\Delta e^{\mathrm{o}}\left(a\Delta \hat{U}_{\mathrm{i}}^{\mathrm{o}}\right) = a\Delta e^{\mathrm{o}}\left(\Delta \hat{U}_{\mathrm{i}}^{\mathrm{o}}\right)$, where a is a scalar.

Hence, when the base vectors and the corresponding $\Delta e^{\mathrm{o}}(\Delta\hat{U}_{\mathrm{i}}^{\mathrm{o}})$ could be obtained, the output error sequence is expressed by

$$\vec{e}(k) = \vec{e}(k-1) - \mathrm{G}\sum_{i=1}^{mL}\lambda_i\Delta\hat{U}_{\mathrm{i}}^{\mathrm{o}} - \mathrm{G}\sum_{i=1}^{mL}\lambda_i\Delta e^{\mathrm{o}}(\Delta\hat{U}_{\mathrm{i}}^{\mathrm{o}})$$

under the updating input sequence $\Delta U(k) = \sum_{i=1}^{mL}\lambda_i\Delta\hat{U}_{\mathrm{i}}^{\mathrm{o}}$ where λ_i $(i = 1, 2, \ldots, mL)$ are the combination coefficients.

However, it is not easy to acquire all base vectors and the corresponding $\Delta e^{\mathrm{o}}(\Delta\hat{U}_{\mathrm{i}}^{\mathrm{o}})$ for implementation in practice. To address this issue, the past input updating sequences are directly chosen as the base vectors $\Delta\hat{U}_{\mathrm{i}}^{\mathrm{o}}$ and $\Delta e^{\mathrm{o}}(\Delta\hat{U}_{\mathrm{i}}^{\mathrm{o}})$. That is, for the batch k, the following matrix can be constructed

$$\Delta\hat{U}^{o}(k) = [\Delta U(k-1), \Delta U(k-2), \ldots, \Delta U(k-\tau)],$$

which corresponds to

$$\Delta E^{o}(k) = \left[\Delta e^{\mathrm{o}}(\Delta U(k-1)), \Delta e^{\mathrm{o}}(\Delta U(k-2)), \ldots, \Delta e^{o}(\Delta U(k-\tau))\right].$$

Based on the above two matrices, a data-driven updating control sequence for batch k is formulated as

$$\Delta U(k) = \Delta\hat{U}(k) + \Delta\hat{U}^{o}(k)l(k), \tag{7.30}$$

where $l(k)$ is the parameter vector denoted by $[\lambda(k-1), \lambda(k-2), \ldots, \lambda(k-\tau)]^T$, $\Delta\hat{U}(k)$ is an auxiliary updating input; a moving window of batches with a length τ is introduced to forget the data far from the current batch. Here, the past updating input sequences in the moving window are directly used as the base vectors, which may be linearly dependent of each other, and therefore, $\Delta\hat{U}^{o}(k)l(k)$ can approximately express $\Delta U(k)$. The auxiliary input $\Delta\hat{U}(k)$ is added to improve the expression accuracy.

With regard to the common input sequence, the main difference in (7.30) is the combination of historical input sequences, as determined by the mapping relationship from $\Delta\hat{U}^{o}(k)$ to $\Delta E^{o}(k)$. Hence, it is called the input mapping scheme. In fact, this scheme can not only be used in ILC-MPC, but also used to design an MPC or a sliding mode controller.

By the input mapping scheme, the tracking error model in (7.28) can be rewritten as

$$\vec{e}(k) = \vec{e}(k-1) - G\left[\Delta\hat{U}(k) + \Delta\hat{U}^{o}(k)l(k)\right] - \Delta E^{o}(k)l(k) - \Delta G\Delta\hat{U}(k), \tag{7.31}$$

which implies that there is no uncertainty for the input increment $\Delta U(k)$ corresponding to the combination of vectors in the moving window, which is completely determined by the parameter vector $l(k)$. Therefore, the input mapping scheme can reduce the effect from the uncertainties and improve the batch control performance. Note that, in the above input mapping scheme, the past data is not used to identify the system model, while neither matrix $\Delta\hat{U}^o(k)$ nor $\Delta E^o(k)$ is required to be full-rank square matrix.

To handle the input constraints and ensure the convergence of tracking error, the two-mode control framework in (7.11) is adopted to combine the input-mapping scheme, i.e.

$$\Delta U(k+i) = \begin{cases} \Xi(k)\begin{bmatrix} l(k) \\ \Delta\hat{U}(k) \end{bmatrix}, & i = 0, \\ \rho^{i-1}\Delta\hat{U}(k+1|k), & i \geq 1, \end{cases} \tag{7.32}$$

where $\Xi(k) = \left[\Delta\hat{U}^o(k)I\right]$. The variables $l(k)$, $\Delta\hat{U}(k)$, and $\Delta\hat{U}(k+1|k)$ are determined by the ILC-MPC controller. Accordingly, the control structure of the data-driven run-to-run ILC-MPC is shown Fig. 7.6.

Note that the historical input sequences are used in the first mode, since only the first updating sequence will be implemented according to the receding horizon optimization of MPC. The length of the moving window, denoted by τ is a parameter of the designed controller, which should be chosen beforehand. With a larger window length, the unknown process dynamics could be described more exactly, thus facilitating better control performance. Meanwhile, the optimization freedom of the ILC-MPC is also increased. Therefore, a larger τ can accelerate the convergent procedure. On the contrary, the moving window can also provide the forgotten function to accommodate for the variation of process dynamics reflected by $\Delta A(t)$ and $\Delta B(t)$. In other words, a larger moving window may degrade the closed-loop system stability, let alone the storage capability of the ILC-MPC controller.

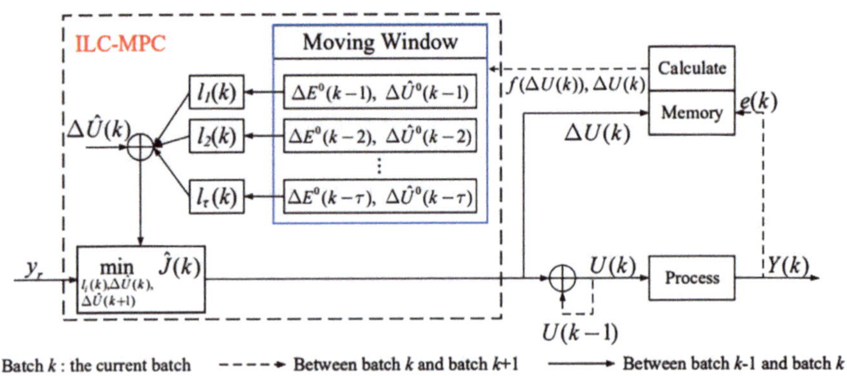

Fig. 7.6 Block diagram of the data-driven run-to-run ILC-MPC

7.3.2 Input Mapping Based Run-To-Run ILC-MPC Design

According to the control updating law in (7.32), the tracking error sequence can be written as

$$\vec{e}(k|k) = \vec{e}(k-1) - G\big[\Delta\hat{U}(k) + \Delta\hat{U}^o(k)l(k)\big] - \Delta E^o(k)l(k) - \Delta G\Delta\hat{U}(k),$$

$$\vec{e}(k+i|k) = \vec{e}(k|k) - G\sum_{j=1}^{i}\rho^{j-1}\Delta\hat{U}(k+1|k) - \Delta G\sum_{j=1}^{i}\rho^{j-1}\Delta\hat{U}(k+1|k), i \geq 1.$$

Correspondingly, the output error should converge to

$$\vec{e}(k+\infty|k) = \vec{e}(k|k) - G\frac{\Delta\hat{U}(k+1|k)}{1-\rho} - \frac{1}{1-\rho}\Delta G\Delta\hat{U}(k+1|k).$$

Hence, the optimization problem of the data-driven run-to-run ILC-MPC at batch k can be formulated as

$$\min_{l(k),\Delta\hat{U}(k),\Delta\hat{U}(k+1|k)} J(k) s.t., (7.16)-(7.18), \tag{7.33}$$

where $J(k) = r_1\|\Delta U(k)\|^2 + r_2\sum_{i=1}^{\infty}\|\Delta U(k+i|k)\|^2 + q_\infty\|\vec{e}(k+\infty|k)\|^2$, q_∞ and $r_2 > r_1 > 0$ are weighting parameters. The input constraints are the same with (7.3)–(7.5) and the above two-mode framework is adopted. Proposition 7.1 is used to guarantee the input constraints from batch k. The control goal is to make the process output sequence converge to the reference, while the convergent output sequence is considered into the objective function along with the control input sequences.

Due to the unknown items $\Delta A(t)$ and $\Delta B(t)$, the optimization problem in (7.33) is not a quadratic problem, which cannot be solved by the standard quadratic programming method. It is therefore proposed to use the upper bound of the objective function to formulate the ILC-MPC for the ease of online computation.

$$\begin{aligned} J(k) \leq\ & r_1\begin{bmatrix} l(k) \\ \Delta\hat{U}(k)\end{bmatrix}^T \Xi^T(k)\Xi(k)\begin{bmatrix} l(k) \\ \Delta\hat{U}(k)\end{bmatrix} + \frac{r_2}{1-\rho^2}\|\Delta\hat{U}(k+1|k)\|^2 \\ & + q_\infty\left\|\vec{e}(k-1) - G\Delta\hat{U}(k) - G\Delta\hat{U}^o(k)l(k) - \Delta E^o(k)l(k) - G\frac{\Delta\hat{U}(k+1|k)}{1-\rho}\right\|^2 \\ & + \frac{q_\infty}{1-\rho^2}\overline{\Delta G}^2\|\Delta\hat{U}(k+1|k)\|^2 + q_\infty\overline{\Delta G}^2\|\Delta\hat{U}(k)\|^2 \\ & := \overline{J}(k). \end{aligned}$$

The above Assumption 7.2 is used to acquire the upper bound of the objective function. Note that the weighting parameter q_∞ in $\overline{J}(k)$ is not the same as that in the above $J(k)$, which is used to tune the optimization problem of ILC-MPC for obtaining a feasible solution.

Considering that the vectors $\Delta\hat{U}^o(k-i)$ in the moving window may be linearly dependent, and the Hessian matrix of $\overline{J}(k)$ may be semi-positive-definite, which may cause difficulty in solving the above optimization problem, the objective function is further modified as

$$\hat{J}(\mathrm{k}) = \overline{J}(k) + \varepsilon l^T(k)l(k),$$

where the parameter $\varepsilon > 0$ is a small scalar specified for the convenience of computation.

Thus, the optimization problem of ILC-MPC is formulated as

$$\min_{l(k),\Delta\hat{U}(k),\Delta\hat{U}(k+1|k)} \hat{J}(k) s.t., (7.16) - (7.18). \tag{7.34}$$

This is a standard quadratic programming (QP) problem and can be efficiently solved by many QP solvers.

For clarity, the run-to-run ILC-MPC algorithm is summarized below.

Algorithm II (Data-driven run-to-run ILC-MPC)

Step 1. Choose parameters $\rho \in (0, 1)$, τ, $\varepsilon > 0$, and weighting parameters q_∞, r_1, r_2. Set $\Xi(0) = [0\, I]$ and $\Delta E^o(0) = 0$;
Step 2. Solve the optimization problem (7.34) at the interval between batch $k-1$ and k;
Step 3. Calculate $\Delta U(k) = \Xi(k)\begin{bmatrix} l(k) \\ \Delta\hat{U}(k) \end{bmatrix}$ and implement $U(k) = U(k-1) + \Delta U(k)$ on the controlled batch process;
Step 4. At the end of batch k, add $\Delta U(k)$ and $\Delta G \Delta U(k)$ from (7.29) into matrices $\Xi(k)$ and $\Delta E^o(k)$. Let $k = k+1$, return to Step 2.

7.3.3 Convergence Analysis

The following theorem is presented for analyzing the convergence of Algorithm II.

Theorem 7.2. *Consider a batch process described by* (7.23) and (7.24) *subjected to the input constraints in* (7.3)–(7.5), *the control Algorithm II is feasible and guarantees the convergence if* Assumption 7.2 *holds, and the optimization problem in* (7.34) *is always feasible if it is feasible at batch k*.

Proof Firstly, the feasibility of Algorithm II at the current batch is clarified herein. Since the optimization problem (7.34) in Algorithm II is feasible at batch k, one can assume that the optimal solution is $\Sigma^*(k) = \{l^*(k), \Delta\hat{U}^*(k), \Delta\hat{U}^*(k+1)\}$, which satisfies all the input constraints and corresponds to $\hat{J}^*(k)$.

When using Algorithm II, the incremental input sequence at batch k is $\Delta U(k) = \Xi(k)\begin{bmatrix} l^*(k) \\ \Delta\hat{U}^*(k) \end{bmatrix}$. At the end of batch k, the matrices $\Delta\hat{U}^o(k)$ and $\Delta E^o(k)$ are updated as $\Delta\hat{U}^o(k+1)$ and $\Delta E^o(k+1)$, where the $\Delta U(k)$ will be inserted into $\Delta\hat{U}^o(k+1)$ and its corresponding $\Delta G\Delta U(k) = \vec{e}(k-1) - \vec{e}(k) - G\Delta U(k)$ will be inserted into $\Delta E^o(k+1)$.

Then, for batch $k+1$, a solution of the optimization problem (7.34) is constructed as $\{0, \Delta\hat{U}^*(k+1|k), \rho\Delta\hat{U}^*(k+1|k)\}$, i.e., $\Delta\hat{U}^*(k+1|k+1) = \Delta\hat{U}^*(k+1|k)$ and $\Delta\hat{U}^*(k+2|k+1) = \rho\Delta\hat{U}^*(k+1|k)$. Note that the solution is assumed to offer the same incremental sequence from batch $k+1$ to ∞ as that given by the optimal solution at batch k. Since the optimal solution at batch k satisfies all the input constraints, it can be concluded that the constructed solution also satisfies these constraints at batch $k+1$, implying that it is feasible for batch $k+1$.

Consider $\hat{J}(k+1|k+1)$ of the constructed solution. Since item $\Delta U(k)$ at batch k is erased at batch $k+1$ and $r_2 > r_1$, the item $r_1||\Delta U(k)||^2 + r_2\sum_{i=1}^{\infty}\left|\left|\rho^{i-1}\Delta U^*(k+1|k)\right|\right|^2$ in the objective function at batch k becomes $r_1||\Delta U^*(k+1|k)||^2 + r_2\sum_{i=2}^{\infty}\left|\left|\rho^{i-1}\Delta U^*(k+1|k)\right|\right|^2$ at batch $k+1$, whose value must decrease.

Then, consider the item $q_\infty\|\vec{e}(k+\infty|k+1)\|^2$. The constructed solution implies that the updated control input sequences from batch $k+1$ to ∞ are the same as that at batch k, that is

$$\vec{e}^*(k) = \vec{e}(k-1) - G\left[\Delta\hat{U}^*(k) + \Delta\hat{U}^o(k)l^*(k)\right] - \Delta E^o(k)l^*(k) - \Delta G\Delta\hat{U}^*(k),$$

and $l(k+1) = 0$ means that $\vec{e}(k+\infty|k+1)$ can be rewritten as

$$\begin{aligned}
\vec{e}(k+\infty|k+1) &= \vec{e}(k) - \mathrm{G}\Delta\hat{U}^*(k+1|k) - \Delta G\frac{\Delta\hat{U}^*(k+1|k)}{1-\rho} \\
&= \vec{e}(k-1) - G\left[\Delta\hat{U}^*(k) + \Delta\hat{U}^o(k)l^*(k)\right] \\
&\quad - \Delta E^o(k)l^*(k) - \Delta G\Delta\hat{U}^*(k) \\
&\quad - \mathrm{G}\Delta\hat{U}^*(k+1|k) - \Delta G\frac{\Delta\hat{U}^*(k+1|k)}{1-\rho}.
\end{aligned}$$

Since $\Delta G\Delta\hat{U}^*(k)$ is exactly known after batch k, it follows from the triangle inequality that

$$\begin{aligned}
&\left\| \vec{e}(k) - \mathrm{G}\Delta\hat{U}^*(k+1|k) - \Delta G\frac{\Delta\hat{U}^*(k+1|k)}{1-\rho} \right\|^2 \\
&\le \Big\| \vec{e}(k-1) - G\big[\Delta\hat{U}^*(k) + \Delta\hat{U}^o(k)l^*(k)\big] - \Delta E^o(k)l^*(k) \\
&-\mathrm{G}\Delta\hat{U}^*(k+1|k) - \Delta G\frac{\Delta\hat{U}^*(k+1|k)}{1-\rho} \Big\|^2 \\
&+\big\| \Delta G\Delta\hat{U}^*(k) \big\|^2.
\end{aligned}$$

Comparing all items of the above inequality with the corresponding items in $\hat{J}(k)$, it can be found that the value of items corresponding to $\vec{e}(k+\infty)$ predicted at batch $k+1$ will decrease from that at batch k. In addition, since $l(k+1)$ is 0, $\varepsilon l^T(k+1)l(k+1)$ must be smaller than $\varepsilon l^T(k)l(k)$\$. That is, $\hat{J}(k+1|k+1) \le \hat{J}^*(k)$. Then, from the optimization theory, there stands $\hat{J}^*(k+1) \le \hat{J}(k+1|k+1) \le \hat{J}^*(k)$, where $\hat{J}^*(k+1)$ is the optimal value of the objective function at batch $k+1$. That is, the value of the cost function in the optimization problem (7.34) decreases along the batch direction until convergence, which is always feasible if it is feasible at batch k. This completes the proof. □

7.3.4 Experiments on a 6-Degree-of-Freedom Manipulator

A 6-degree-of-freedom manipulator is used for verifying Algorithm II to track a predefined reference periodically, which is shown in Fig. 7.7 together with its coordinate system of joints. For this manipulator, the repeated position accuracy of joint motors is 0.01 degree, and the limit of each joint angle is $\pm[180^o, 135^o, 115^o, 180^o, 115^o, 180^o]$, the limit of each joint angle speed is $\pm[180^o, 180^o, 180^o, 180^o, 180^o, 180^o]$. The corresponding Denavit-Hartenberg model is established with parameters listed in Table 7.1.

The model is in the form of

$$\dot{s} = J(\boldsymbol{\theta})\dot{\theta},$$

where $s = \left[x, y, z, \boldsymbol{a}, \boldsymbol{o}, \boldsymbol{p}\right]^T$ indicates the terminal position $\{x, y, z\}$ and posture $\{\boldsymbol{a}, \boldsymbol{o}, \boldsymbol{p}\}$ of the manipulator, and $\boldsymbol{\theta}$ is the angular vector of the manipulator's joints. Considering the sampling time and the periodical operation, the above model can be discretized as

$$s(t+1, k) = s(t, k) + B(t, k)\Delta\theta(t, k),$$

where $B(t, k) = TJ(\theta(t, k))$ and T is the cycle time of the system reference. The tracking task starts from the same initial position. For each batch run, $\theta(0, k)$, $B(0, k)$, and $s(0, k)$ are fixed for initialization.

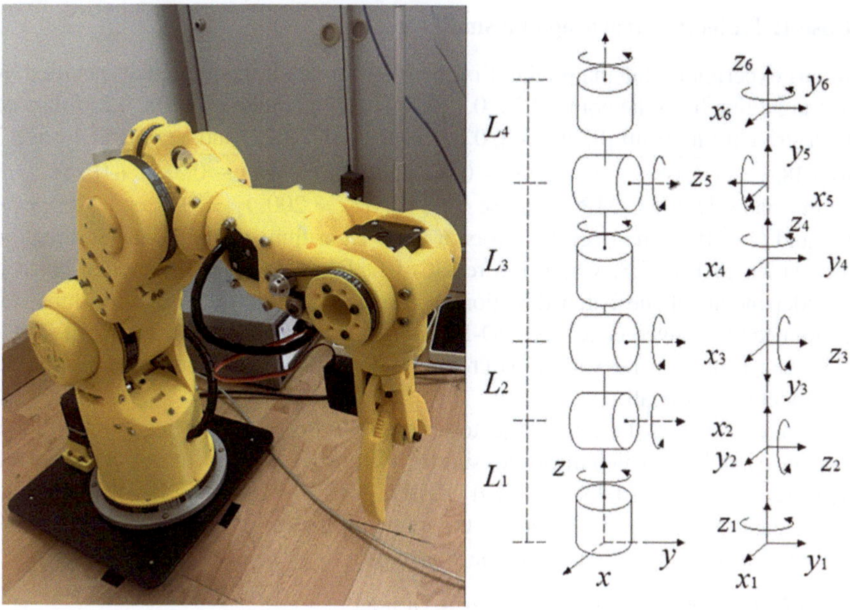

Fig. 7.7 A 6-degree-of-freedom manipulator

Table 7.1 D-H model of manipulators

i	α_{i-1}	$a_{i-1}(mm)$	θ_{i-1}	$d_{i-1}(mm)$
1	0°	0	$\boldsymbol{\theta}_1$	$L_1 = 242$
2	-90°	0	$\boldsymbol{\theta}_2 - 90^o$	0
3	0°	$L_2 = 225$	$\boldsymbol{\theta}_3 + 90^o$	0
4	90°	0	$\boldsymbol{\theta}_4$	$L_3 = 228.86$
5	90°	0	$-\boldsymbol{\theta}_5$	0
6	-90°	0	$\boldsymbol{\theta}_6$	$L_3 = 50$

Since $J(\boldsymbol{\theta})$ is nonlinear with respect to θ, the variation of θ during the manipulator operation will result in the change of its mode, which is not exactly known before the tracking error is eliminated.

In all experiments, the initial joint angles are set as $\theta(0, k) = [14.74, 66.37, 58.92, 0.00, 54.71, 14.74]$ and $s(0, k) = [0.38, 0.1, 0.15, -1, 0, 0, 0, 1, 0, 0, 0, -1]$. The nominal model is established under the initial condition. To make a comparison with the existing ILC methods, these experiments also include: Algorithm II without data-driven (named as ILC-MPC), the design in (Chi et al. 2015) (named as DDO-ILC), and the MPC algorithm with online discretized model (Online-MPC). The above Algorithm II is denoted by DD-ILC-MPC. Note that the model of Online-MPC is updated according to the current angles of joints.

Case 1: Trajectory tracking in a small range

In this experiment, the manipulator is conducted to track the reference specified by a straight line from the point [0.38, 0.1, 0.15] to the point [0.34, 0.08, 0.2] followed by a straight line from [0.34, 0.08, 0.2] to [0.3, 0.02, 0.25] and a curve (x from 0.3 to 0.38, $y = \sqrt{0.08^2 - (x - 0.3)^2} - 0.06$, $z = \sqrt{0.08^2 - (x - 0.38)^2} + 0.25$).

For using DD-ILC-MPC, choose $\tau = 10$, $q_\infty = 200$, $r_1 = 0.1$, $r_2 = 200$, $\rho = 0.1$ and $\overline{G} = 0.1$. The tracking errors of these algorithms along the batch direction are shown in Fig. 7.8, where the result of Online-MPC is not included since it is independent of the batch direction. It can be seen that all ILC algorithms can guarantee the convergence, and DD-ILC-MPC is almost the same as that of DDO-ILC. This is because the variation of reference is within a small range and the change of model is also small.

Figure 7.9 shows the actual trajectories of different algorithms in a 3D coordinate system at batch 49, which illustrates similar results to in Fig. 7.8. It is obvious that the convergence speed of DDO-ILC and DD-ILC-MPC is better than ILC-MPC owing to the data-driven approach. Due to the model of Online-MPC not being accurate, there exists a steady state error in the control result.

Case 2: Trajectory tracking in a large range

In this experiment, the manipulator is conducted to track the reference specified by a straight line from [0.38, 0.1, 0.15] to [0.34, −0.2, 0.2] followed by a straight line from [0.34, −0.2, 0.2] to [0.2, −0.2, 0.15] and a straight line from [0.2, −0.2, 0, 15] to [0.22, −0.3, 0.2]. The variation of this reference is obviously larger than Case 1.

Choose the same parameters for DD-ILC-MPC as in Case 1. Figure 7.10 shows the tracking errors of different algorithms. Unlike Case 1, the result of DDO-ILC is not convergent. By checking the model variation along the reference, it can be found

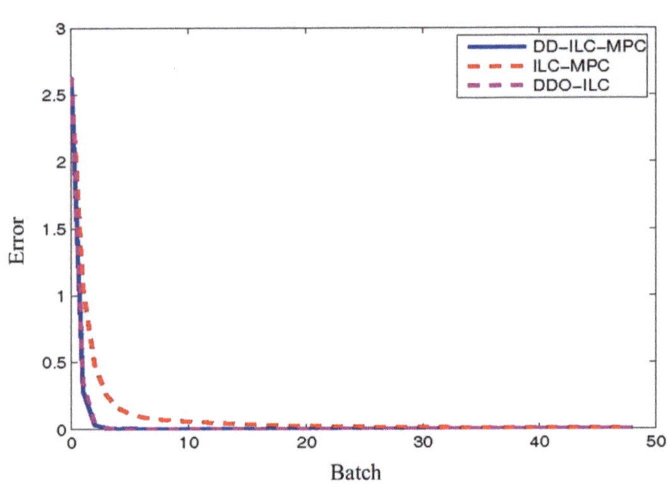

Fig. 7.8 The error variation of Case 1

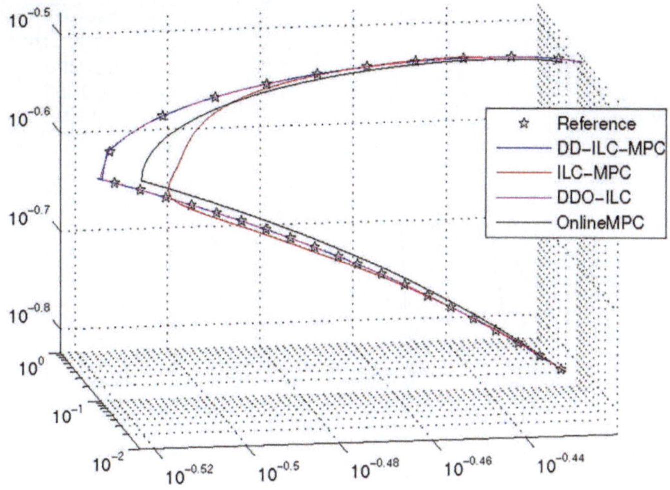

Fig. 7.9 The tracking trajectories in 3D space at batch 49

that the real system does not satisfy Assumption 3 in (Chi et al. 2015), which leads to the divergent result.

Figure 7.11 shows the actual trajectories of different algorithms in a 3D coordinate system at batch 99. It can be seen that DD-ILC-MPC offers better control results than other algorithms.

The above two experiments illustrate that if the model change is small, all the ILC algorithms can acquire good tracking performance, especially for DD-ILC-MPC and

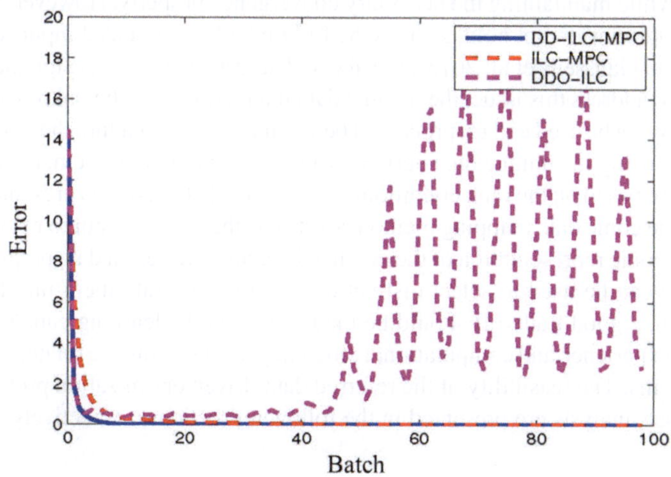

Fig. 7.10 The error variation curves of Case 2

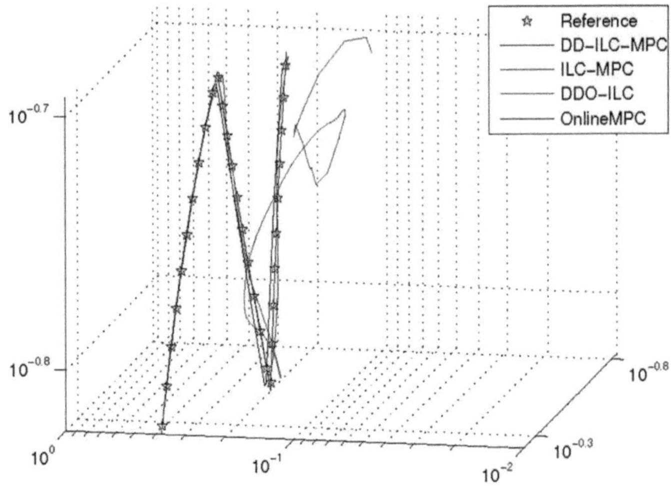

Fig. 7.11 The tracking trajectories in 3D space at batch 99

DDO-ILC. However, when the change of the model becomes larger, the proposed design, i.e., DD-ILC-MPC, has an obvious advantage over other designs.

7.4 Conic Data-Driven Optimization-Based ILC Scheme

Based on the selected data in the above section, a data-driven control scheme can be designed while maintaining the necessary convergence property. However, when the related conditions do not hold, such as the full rank of the past trial input sequences as studied in (Janssens et al. 2013), the above data-driven control design may not be feasible. To address this issue, the distinct data interacting with the unknown process uncertainty can be used and preprocessed before incorporating it into the optimal ILC design. By using the unique data derived from historical input increment sequences and selected based on the rank conditions, a new conic ILC design is presented in this section. The conic data mapping method constructs the control input to be optimized by using a conic representation of the distinct data. Since different data preprocessing techniques can be used to select distinct data and incorporate them into the conic ILC design, a good trade-off should be made between the learning capability from historical experience and computational efficiency from solving the related optimization problems. The feasibility of the referred data-driven optimization problems and convergence analysis are presented in the following sections, respectively.

7.4.1 Data Selection and Conic Data Mapping

For the batch process described by (7.1) and (7.2), if the nonlinear part is not considered, the above optimization-based ILC can be used to handle the system uncertainty together with the input constraints in (7.3)–(7.5). By ignoring the state-dependent uncertainty $f(\cdot)$ and defining $\hat{x}(t,k) = \mathrm{x}(t,k) - x_d(t)$ and $\hat{u}(t,k) = \mathrm{u}(t,k) - u_d(t)$, one obtains

$$\hat{x}(t+1,k) = A\hat{x}(t,k) + B\hat{u}(t,k), \tag{7.35}$$

$$e(t,k) = C\hat{x}(t,k), \tag{7.36}$$

where the tracking error is $e(t,k) = y(t,k) - y_d(t)$.

Then, using $\delta u(t,k) = u(t,k) - u(t-1,k) - \delta u_d(t)$, the nominal control system based on the above process described by (7.1) and (7.2) can be written in the incremental state-space form

$$\delta x(t+1,k) = A\delta x(t,k) + B\delta u(t,k), \tag{7.37}$$

where $\delta x(t,k) = \hat{x}(t,k) - \hat{x}(t-1,k)$. Accordingly, there is

$$e(t+1,k) = e(t,k) + CA\delta x(t,k) + CBu(t,k). \tag{7.38}$$

Define an augmented state by $x\ (t,k) = [\delta x(t,k), e(t,k)]^T$, one obtains the augmented incremental state-space model as

$$x(t+1,k) = \boldsymbol{A}x(t,k) + \boldsymbol{B}\delta u(t,k), \tag{7.39}$$

where the augmented system matrices are given by

$$\boldsymbol{A} = \begin{bmatrix} A & 0 \\ CA & I \end{bmatrix} \text{and}\, \boldsymbol{B} = \begin{bmatrix} B \\ CB \end{bmatrix}.$$

To realize the optimal ILC design for the above batch process, a stage cost function denoted by $L_k(\mathrm{x}(t,k), \delta u(t,k))$ is introduced in terms of the state $f(x(t,k))$ and the control input increment $\delta u(t,k)$, as detailed in the next section.

7.4.2 Optimization-Based Conic ILC Algorithm

A basic form of the data-driven ILC is

$$\delta u(t,k) = h(D), \tag{7.40}$$

where $h(\cdot)$ denotes a mapping function with respect to a data set D, including all data collected by the real system, as defined by

$$\begin{aligned} D = \{ & x(0,1), \ \ldots, x(N,1), \ \delta u(0,1), \ \ldots, \delta u(N-1,1), \\ & x(0,k-1), \ldots, x(N,k-1), \delta u(0,k-1), \ \ldots, \\ & \delta u(N-1,k-1), \ \ldots, \ e(t,k), \ \ \delta u(0,k), \ \ldots, \delta u(N-1,k).\} \end{aligned} \tag{7.41}$$

However, it could be computationally expensive to design the mapping function $h(\cdot)$ with all data from the set D, causing a large computational burden on the control system. To overcome the deficiency, the following two preprocessing steps are proposed. The first step is a data processing technique for the set D, which provides a subset of distinct data to be incorporated into the ILC design. The second step is a conic design on $h(\cdot)$, formulating a new conic data mapping based on the selected distinct data.

Firstly, a subset of the input data from D is selected as

$$\begin{aligned} D_0 = \{ & \delta u(0,1), \ldots, \delta u(N,1), \ldots, \\ & \delta u(0,k-1), \ldots, \delta u(N,k-1), \delta u(0,k), \ldots, \delta u(t-1,k)\}. \end{aligned} \tag{7.42}$$

By using linear algebra, only a finite family of input vectors in D_0, that is

$$D_1 = \{s_1, \ldots, s_m\} \tag{7.43}$$

is needed, where s_i $(i = 1, \ldots, m.)$ denotes a vector selected from D_0. Thus, all admissible inputs in R_m can be represented as a linear combination of the vectors in the set D_1, which can be selected by using the rank conditions. Note that it should be guaranteed that the distinct data in D_1 contain at least one basis of R_m, such that there stands rank $([s_1, \ldots, s_m]) = m$ for all vectors in D_1.

Correspondingly, the data mapping $h(D)$ can be simplified as

$$\delta u(t,k) = h(D_1), \tag{7.44}$$

which indicates that only m data points are needed for the optimization-based ILC design.

Next, using the distinct data set D_1, a conic representation of the data mapping $h(\cdot)$ is can be obtained. To incorporate the distinct data points in D_1 into the optimal ILC design, the data mapping $h(D_1)$ is designed by using the conic hull theory, i.e.

$$h(D_1) = CH(D_1) + \upsilon_{t,k} \tag{7.45}$$

where D_1 denotes the set of data points $s_1, \ldots, s_m$ and the vector $\upsilon_{t,k}$ denotes an auxiliary control input vector, which can be seen as a projection of $h(D_1)$ on the conic hull $CH(D_1)$.

Hence, the control input $\delta u(t, k)$ can be rewritten as

$$\delta u(t,k) = h(D_1) = \sum_{i=1}^{m} \alpha_i s_i + \upsilon_{t,k}. \tag{7.46}$$

Based on the above conic data mapping, a quadratic cost function is defined as

$$L_k(\cdot,\cdot) = ||x(t+1,k)||_Q^2 + ||\delta u(t,k)||_R^2, \tag{7.47}$$

and correspondingly, an optimization problem is formulated by

$$\min_{\alpha_1,\alpha_2,\ldots,\alpha_m,\upsilon\{k,t\}} \quad L_k(x(t,k), \delta u(t,k)) \tag{7.48}$$

$$s.t. x(t+1,k) = \boldsymbol{A}x(t,k) + \boldsymbol{B}\delta u(t,k),\ \delta u(t,k) = CH(D_1) + \upsilon_{t,k},$$
$$\textit{Constraints}\ (7.3) - (7.5).$$

By solving the above QP optimization problem, the control updating law is obtained as

$$\delta u^\star(t,k) = \sum_{i=1}^{m} \alpha_i^\star s_i + \upsilon_{t,k}^\star + u(t,k-1) + \delta u_d(t) \tag{7.49}$$

for implementation at time t of batch k. The procedure can be easily repeated at the next time instant $t + 1$.

7.4.3 Convergence Analysis

For analyzing the feasibility and robust convergence of the above conic ILC design, the following lemma is presented.

Lemma 7.2. Consider a batch process described by (7.1) and (7.2) subject to the constraints (7.3)–(7.5), the optimal solution $\delta u^\star(t, k)$ obtained by solving the optimization problem in (7.48) is always feasible if the optimization problem is feasible at any time t of the first batch.

Proof Since for any time t of the first batch, the optimization problem in (7.48) is feasible, so that the optimal solution $\delta u^\star(t, 1)$ can be executed by constructing the distinct data set D_1. Hence, the proof follows by the mathematical induction approach:

(1) For the second batch, denoted by $k = 2$, the same control input can be adopted as the optimal input for the first batch, that is, $u(t, 2) = u^{\star}(t, 1)$, which means that $\delta u(t, 2) = 0$. In this case, one can set $\alpha_i = 0, i = 1, \ldots, m$, and $\upsilon_{t,k} = 0$. Thus, it can be easily verified that this is a feasible solution for the above optimization problem at any time t with $k = 2$.
(2) Suppose that at any time t of batch $k \geq 2$, the above optimization problem has a feasible solution, denoted by $\{\alpha_1^{\star}, \ldots, \alpha_m^{\star}\}and\upsilon_{\mathrm{t,k}}^{\star}$. Then $\delta u_k^{\star}(t) = \sum_{i=1}^{m} \alpha_i^{\star} s_i + \upsilon_{\{k,t\}}^{\star}$ and $\delta u_{\mathrm{k}}^{\star}(\mathrm{t})$ satisfies the above conic representation and the input constraints in (7.3)–(7.5).
(3) For the batch $k + 1$, since the vectors $\{s_1, \ldots, s_m\}$ in D_1 are a basis of the Euclidean space R_m, there exist m parameters denoted by $\{\beta_1, \ldots, \beta_m\}$ satisfying $\xi_k(t) = \delta u_k^{\star}(t) - \upsilon_{t,k+1}$ and $\xi_k(t) = \sum_{\{i=1\}}^{\{m\}} \beta_i s_i$. To prove the feasibility of the above optimization problem at any time t of batch $k + 1$, an admissible input vector $\delta u(t, k + 1)$ is constructed by

$$\delta u(t, k + 1) = \sum_{i=1}^{m} \gamma_i s_i + \upsilon_{t,k+1}, \tag{7.50}$$

where $\gamma_i = \beta_i$, $i = 1, \ldots, m$, if $\beta_i \geq 0$; otherwise, $\gamma_i = \alpha_i^{\star}$. In addition, let $\upsilon_{t,k+1} = \delta u_k^{\star}(t) - \xi_k(t) = \sum_{i=1}^{m} \left(\alpha_i^{\star} - \beta_i\right) s_i + \upsilon_{t,k}^{\star}$. Thus, the non-negative coefficients $\{\gamma_1, \ldots, \gamma_m\}$ are obtained at time t of batch $k + 1$. By taking all parameters γ_i, $(i = 1, \ldots, m)$ into the conic representation, there follows $\delta u(t, k + 1) = \delta u_k^{\star}(t)$, indicating that the dynamics and the input constraints are both satisfied for batch $k + 1$. Thus, there is at least one admissible solution constructed as above for the optimization problem at batch $k + 1$. Hence, the optimization problem is feasible at batch $k + 1$.

Based on the above mathematical induction, it is concluded that the above optimization problem is feasible. This completes the proof. □

Theorem 7.3 *Consider a batch process described by* (7.23) and (7.24) *subjected to the input constraints in* (7.3)–(7.5) *under the control law* $u^{\star}(t, k)$ *determined by solving the optimization problem in* (7.48) *with the cost function defined by an extended quadratic form of* $L_k(\cdot, \cdot) = ||x(t + 1, k)||_Q^2 + ||\delta u(t, k)||_R^2 + ||\Delta u(t, k)||_{R_u}^2$ *where* $\Delta u(t, k) = \delta u(t, k) - \delta u(t, k - 1)$. Assume that the pair $(\overline{A}, \boldsymbol{B})$ with $\overline{A} = \boldsymbol{A} + (1 + ||\mathrm{C}||) l_f \mathbf{I}$ can be readily stabilized via feedback. Then robust asymptotic convergence of the controlled system along the batch direction is guaranteed as $k \to \infty$ if $\rho\left(\left(\boldsymbol{B}^T Q \boldsymbol{B} + R + R_u\right)^{-1} R\right) < 1$, where $\rho(\cdot)$ denotes the spectral radius.

Proof Firstly, it should prove that the controlled system is subject to the input constraints (7.3)–(7.5) under the control law $u_k^{\circ}(t)$ from solving the optimization problem in (7.48) holds the robust asymptotic convergence property, if the cost function $L_k(\cdot,\cdot)$ is defined by the above form. To this end, a simplified and augmented system model is formulated by linearizing the system uncertainty as

$$x^1(t+1,k) = \left[\boldsymbol{A} + (1+||C||)l_f\boldsymbol{I}\right]x^1(t,k) + \boldsymbol{B}\delta u^\circ(t,k), \tag{7.51}$$

where the control law $\delta u_k^\circ(t)$ is adopted. Note that the stationary condition $\frac{\partial L_k}{\delta u_k^\circ} = 0$ for optimality of the output trajectory is the same as that of (7.1) and (7.2).

By taking the similar arguments defined in the proof of Zhou et al. (2022), it is straightforward that robust asymptotic convergence of the controlled system along the batch direction is guaranteed.

Next, consider the process model (7.1) and (7.2) with unknown $f(\cdot) \neq 0$. For every $x(t,k) = x^1(t,k)$ with the same input $\delta u^\circ(t,k)$, an augmented system model is

$$x(t+1,k) = \boldsymbol{A}x(t,k) + \boldsymbol{B}\delta u_k^\circ(t) + \boldsymbol{w}(t,k), \tag{7.52}$$

where $\boldsymbol{w}(t,k) = \left[\overline{f}^T(x(t,k)), C\overline{f}^T(x(t,k))\right]^T$ with $\overline{f}(x(t,k)) = f(x(t,k)) - f(x(t-1,k))$. Then, it follows that $\left|\overline{f}(x(t,k))\right| \leq l_f|\delta x(t,k)|$, and

$$\begin{aligned}
&\|x(t+1,k)\| - \left\|x^1(t+1,k)\right\| \leq \|\boldsymbol{w}(t,k)\| - \left\|(1+||C||)l_f x^1(t,k)\right\| \\
&\leq \left(1+||C||l_f\right)(\|\delta x_k(t)\| - \|x(t,k)\|).
\end{aligned}$$

By the definition of $x(t,k) = \left[\delta x^T(t,k), e^T(t,k)\right]^T$, there is $||\delta x(t,k)|| \leq ||x(t,k)||$. Accordingly,

$$0 \leq \|x(t+1,k)\| \leq \left\|x^1(t+1,k)\right\|. \tag{7.53}$$

Then, by iteration, it can be easily verified that robust asymptotic convergence of $x(t+1,k)$ is guaranteed if the robust convergence property of $x^1(t+1,k)$ holds true. Correspondingly, the robust convergence of the system $x(t+1,k)$ is ensured. Hence, robust asymptotic convergence of the above system under the control law $u^\circ(t,k)$ is guaranteed.

Subsequently, consider the control law $u^\star(t,k)$ from solving the above optimization problem with the same cost function. To prove the robust asymptotic convergence of the above system controlled by $u^\star(t,k)$, it needs to clarify that $\delta u^\circ(t,k)$ is a feasible solution to the conic ILC optimization problem. To this end, let $u^\star(t,k) = u(t-1,k) + \delta u^\circ(t,k)$. Owing to that the control law $\delta u^\circ(t,k)$ from solving the above optimization problem satisfies the input constraints in (7.3)–(7.5), it also satisfies the reformulated constraints in the conic ILC optimization problem. At this point, it shall be proved that there exists a family of nonnegative coefficients

$\{\gamma_1, \ldots, \gamma_m\}$ and vector $\upsilon_{t,k}$ such that $\delta u^{\circ}(t,k) = \sum_{i=1}^{m} \gamma_i s_i + \upsilon_{t,k}$. This can be guaranteed by using similar arguments as the construction of feasible solution in the proof of Lemma 7.2. Thus, there exists at least one admissible solution, taking $\delta u^{\circ}(t,k)$ as an example, for the optimization problem satisfying all the input constraints.

Next, since every solution of $\delta u^{\circ}(t,k)$ from solving the above optimization problem is feasible to the conic optimization problem, it is sure that for batch k, $0 \leq L_k(x(t,k), 0) \leq L_k(x(t,k), \delta u^{\star}(t,k) \leq L_k(x(t,k), \delta u^{\circ}(t,k))$, which means that the control performance of the system controlled by $u^{\star}(t,k)$ will not be inferior to that controlled by $u^{\circ}(t,k)$. This indicates that the robust asymptotic convergence property of the above system under the control law $u^{\star}(t,k)$ still holds as it is controlled by the law $u^{\circ}(t,k)$. This completes the proof. □

Theorem 7.4 *Consider a batch process described by* (7.23) and (7.24) *subjected to the input constraints* in (7.3)–(7.5) *under the control law* $u^{\star}(t,k)$ *determined by solving the optimization problem in* (7.48), *where* $L_k(\cdot,\cdot) = \|x(t+1,k)\|_Q^2 + \|\delta u(t,k)\|_R^2$. *Assume that the pair* $(\overline{A}, \boldsymbol{B})$ *with* $\overline{A} = \boldsymbol{A} + (1 + ||\mathrm{C}||)\mathrm{l_f}1$ *can be readily stabilized via feedback. Then robust asymptotic convergence of the controlled system along the batch direction is guaranteed as* $k \to \infty$ *if* $\rho\left(\left(\boldsymbol{B}^T Q \boldsymbol{B} + R\right)^{-1} R\right) < 1$.

Proof By letting $R_u = 0$ in the sufficient condition of Theorem 7.3, the premise of Theorem 7.4 is the same as that of Theorem 7.3. Hence, the same conclusion can be obtained by using similar arguments for analysis. This completes the proof. □

7.4.4 An Illustrative Example

To demonstrate the effectiveness of the above conic ILC design, the packing phase of an injection molding machine shown in Fig. 7.1 is considered here for illustration. The nozzle packing pressure response to the hydraulic control valve opening was identified by the least-squares regression method (Shi et al. 2006), as described by a state-space model,

$$x(t+1,k) = \begin{bmatrix} 1.6070 & 1 \\ -0.6086 & 0 \end{bmatrix} x(t,k) + \begin{bmatrix} 1.2390 \\ -0.9282 \end{bmatrix} u(t,k) + f(x(t,k)),$$

$$y(t,k) = \begin{bmatrix} 1 & 0 \end{bmatrix} x(t,k),$$

where $0 \leq t \leq 100; f(x(t,k)) = \Delta_a(t) x(t,k)$ with

$$\Delta_\mathrm{a}(\mathrm{t}) = \begin{bmatrix} 0.04\delta_1(t) & 0 \\ -0.04\delta_2(t) & 0 \end{bmatrix}$$

and $\delta_i(t)$, $i \in \{1, 2\}$, denote two independently and uniformly distributed random variables in the interval (0, 1). The desired reference trajectory $y_d(t)$ is herein set as

$$y_d(t) = \begin{cases} 0, 0 \le t < 20, \\ 15(t-20), 20 \le t < 40, \\ 300, 40 \le t < 60, \\ 300 - 15(t-60),\ 40 \le t < 60, \\ 0, 80 \le t \le 100. \end{cases}$$

For simulation tests, the sampling time is taken as one second, and the trail length of each batch is $N = 100$. Besides, the input constraints are assumed to be $\overline{u}_i = -\underline{u}_i = 20$ and $\overline{\delta u}_i = -\underline{\delta u}_i = 10$.

***Case 1*: *Conic ILC design with a quadratic cost function in* (7.47)**

To illustrate the control performance, the weighting matrices are taken as $Q = \boldsymbol{I}$ and $R = 0.1\boldsymbol{I}$ to track the reference $y_d(t)$ for the batch packing process of injection molding in the presence of process uncertainties. It can be easily verified that the convergence condition in Theorem 7.4 holds with $0.0248 < 1$. The output trajectories and control inputs of the packing process at the 1st, 2nd, 10th, and 30th cycles are depicted in Fig. 7.12. It can be seen that the output trajectories converge to the reference with good accuracy, while all the input constraints are satisfied, despite that the input signals have multiple vibrations due to the state-dependent uncertainty.

To demonstrate the advantage of the above conic ILC design with the quadratic cost function, a comparison with the optimal ILC design given by Xu et al. (2013) is made herein. The ILC method in (Xu et al. 2013) adopted a lift technique to design the batch-wise control input, which could be redesigned by solving the optimization control problem at each sampling time. Meanwhile, the cost function can be modified to include the total tracking errors of a batch run. For performance evaluation, the following index is adopted

$$RMSE(k) = \sqrt{\frac{1}{N}\sum_{i=1}^{N} e^2(t, k)},$$

which represents the squared root of the averaged tracking errors.

The comparison results with respect to batch k are shown in Fig. 7.13. It is seen that the above conic ILC design with the quadratic cost function converges evidently faster than the optimal ILC method given in (Xu et al. 2013) with the same cost function.

Case 2*: *Conic ILC design with an extended quadratic cost function

By using the extended quadratic cost function $L_k(\cdot,\cdot) = \|x(t+1,k)\|_Q^2 + \|\delta u(t,k)\|_R^2 + \|\Delta u(t,k)\|_{R_u}^2$ for the conic ILC design, the weighting matrices are taken as $Q = \boldsymbol{I}$ and $R = 0.1\boldsymbol{I}$, identical with the above Case 1 for comparison. Besides, two different values are taken to demonstrate the effects of the weighting term R_u, i.e., $R_u = 0.1$ and $R_u = 1$. It can be easily verified that if $R_u = 0.1$, the

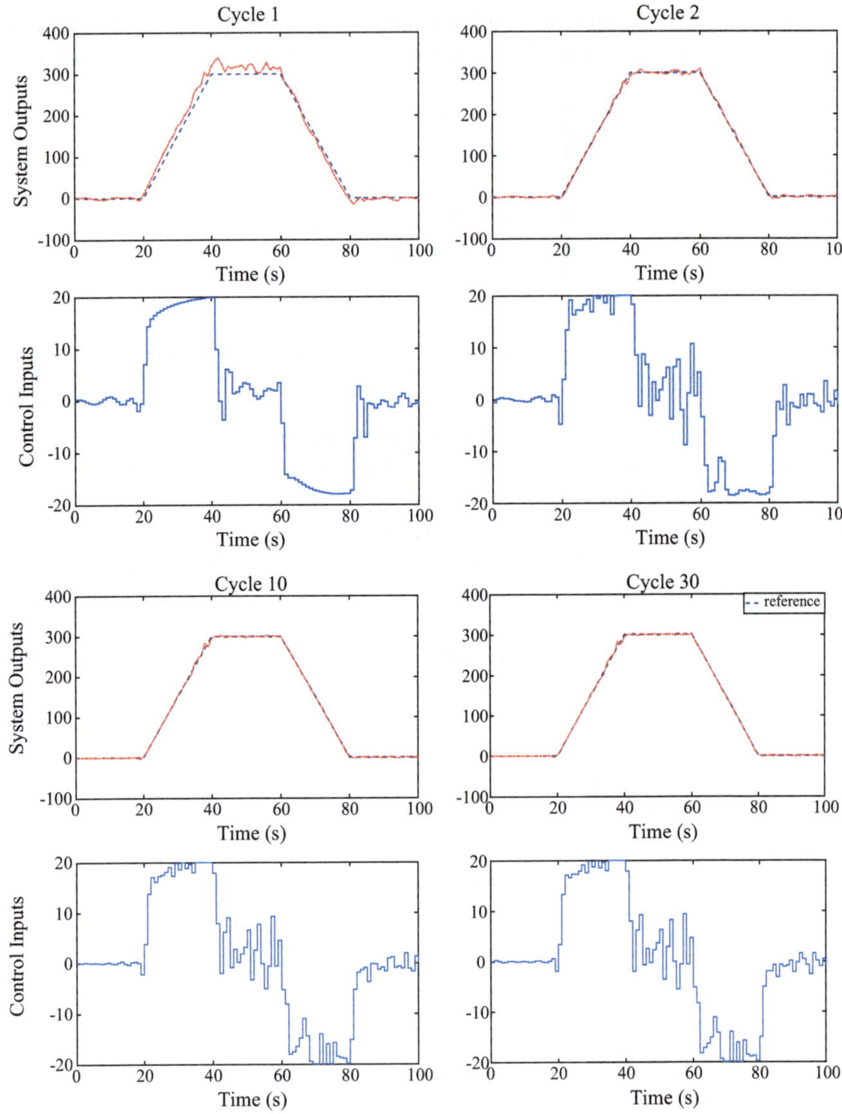

Fig. 7.12 The output trajectories and control inputs at the 1st, 2nd, 10th, and 30th cycles

convergence condition in Theorem 7.3 holds with $0.0242 < 1$; if $R_u = 1$, it holds with $0.0199 < 1$.

Figures 7.14 and 7.15 show the control input signals at the 1st, 2nd, 5th, 10th, 20th, and 30th iterations, by taking $R_u = 0.1$ and $R_u = 1$, respectively.

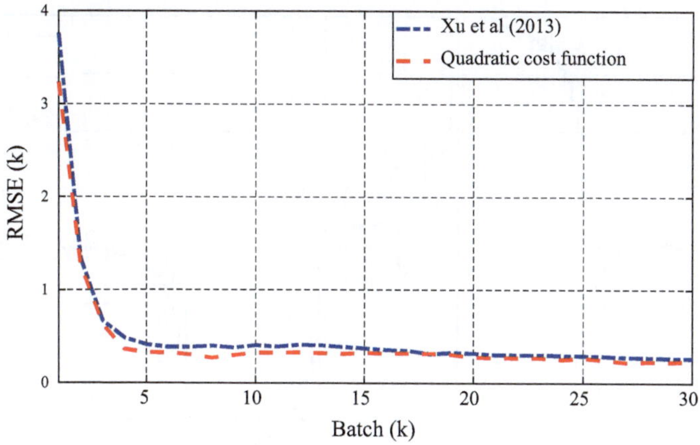

Fig. 7.13 Comparisons of *RMSE* with respect to batch *k*

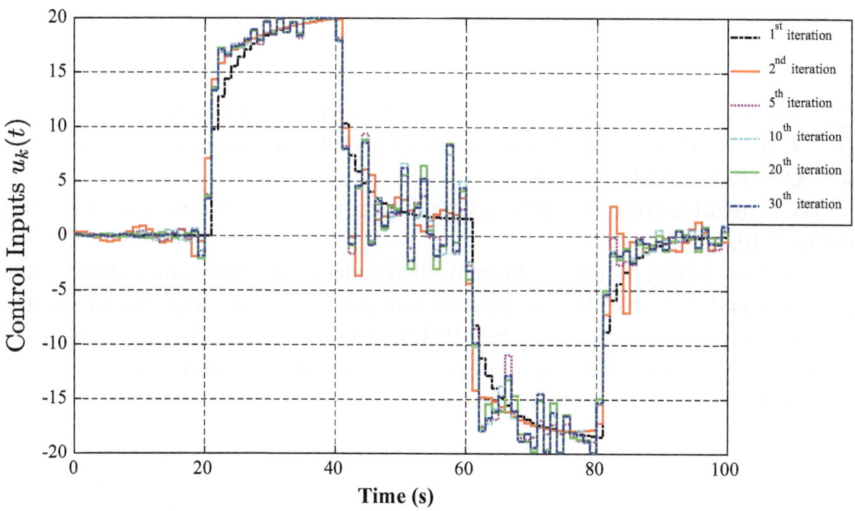

Fig. 7.14 The control inputs with $R_u = 0.1$

It can be seen that all the input constraints are satisfied for all cycles, though the control input signals have multiple vibrations due to the state-dependent uncertain term $f(x(t, k))$.

For comparison, the ILC method given by Sun and Alleyne (2014) is also performed. Note that the cost function in the cited reference merely contains information at each individual sampling time within a batch, which took the same cost function with the above conic ILC design. Four cases with different weighting matrices $[Q, R, R_u]$ are considered here, that is, Case 1: the quadratic cost function is used with $Q = \boldsymbol{I}, R = 0.1\boldsymbol{I}, R_u = 0.1\boldsymbol{I}$; Case 2: the quadratic cost function is used with

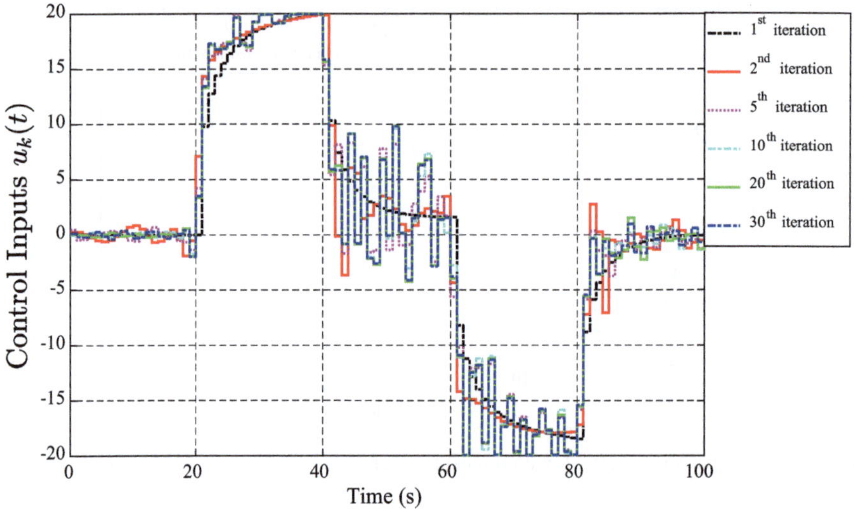

Fig. 7.15 The control inputs with $R_u = 1$

$Q = \boldsymbol{I}, R = 0.1\boldsymbol{I}, R_u = 1\boldsymbol{I}$; Case 3: the extended quadratic cost function is used with $Q = \boldsymbol{I}, R = 0.1\boldsymbol{I}, R_u = 0.1\boldsymbol{I}$; Case 4: the extended quadratic cost function is used with $Q = \boldsymbol{I}, R = 0.1\boldsymbol{I}, R_u = 1\boldsymbol{I}$.

The comparison results of *RMSE* with respect to batch *k* in four cases are shown in Fig. 7.16.

It can be seen that for the system controlled under cases 3 and 4, the control results are stable and desirable. By comparison, a larger value of R_u degrades the control performance corresponding to a larger RMSE. The above conic ILC algorithm under cases 3 and 4 acquires better control performance than the optimal ILC method in (Sun and Alleyne 2014) under cases 1 and 2.

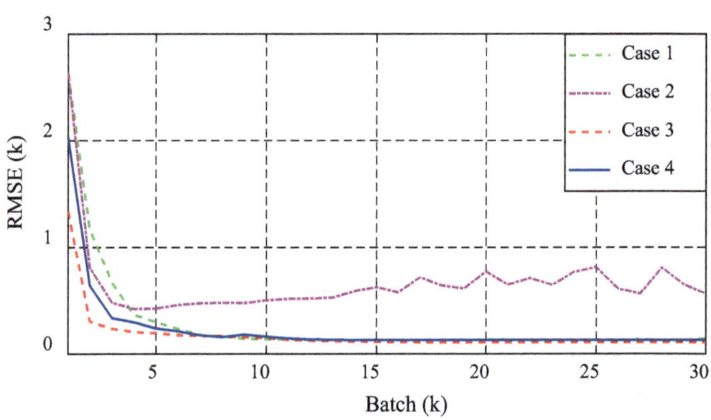

Fig. 7.16 Comparison of *RMSE* in different cases

7.5 Summary

To deal with the input constraints of batch run, the design approaches of ILC combined with MPC (Li et al. 2016) and optimal control (Zhou et al. 2022) can provide good tracking performance with convergence. Meanwhile, the influence of process uncertainties could be effectively suppressed by these data-driven ILC methods. Moreover, the input mapping strategy (Li et al. 2019) can be used to accelerate the convergence speed, which can be efficiently implemented online in practice.

To guarantee the convergence of constrained ILC-MPC, the chapter presents the two-mode control framework, which could address the convergence from the current batch to the infinite batch, by designing different control laws in each mode. Accordingly, the real-time-feedback ILC-MPC and run-to-run ILC-MPC are presented to guarantee convergence of the tracking error, respectively.

The control tests on a batch run of an injection molding machine and a 6-freedom manipulator well demonstrate the effectiveness and advantages of the above ILC designs, in contrast with previously developed ILC methods.

References

Chi R, Hou Z, Huang B, Jin S (2015) A unified data-driven design framework of optimality-based generalized iterative learning control. Comput Chem Eng 77:10–23

He S, Chen W, Li D, Xi Y, Xu Y, Zheng P (2022) Iterative learning control with data-driven-based compensation. IEEE Trans Cybern 52(8):7492–7503

Janssens P, Pipeleers G, Swevers J (2013) A data-driven constrained norm-optimal iterative learning control framework for LTI systems. IEEE Trans Control Syst Technol 21(2):546–551

Lee K, Chin I, Lee H, Lee J (1999) Model predictive control technique combined with iterative learning for batch processes. AIChE J 45(10):2175–2187

Li D, He S, Xi Y, Liu T, Gao F, Wang Y, Lu J (2019) Synthesis of ILC–MPC controller with data-driven approach for constrained batch processes. IEEE Trans Industr Electron 67(4):3116–3125

Li D, Xi Y, Lu J, Gao F (2016) Synthesis of real-time-feedback-based 2D iterative learning control–model predictive control for constrained batch processes with unknown input nonlinearity. Ind Eng Chem Res 55(51):13074–13084

Li Q, Wan HY, Zhao Z, Liu F (2024a) Robust batch-to-batch optimization with global sensitivity analysis for microbial fermentation processes under model-plant mismatch. Chem Eng Sci 301:120658

Li Z, Han M, Vo D, Yin X (2024b) Machine learning-based input-augmented Koopman modeling and predictive control of nonlinear processes. Comput Chem Eng 191:108854

Liu J, Ruan X, Zheng Y (2022) Iterative learning control for discrete-time systems with full learnability. IEEE Trans Neural Netw Learn Syst 33(2):629–643

Lu J, Cao Z, Zhao C, Gao F (2019) 110th anniversary: An overview on learning-based model predictive control for batch processes. Ind Eng Chem Res 58(37):17164–17173

Mayne D, Rawlings J, Rao C, Scokaert P (2000) Constrained model predictive control: Stability and optimality. Automatica 36:789–814

Qin S, Badgwell T (2003) A survey of industrial model predictive control technology. Control Eng Pract 11(7):733–764

Shi J, Gao F, Wu T-J (2006) Robust iterative learning control design for batch processes with uncertain perturbations and initialization. AIChE J 52(6):2171–2187

Shi J, Gao F, Wu T (2007) Single-cycle and multi-cycle generalized 2D model predictive iterative learning control (2D-GPILC) schemes for batch processes. J Process Control 17(9):715–727

Sun H, Alleyne AG (2014) A computationally efficient norm optimal iterative learning control approach for LTV systems. Automatica 50(1):141–148

Wang Y, Zhou D, Gao F (2008) Iterative learning model predictive control for multi-phase batch processes. J Process Control 18(6):543–557

Xu Z, Zhao J, Yang Y, Shao Z, Gao F (2013) Optimal iterative learning control based on a time-parametrized linear time-varying model for batch processes. Ind Eng Chem Res 52(18):6182–6192

Zhang G, Li Z, Li J, Zhang X (2024) Robust iterative learning fault-tolerant control for unsupervised path following of an ASV via the AIS data-based guidance. Ocean Eng 310:118616

Zhou Y, Li D, Gao F (2022) Conic iterative learning control using distinct data for constrained systems with state-dependent uncertainty. IEEE Trans Industr Inf 18(5):3095–3104

Chapter 8
2D State Predictor Based ILC Design Under Input Delay

8.1 Introduction

The input delay is often encountered in industrial process operation, e.g., the heat exchange of a heating barrel for heating an injection molding machine (Liu et al. 2018) or a thermocirculator for cooling down a glass crystallizer (Liu et al. 2017), due to the limited surfaces for thermo-energy exchange. It may provoke performance degradation of a closed-loop control system or even its instability. The survey literature (Bonvin et al. 2006) pointed out that the input delay had become a notable challenge to design ILC for batch process optimization, especially when there exist time-varying process uncertainties (Wang et al. 2009). An increasing quantity of ILC methods were devoted to addressing this challenge in the past two decades (Shokri-Ghaleh et al. 2021 and 2023). Concerning the developed ILC methods in the frequency domain, a Smith predictor (SP) based ILC design was presented by Xu et al. (2001) to enhance the tracking of the set-point profile under the process input delay. In contrast, an internal model based ILC design (Liu et al. 2010) was proposed to accommodate the process input delay with uncertainty in practice. A double-loop ILC structure based on the reference shift (Cai et al. 2008) was presented for application to a non-minimum phase batch process with delay response, which could obtain evidently improved performance for repetitive operation. Concerning the other ILC methods developed in the time domain, a so-called holding mechanism was introduced for the ILC design (Park et al. 1998) to conduct batch optimization under the input delay. Another dead-time compensator-based ILC design (Xu et al. 2005) was provided to realize perfect tracking for industrial batch process operation subject to a constant input delay. In contrast, a phase lag compensation strategy was adopted to design ILC so as to accommodate the process input delay (Tan et al. 2009). By incorporating a state observer-based predictor into the real-time feedback control loop, an indirect-type ILC scheme was developed in Tao et al. (2023) for batch processes with model uncertainties and input delay. Based on an extended non-minimum state space model of the batch process, a 2D model predictive ILC strategy based on the

T. Liu et al., *Robust Iterative Learning Control of Industrial Batch Systems*, Intelligent Control and Learning Systems 22, https://doi.org/10.1007/978-981-96-9778-6_8

set point learning was presented in Li et al. (2024) to reduce the impact of time delay so as to improve the tracking performance. It should be noted that although a lot of robust control methods had been developed for time delay systems in the existing references, it is not straightforward to extend these methods for application to industrial batch processes subject to input delays, especially under time-varying process uncertainties from batch to batch (Wang et al., 2009; Shen et al. 2014).

To address the above issues related to time-varying process uncertainties and the input delay, a robust ILC design is therefore presented in this chapter for practical applications. For the convenience of ILC design, the batch operation is described by a 2D Fornasini-Marchesini (FM) system form. To compensate for the process input delay, a 2D predictor is designed by using the measured output errors with regard to the current and previous batches, so as to predict the augmented FM system states. Accordingly, the controller tuning could be simply designed for the resulting delay-free system. Moreover, a synthetic 2D Lyapunov–Krasovskii function is constructed to establish the robust stability conditions for the corresponding closed-loop control system. The ILC controller is determined by solving the related matrix inequalities with free-weighting matrices and a modified cone complementarity linearization (CCL) method. In addition, it is suggested to take the integral of tracking error as an augmented system state for the design of closed-loop ILC, in order to remove the steady-state output tracking error for each batch run. For clarity, the proposed robust ILC design for a batch process with input delay is detailed in the following sections.

8.2 Design of 2D State Predictor

Generally, an observable canonical state-space model in the discrete time domain may be used to describe an industrial batch process with input delay and time-varying uncertainties

$$P_\Delta : \quad \begin{cases} x(t+1,k) = [A + \Delta A(t,k)]x(t,k) + [B + \Delta B(t,k)]u(t-h,k), \\ y(t,k) = Cx(t,k), \ t \in \mathbb{Z}_N, \\ x(0,k) = x(0), \ k \in \mathbb{Z}_+, \end{cases} \tag{8.1}$$

where t and k denote the time index and batch number, respectively; Denote the process input, state variables, and output by $u(t,k) \in \Re^{n_u}$, $x(t,k) \in \Re^{n_x}$ and $y(t,k) \in \Re^{n_y}$, respectively; The initial resetting condition of each batch is denoted by $x(0)$; The system matrices are indicated by $\{A, B, C\}$ with appropriate dimensions, while h denotes the process input delay; The overall batch time (or period) is denoted by N.

The process dynamic uncertainties are denoted by $\Delta A(t,k)$ and $\Delta B(t,k)$. The matrix uncertainties are expressed by the norm-bounded form

$$[\Delta A(t,k), \quad \Delta B(t,k)] = E\Delta(t,k)[F_\text{A}, \quad F_\text{B}] \tag{8.2}$$

with

$$\Delta^{\mathrm{T}}(t,k)\Delta(t,k) \le \mathbf{I},$$

where E,F_{A}, F_{B} are prespecified matrices based on a prior knowledge of process operation, and $\Delta(t,k)$ indicates a time-varying function with a magnitude bound of unity during each batch run.

For analysis, the output tracking error in the current batch is expressed by

$$e(t,k) \triangleq y(t,k) - y_{\mathrm{d}}(t), \tag{8.3}$$

where $y_{\mathrm{d}}(t)$ indicates the target output profile for an SISO system, and $y(t,k)$ denotes the real system output in the current batch.

Taking into account the output response delay, the above tracking error is written as

$$e(t,k) \triangleq y(t,k) - y_{\mathrm{d}}(t-h). \tag{8.4}$$

Define a batchwise-direction incremental function by

$$\delta f(t,k) \triangleq f(t,k) - f(t,k-1), \tag{8.5}$$

where f could be the process input, state, or output variable.

For a batch process P_{Δ} expressed in (8.1), the ILC control law is generally in the form

$$\Sigma_{\mathrm{ILC}}: \quad \begin{cases} u(t,k) = u(t,k-1) + \delta u(t,k), \\ u(t,0) = 0, \quad t \in \mathbb{Z}_N, \quad k \in \mathbb{Z}_+, \end{cases} \tag{8.6}$$

where $\delta u(t,k)$ denotes the control increment to be designed, $u(t,0)$ indicates the initial control input that is generally reset to zero (or a constant value that may be scaled to zero) for the convenience of batch run.

Using (8.1) and (8.6) along with the definitions in (8.4) and (8.5), it follows that

$$e(t+1,k) = e(t+1,k-1) + C\delta x(t+1,k), \tag{8.7}$$

$$\delta x(t+1,k) = [A + \Delta A(t,k)]\delta x(t,k) + [B + \Delta B(t,k)]\delta u(t-h,k) + \varpi(t,k), \tag{8.8}$$

where

$$\varpi(t,k) \triangleq [\Delta A(t,k) - \Delta A(t,k-1)]x(t,k-1) + [\Delta B(t,k) - \Delta B(t,k-1)]u(t-h,k-1). \tag{8.9}$$

It is evident that $\varpi(t,k) \neq 0$ corresponds to nonrepeatable process uncertainties, which is therefore regarded as nonrepeatable disturbance to treat with. For the simplicity of analysis, it is used to represent all kinds of disturbances to evaluate the control system stability and performance.

According to the 2D system control theory (Kaczorek 1985), the above batch process operation is herein described by a 2D FM system form for the convenience of designing the ILC updating law

$$\Sigma_{2D\text{-}P_\Delta}: \begin{cases} \begin{bmatrix} \delta x(t+1,k) \\ e(t+1,k) \end{bmatrix} = (A_1 + \Delta A_1)\begin{bmatrix} \delta x(t,k) \\ e(t,k) \end{bmatrix} + A_2\begin{bmatrix} \delta x(t+1,k-1) \\ e(t+1,k-1) \end{bmatrix} \\ \qquad + \left(B_g + \Delta B_g\right)\delta u(t-h,k) + D_g\varpi(t,k), \\ y(t,k) = \begin{bmatrix} e(t,k-1) \\ e(t,k) \end{bmatrix} = G\begin{bmatrix} \delta x(t,k) \\ e(t,k) \end{bmatrix}, \\ z(t,k) = e(t,k) = H\begin{bmatrix} \delta x(t,k) \\ e(t,k) \end{bmatrix}, \end{cases} \tag{8.10}$$

where

$$H = \begin{bmatrix} \mathbf{0} & 1 \end{bmatrix}, G = \begin{bmatrix} -C & 1 \\ \mathbf{0} & 1 \end{bmatrix}, D_g = \begin{bmatrix} \boldsymbol{I} \\ C \end{bmatrix}.$$

$$A_1 = \begin{bmatrix} A & 0 \\ CA & 0 \end{bmatrix}, \Delta A_1 = \begin{bmatrix} \Delta A(t,k) & 0 \\ C\Delta A(t,k) & 0 \end{bmatrix} = \begin{bmatrix} E \\ CE \end{bmatrix}\Delta(t,k)\begin{bmatrix} F_A & 0 \end{bmatrix} \triangleq \overline{E}\Delta(t,k)\overline{F}_A,$$

$$A_2 = \begin{bmatrix} \mathbf{0} & \mathbf{0} \\ \mathbf{0} & 1 \end{bmatrix}, B_g = \begin{bmatrix} B \\ CB \end{bmatrix}, \Delta B_g = \begin{bmatrix} \Delta B(t,k) \\ C\Delta B(t,k) \end{bmatrix} = \begin{bmatrix} E \\ CE \end{bmatrix}\Delta(t,k)F_B \triangleq \overline{E}\Delta(t,k)F_B,$$

For the purpose of designing the control law, it is desired to stabilize the system state variables in (8.10) to the origin in the state space. Considering that the above state variables could not be measured or are difficult to acquire in engineering practice, only the measured outputs in the current and last batches are adopted to estimate these state variables. Correspondingly, an ILC updating law is designed by using these estimated states for implementation. Note that the output tracking error $e(t,k-1)$ in the last batch is adopted for designing the feedforward control along the batch direction, while $e(t,k)$ in the current batch is used for designing the feedback control along the time direction in the current batch. Denote by $z(t,k)$ the output tracking error in the current batch. The robust control objective for batch run can be formulated as

$$\|z(t,k)\|_2 \leq \gamma\|\varpi(t,k)\|_2, \tag{8.11}$$

where γ indicates a prescribed $\mathcal{H}_\infty$ control performance index for disturbance rejection.

Accordingly, an augmented 2D system state vector is denoted by $x_g(t,k) = [\,\delta x^T(t,k)\ e^T(t,k)\,]^T$, which yields an equivalent 2D system description of (8.10)

$$\begin{cases} x_g(t+1,k) = (A_1 + \Delta A_1)x_g(t,k) + A_2 x_g(t+1,k-1) \\ \qquad\qquad + \big(B_g + \Delta B_g\big)\delta u(t-h,k) + D_g\varpi(t,k), \\ y(t,k) = Gx_g(t,k), \\ z(t,k) = Hx_g(t,k). \end{cases} \tag{8.12}$$

To make compensation for the input delay, a 2D predictor is constructed by

$$\begin{aligned} \hat{x}_g(t+1,k) = {} & A_1\hat{x}_g(t,k) + A_2\hat{x}_g(t+1,k-1) + B_g\delta u(t,k) + L_1\big[G\hat{x}_g(t-h,k) - y(t,k)\big] \\ & + L_2\big[G\hat{x}_g(t+1,k-1) - y(t+1+h,k-1)\big], \end{aligned} \tag{8.13}$$

where $\hat{x}_g(t,k)$ denotes the estimate of $x_g(t+h,k)$. Thus, the prediction error is denoted by

$$\tilde{x}_g(t,k) = \hat{x}_g(t-h,k) - x_g(t,k). \tag{8.14}$$

With the above state prediction by h steps ahead, the 2D system error dynamics can be formulated as

$$\begin{aligned} \tilde{x}_g(t+1,k) = {} & A_1\tilde{x}_g(t,k) + (A_2 + L_2G)\tilde{x}_g(t+1,k-1) + L_1G\tilde{x}_g(t-h,k) \\ & - \Delta A_1 x_g(t,k) - \Delta B_g\delta u(t-h,k) - D_g\varpi(t,k). \end{aligned} \tag{8.15}$$

Hence, the ILC updating law based on the above state prediction is designed as

$$\delta u(t,k) = K_1\hat{x}_g(t,k) + K_2\hat{x}_g(t+1,k-1), \tag{8.16}$$

where K_1 is a feedback controller for stabilizing the closed-loop system by using the predicted state of $x_g(t+h,k)$, and K_2 is a feedforward controller for output tracking by using the one-step ahead state information in the last batch.

8.3 Robust 2D ILC Design

By substituting (8.16) into (8.12) and introducing a new augmented state

$$X(t,k) \triangleq \begin{bmatrix} x_g(t,k) + \tilde{x}_g(t,k) \\ \tilde{x}_g(t,k) \end{bmatrix}, \tag{8.17}$$

the following closed-loop 2D system description is obtained

$$\begin{cases} X(t+1,k) = \left(\widehat{A}_1 + \Delta\widehat{A}_1\right)X(t,k) + \left(\widehat{A}_2 + \Delta\widehat{A}_2\right)X(t+1,k-1) + \widehat{A}_{\rm h}X(t-h,k) \\ \qquad\qquad + \widehat{D}_{\rm g}\varpi(t,k), \\ z(t,k) = \widehat{H}X(t,k), \end{cases} \tag{8.18}$$

where $\widehat{H} = \begin{bmatrix} H & -H \end{bmatrix}$,

$$\widehat{A}_1 = \begin{bmatrix} A_1 + B_{\rm g}K_1 & \mathbf{0} \\ \mathbf{0} & A_1 \end{bmatrix}, \widehat{A}_2 = \begin{bmatrix} A_2 + B_{\rm g}K_2 & L_2G \\ \mathbf{0} & A_2 + L_2G \end{bmatrix}, \widehat{A}_{\rm d} = \begin{bmatrix} \mathbf{0} \; L_1G \\ \mathbf{0} \; L_1G \end{bmatrix}, \widehat{D}_{\rm g} = \begin{bmatrix} \mathbf{0} \\ -D_{\rm g} \end{bmatrix},$$

$$\Delta\widehat{A}_1 = \begin{bmatrix} \mathbf{0} & \mathbf{0} \\ -\Delta A_1 - \Delta B_{\rm g}K_1 & \Delta A_1 \end{bmatrix} = \begin{bmatrix} \mathbf{0} \\ \overline{E} \end{bmatrix}\Delta(t,k)\begin{bmatrix} -\overline{F}_{\rm A} - F_{\rm B}K_1 \; \overline{F}_{\rm A} \end{bmatrix} \triangleq \widehat{E}\Delta(t,k)\widehat{F}_{\rm A},$$

$$\Delta\widehat{A}_2 = \begin{bmatrix} \mathbf{0} & \mathbf{0} \\ -\Delta B_{\rm g}K_2 & \mathbf{0} \end{bmatrix} = \begin{bmatrix} \mathbf{0} \\ \overline{E} \end{bmatrix}\Delta(t,k)\begin{bmatrix} -F_{\rm B}K_2 \; \mathbf{0} \end{bmatrix} \triangleq \widehat{E}\Delta(t,k)\widehat{F}_{\rm B}.$$

To establish an ILC updating law for the above closed-loop system, the following lemma is briefly introduced for the convenience of analysis.

Lemma 8.1 (Shao et al. 2011) Given a matrix $M \in \Re^{m\times m}$ satisfying $M > 0$, two integers satisfying $l_2 > l_1$, and a vector function $\chi : \{l_1, \cdots, l_2\} \rightarrow \Re^m$, there holds

$$(l_2 - l_1 + 1)\sum_{i=l_1}^{l_2} \chi^{\rm T}(i)M\chi(i) \geq \left(\sum_{i=l_1}^{l_2}\chi(i)\right)^{\rm T} M \left(\sum_{i=l_1}^{l_2}\chi(i)\right). \tag{8.19}$$

Based on the above lemma, the following theorem is presented for evaluating the robust stability of the above closed-loop 2D system in (8.18).

Theorem 8.1 The closed-loop 2D system in (8.18) with the input delay indicated by h is asymptotically stable while satisfying (8.11), if there exist matrices $P > P_0 > 0$, $R > 0$, $Q = Q^{\mathrm{T}}$, $N \geq 0$, M, K_1, K_2, L_1, L_2, and scalar $\varepsilon > 0$ such that the following matrix inequalities hold

$$\begin{bmatrix} \Omega_{11} & M_2^{\mathrm{T}} + hN_{12} & \Omega_{13} & M_4^{\mathrm{T}} + hN_{14} & \widehat{H}^{\mathrm{T}} & \widehat{A}_1^{\mathrm{T}} & h(\widehat{A}_1 - \boldsymbol{I})^{\mathrm{T}} & \widehat{F}_{\mathrm{A}}^{\mathrm{T}} \\ * & \Omega_{22} & -M_2 + hN_{23} & hN_{24} & \mathbf{0} & \widehat{A}_2^{\mathrm{T}} & h\widehat{A}_2^{\mathrm{T}} & \widehat{F}_{\mathrm{B}}^{\mathrm{T}} \\ * & * & \Omega_{33} & -M_4^{\mathrm{T}} + hN_{34} & \mathbf{0} & \widehat{A}_{\mathrm{h}}^{\mathrm{T}} & h\widehat{A}_{\mathrm{h}}^{\mathrm{T}} & \mathbf{0} \\ * & * & * & -\gamma \boldsymbol{I} + hN_{44} & \mathbf{0} & \widehat{D}_{\mathrm{g}}^{\mathrm{T}} & h\widehat{D}_{\mathrm{g}}^{\mathrm{T}} & \mathbf{0} \\ * & * & * & * & -\gamma \boldsymbol{I} & \mathbf{0} & \mathbf{0} & \mathbf{0} \\ * & * & * & * & * & -P^{-1} + \varepsilon \widehat{E}\widehat{E}^{\mathrm{T}} & \varepsilon h \widehat{E}\widehat{E}^{\mathrm{T}} & \mathbf{0} \\ * & * & * & * & * & * & -hR^{-1} + \varepsilon h^2 \widehat{E}\widehat{E}^{\mathrm{T}} & \mathbf{0} \\ * & * & * & * & * & * & * & -\varepsilon \boldsymbol{I} \end{bmatrix} < 0, \tag{8.20}$$

$$\begin{bmatrix} N_{11} & N_{12} & N_{13} & N_{14} & M_1 \\ * & N_{22} & N_{23} & N_{24} & M_2 \\ * & * & N_{33} & N_{34} & M_3 \\ * & * & * & N_{44} & M_4 \\ * & * & * & * & R \end{bmatrix} \geq 0, \tag{8.21}$$

$$\begin{bmatrix} P + R & -R \\ * & hQ + R \end{bmatrix} > 0, \tag{8.22}$$

where

$$\Omega_{11} = -P_0 + Q + M_1 + M_1^{\mathrm{T}} + hN_{11},$$
$$\Omega_{13} = -M_1 + M_3^{\mathrm{T}} + hN_{13},$$
$$\Omega_{22} = -P + P_0 + hN_{22},$$
$$\Omega_{33} = -Q - M_3 - M_3^{\mathrm{T}} + hN_{33},$$

$$N = \begin{bmatrix} N_{11} & N_{12} & N_{13} & N_{14} \\ * & N_{22} & N_{23} & N_{24} \\ * & * & N_{33} & N_{34} \\ * & * & * & N_{44} \end{bmatrix}, M = \begin{bmatrix} M_1 \\ M_2 \\ M_3 \\ M_4 \end{bmatrix}.$$

Proof Take the following Lyapunov–Krasovskii function for analyzing the stability of the above closed-loop 2D system in (8.18)

$$V(X(t,k)) = V_h(X(t,k)) + V_v(X(t,k)),$$

$$V_h(X(t,k)) = V_1(X(t,k)) + V_2(X(t,k)) + V_3(X(t,k)),$$

$$V_1(X(t,k)) = X^{\mathrm{T}}(t,k)P_0X(t,k),$$

$$V_2(X(t,k)) = \sum_{i=t-h}^{t-1} X^{\mathrm{T}}(i,k)QX(i,k),$$

$$V_3(X(t,k)) = \sum_{j=-h}^{-1}\sum_{i=t+j}^{t-1} \delta^{\mathrm{T}}(i,k)R\delta(i,k),$$

$$V_v(X(t,k)) = X^{\mathrm{T}}(t,k)(P-P_0)X(t,k),$$

$$\delta(i,k) = X(i+1,k) - X(i,k).$$

Firstly, it should make sure that there stands $V_h(X(t,k)) > 0$ by taking $P > P_0 > 0$, $R > 0$, $Q = Q^{\mathrm{T}}$. In other words, there exists a scalar $\sigma > 0$, ensuring

$$V_h(X(t,k)) \geq \sigma \|X(t,k)\|^2. \tag{8.23}$$

By using Lemma 8.1 and the premise $R > 0$, it can be verified that

$$\begin{aligned}
\sum_{j=-h}^{-1}\sum_{i=t+j}^{t-1} \delta^{\mathrm{T}}(t,k)R\delta(i,k) &\geq \sum_{j=-h}^{-1} -\frac{1}{j}\left(\sum_{i=t+j}^{t-1}\delta(i,k)\right)^{\mathrm{T}} R\left(\sum_{i=t+j}^{t-1}\delta(i,k)\right)\\
&= \sum_{j=-h}^{-1} -\frac{1}{j}\big[X(t,k)-X(t+j,k)\big]^{\mathrm{T}}R\big[X(t,k)-X(t+j,k)\big]\\
&\geq \sum_{j=-h}^{-1} -\frac{1}{h}\big[X(t,k)-X(t+j,k)\big]^{\mathrm{T}}R\big[X(t,k)-X(t+j,k)\big]\\
&= \frac{1}{h}\sum_{i=t-h}^{t-1} [X(t,k)-X(i,k)]^{\mathrm{T}}R[X(t,k)-X(i,k)].
\end{aligned}$$

Thus, there is

$$\begin{aligned}
V_h(X(t,k)) &= X^{\mathrm{T}}(t,k)PX(t,k) + \sum_{i=t-h}^{t-1} X^{\mathrm{T}}(i,k)QX(i,k) + \sum_{j=-h}^{-1}\sum_{i=t+j}^{t-1} \delta^{\mathrm{T}}(i,k)R\delta(i,k)\\
&\geq X^{\mathrm{T}}(t,k)PX(t,k) + \sum_{i=t-h}^{t-1} X^{\mathrm{T}}(i,k)QX(i,k)\\
&+ \sum_{i=t-h}^{t-1} \frac{1}{h}[X(t,k)-X(i,k)]^{\mathrm{T}}R[X(t,k)-X(i,k)]
\end{aligned}$$

$$= \frac{1}{h} \sum_{i=t-h}^{t-1} \left[X^{\mathrm{T}}(t,k), X^{\mathrm{T}}(i,k) \right] \begin{bmatrix} P+R & -R \\ * & hQ+R \end{bmatrix} \begin{bmatrix} X(t,k) \\ X(i,k) \end{bmatrix}.$$

Using the inequality in (8.22), there follows for a scalar $\sigma > 0$ that

$$\begin{bmatrix} P+R & -R \\ * & hQ+R \end{bmatrix} - \begin{bmatrix} \sigma \boldsymbol{I} & \boldsymbol{0} \\ * & \boldsymbol{0} \end{bmatrix} > 0,$$

which results in the inequality in (8.23).

In the sequel, it needs to clarify that the above closed-loop 2D system under $\varpi(t,k) = 0$ could maintain asymptotical stability. For this purpose, define

$$\Delta V(X(t,k)) = \sum_{i=1}^{3} \Delta V_i(X(t,k)) + \Delta V_{\mathrm{v}}(X(t,k)), \tag{8.24}$$

where

$$\Delta V_1(X(t,k)) = X^{\mathrm{T}}(t+1,k) P_0 X(t+1,k) - X^{\mathrm{T}}(t,k) P_0 X(t,k),$$

$$\Delta V_2(X(t,k)) = X^{\mathrm{T}}(t,k) Q X(t,k) - X^{\mathrm{T}}(t-h,k) Q X(t-h,k),$$

$$\Delta V_3(X(t,k)) = h\delta^{\mathrm{T}}(t,k) R \delta(t,k) - \sum_{i=t-h}^{t-1} \delta^{\mathrm{T}}(i,k) R \delta(i,k),$$

$$\Delta V_{\mathrm{v}}(X(t,k)) = X^{\mathrm{T}}(t+1,k)(P-P_0)X(t+1,k) - X^{\mathrm{T}}(t+1,k-1)(P-P_0)X(t+1,k-1).$$

Note that the following two identities stand true

$$2\widehat{\phi}^{\mathrm{T}}(t,k)\widehat{M} \left[X(t,k) - X(t-h,k) - \sum_{i=t-h}^{t-1} \delta(i,k) \right] = 0,$$

$$h\widehat{\phi}^{\mathrm{T}}(t,k)\widehat{N}\widehat{\phi}(t,k) - \sum_{i=t-h}^{t-1} \widehat{\phi}^{\mathrm{T}}(t,k)\widehat{N}\widehat{\phi}(t,k) = 0,$$

where

$$\widehat{\phi}(t,k) = \left[X^{\mathrm{T}}(t,k)\ X^{\mathrm{T}}(t+1,k-1)\ X^{\mathrm{T}}(t-h,k) \right]^{\mathrm{T}},$$

$$\widehat{M} = \begin{bmatrix} M_1 \\ M_2 \\ M_3 \end{bmatrix}, \widehat{N} = \begin{bmatrix} N_{11} & N_{12} & N_{13} \\ * & N_{22} & N_{23} \\ * & * & N_{33} \end{bmatrix}.$$

Using the above identities, one obtains

$$\begin{aligned}\Delta V &= \widehat{\phi}^{\mathrm{T}}(t,k)\left(\widehat{\Gamma}_1^{\mathrm{T}} P\widehat{\Gamma}_1 + h\widehat{\Gamma}_2^{\mathrm{T}} R\widehat{\Gamma}_2 + \widehat{T}\right)\widehat{\phi}(t,k) - \sum_{i=t-h}^{t-1}\delta^{\mathrm{T}}(i,k)R\delta(i,k)\\ &\quad - 2\widehat{\phi}^{\mathrm{T}}(t,k)\widehat{M}\sum_{i=t-h}^{t-1}\delta(i,k) - \sum_{i=t-h}^{t-1}\widehat{\phi}^{\mathrm{T}}(t,k)\widehat{N}\widehat{\phi}(t,k)\\ &= \widehat{\phi}^{\mathrm{T}}(t,k)\left(\widehat{\Gamma}_1^{\mathrm{T}} P\widehat{\Gamma}_1 + h\widehat{\Gamma}_2^{\mathrm{T}} R\widehat{\Gamma}_2 + \widehat{T}\right)\widehat{\phi}(t,k) - \sum_{i=t-h}^{t-1}\widehat{\xi}^{\mathrm{T}}(t,i)\begin{bmatrix}\widehat{N} & \widehat{M}\\ * & R\end{bmatrix}\widehat{\xi}(t,i),\end{aligned}$$

where $\widehat{\xi}(t,i) = \left[\widehat{\phi}^{\mathrm{T}}(t,k)\ \delta^{\mathrm{T}}(i,k)\right]^{\mathrm{T}}$ and

$$X(t+1,k) = \left[\widehat{A}_1 + \Delta\widehat{A}_1\ \widehat{A}_2 + \Delta\widehat{A}_2\ \widehat{A}_{\mathrm{h}}\right]\widehat{\phi}(t,k) = \widehat{\Gamma}_1\widehat{\phi}(t,k),$$

$$\delta(t,k) = \left[\widehat{A}_1 + \Delta\widehat{A}_1 - \boldsymbol{I}\ \widehat{A}_2 + \Delta\widehat{A}_2\ \widehat{A}_{\mathrm{h}}\right]\widehat{\phi}(t,k) = \widehat{\Gamma}_2\widehat{\phi}(t,k),$$

$$\widehat{T} = \begin{bmatrix} -P_0 + Q + M_1 + M_1^{\mathrm{T}} + hN_{11} & M_2^{\mathrm{T}} + hN_{12} & -M_1 + M_3^{\mathrm{T}} + hN_{13}\\ * & -P + P_0 + hN_{22} & -M_2 + hN_{23}\\ * & * & -Q - M_3 - M_3^{\mathrm{T}} + hN_{33}\end{bmatrix}.$$

Hence, there stands $\Delta V(X(t,k)) < 0$ based on the above conditions in (8.20) and (8.21) along with the Schur complement and Lemma 2.1. Accordingly, it can be derived from (8.24) that

$$V_h(X(t+1,k)) + V_v(X(t+1,k)) \le V_h(X(t,k)) + V_v(X(t+1,k-1)). \quad (8.25)$$

By using (8.25) and taking a similar proof with that given in (Liu & Gao 2010), it can be clarified that

$$\sum_{t+k\le N+1} V(X(t,k)) \le \sum_{t+k\le N} V(X(t,k)).$$

Therefore, it is guaranteed that $\lim_{t+k\to\infty} ||X(t,k)|| = 0$ based on the 2D control theory (Kaczorek, 1985), which indicates that the closed-loop 2D system in (8.18) under $\varpi(t,k) = 0$ could maintain asymptotical stability.

In the next, it needs to clarify that the performance index γ in (8.11) is satisfied when $\varpi(t,k) \neq 0$. For this purpose, another control objective function is defined below

$$J_{\mathrm{BP}} = \sum_{t=0}^{N}\sum_{k=0}^{\infty}\left[\gamma^{-1}z^{\mathrm{T}}(t,k)z(t,k) - \gamma\varpi^{\mathrm{T}}(t,k)\varpi(t,k)\right]. \tag{8.26}$$

By resetting the zero initial condition for batch run, one ensures

$$J_{\mathrm{BP}} \le \sum_{t=0}^{N}\sum_{k=0}^{\infty}\left[\gamma^{-1}z^{\mathrm{T}}(t,k)z(t,k) - \gamma\varpi^{\mathrm{T}}(t,k)\varpi(t,k) + \Delta V(X(t,k))\right].$$

Denote the following shorthands

$$X(t+1,k) = \left[\widehat{A}_1 + \Delta\widehat{A}_1\ \widehat{A}_2 + \Delta\widehat{A}_2\ \widehat{A}_{\mathrm{h}}\ \widehat{D}_{\mathrm{g}}\right]\phi(t,k) = \Gamma_1\phi(t,k),$$

$$\delta(t,k) = \left[\widehat{A}_1 + \Delta\widehat{A}_1 - \mathbf{I}\ \widehat{A}_2 + \Delta\widehat{A}_2\ \widehat{A}_{\mathrm{h}}\ \widehat{D}_{\mathrm{g}}\right]\phi(t,k) = \Gamma_2\phi(t,k),$$

$$\phi(t,k) = \left[X^{\mathrm{T}}(t,k)\ X^{\mathrm{T}}(t+1,k-1)\ X^{\mathrm{T}}(t-h,k)\ \varpi^{\mathrm{T}}(t,k)\right]^{\mathrm{T}},$$

$$M = \begin{bmatrix} M_1 \\ M_2 \\ M_3 \\ M_4 \end{bmatrix}, N = \begin{bmatrix} N_{11} & N_{12} & N_{13} & N_{14} \\ * & N_{22} & N_{23} & N_{24} \\ * & * & N_{33} & N_{34} \\ * & * & * & N_{44} \end{bmatrix}.$$

It can be easily clarified that

$$\begin{aligned} J_{\mathrm{BP}} &\le \sum_{t=0}^{N}\sum_{k=0}^{\infty}\left[\gamma^{-1}z^{\mathrm{T}}(t,k)z(t,k) - \gamma\varpi^{\mathrm{T}}(t,k)\varpi(t,k) + \Delta V(X(t,k))\right] \\ &= \sum_{t=0}^{N}\sum_{k=0}^{\infty}\left\{\phi^{\mathrm{T}}(t,k)\left(\Gamma_1^{\mathrm{T}}P\Gamma_1 + h\Gamma_2^{\mathrm{T}}R\Gamma_2 + T\right)\phi(t,k) - \sum_{i=t-h}^{t-1}\xi^{\mathrm{T}}(t,i)\begin{bmatrix} N & M \\ * & R \end{bmatrix}\xi(t,i)\right\}, \end{aligned}$$

where $\xi(t,i) = \left[\phi^{\mathrm{T}}(t,k)\ \delta^{\mathrm{T}}(i,k)\right]^{\mathrm{T}}$ and

$$T = \begin{bmatrix} -P_0 + Q + M_1 + M_1^{\mathrm{T}} + hN_{11} + \gamma^{-1}\widehat{H}^{\mathrm{T}}\widehat{H} & M_2^{\mathrm{T}} + hN_{12} & -M_1 + M_3^{\mathrm{T}} + hN_{13} & M_4^{\mathrm{T}} + hN_{14} \\ * & -P + P_0 + hN_{22} & -M_2 + hN_{23} & hN_{24} \\ * & * & -Q - M_3 - M_3^{\mathrm{T}} + hN_{33} & -M_4^{\mathrm{T}} + hN_{34} \\ * & * & * & -\gamma\mathbf{I} + hN_{44} \end{bmatrix}.$$

By using Lemma 2.1 and the Schur complement together with the constraints in (8.20) and (8.22), it is guaranteed that $J_{\mathrm{BP}} \le 0$, therefore leading to the robust control performance index in (8.11). This completes the proof. □

It should be noted that the condition in (8.20) includes P^{-1} and R^{-1}, and therefore is a bilinear matrix inequality. To resolve the issue, the CCL method (Ghaoui et al., 1997) is adopted here by introducing two slack matrices T and S satisfying

$$PT = \boldsymbol{I}, RS = \boldsymbol{I}.$$

Solving the matrix inequalities in (8.20–8.22) is therefore converted into a nonlinear optimization problem (Op), i.e.

$$\text{(Op) Minimize Tr}(PT + RS) \tag{8.27}$$

s. t. (8.21), (8.22) and

$$\begin{bmatrix} \Omega_{11} & M_2^{\mathrm{T}} + hN_{12} & \Omega_{13} & M_4^{\mathrm{T}} + hN_{14} & \widehat{H}^{\mathrm{T}} & \widehat{A}_1^{\mathrm{T}} & h(\widehat{A}_1 - \mathbf{I})^{\mathrm{T}} & \widehat{F}_{\mathrm{A}}^{\mathrm{T}} \\ * & \Omega_{22} & -M_2 + hN_{23} & hN_{24} & \mathbf{0} & \widehat{A}_2^{\mathrm{T}} & h\widehat{A}_2^{\mathrm{T}} & \widehat{F}_{\mathrm{B}}^{\mathrm{T}} \\ * & * & \Omega_{33} & -M_4^{\mathrm{T}} + hN_{34} & \mathbf{0} & \widehat{A}_{\mathrm{h}}^{\mathrm{T}} & h\widehat{A}_{\mathrm{h}}^{\mathrm{T}} & \mathbf{0} \\ * & * & * & -\gamma\mathbf{I} + hN_{44} & \mathbf{0} & \widehat{D}_{\mathrm{g}}^{\mathrm{T}} & h\widehat{D}_{\mathrm{g}}^{\mathrm{T}} & \mathbf{0} \\ * & * & * & * & -\gamma\mathbf{I} & \mathbf{0} & \mathbf{0} & \mathbf{0} \\ * & * & * & * & * & -T + \varepsilon\widehat{E}\widehat{E}^{\mathrm{T}} & \varepsilon h\widehat{E}\widehat{E}^{\mathrm{T}} & \mathbf{0} \\ * & * & * & * & * & * & -hS + \varepsilon h^2\widehat{E}\widehat{E}^{\mathrm{T}} & \mathbf{0} \\ * & * & * & * & * & * & * & -\varepsilon\mathbf{I} \end{bmatrix} < 0, \tag{8.28}$$

$$\begin{bmatrix} P & \mathbf{I} \\ * & T \end{bmatrix} > 0, \tag{8.29}$$

$$\begin{bmatrix} R & \mathbf{I} \\ * & S \end{bmatrix} > 0. \tag{8.30}$$

Accordingly, a numerical solution is presented in the following algorithm for the above optimization:

Algorithm I

Step 1. With an initially user-specified γ, solve (8.21), (8.22), (8.28)-(8.30) to obtain a feasible solution of $\{P_0^0, Q^0, P^0, T^0, R^0, S^0, M^0, N^0, K_1^0, K_2^0, L_1^0, L_2^0\}$. Meanwhile, take $k=0$ for the initialization.

Step 2. Solve the minimization program subject to the above LMI constraints,

$$\text{Minimize } \operatorname{Tr}(P^kT + PT^k + R^kS + RS^k)$$

$$\text{s. t. } (8.21),\ (8.22),\ (8.28)\text{-}(8.30)$$

and then take $P^{k+1} = P$, $T^{k+1} = T$, $R^{k+1} = R$ and $S^{k+1} = S$.

Step 3. The acquired K_1, K_2, L_1, L_2 in the above step are regarded as the proper solution if the LMI conditions in (8.21) and (8.22) are satisfied together with the following matrix inequality,

$$\begin{bmatrix} \Lambda_{11} & \Lambda_{12} & \Lambda_{13} & M_4^{\mathrm{T}} + hN_{14} & \widehat{H}^{\mathrm{T}} & \widehat{A}_1^{\mathrm{T}}P & h(\widehat{A}_1^{\mathrm{T}} - \mathbf{I})R & \mathbf{0} \\ * & \Lambda_{22} & -M_2 + hN_{23} & hN_{24} & \mathbf{0} & \widehat{A}_2^{\mathrm{T}}P & h\widehat{A}_2^{\mathrm{T}}R & \mathbf{0} \\ * & * & \Lambda_{33} & -M_4^{\mathrm{T}} + hN_{34} & \mathbf{0} & \widehat{A}_h^{\mathrm{T}}P & h\widehat{A}_h^{\mathrm{T}}R & \mathbf{0} \\ * & * & * & -\gamma\mathbf{I} + hN_{44} & \mathbf{0} & \widehat{D}_g^{\mathrm{T}}P & h\widehat{D}_g^{\mathrm{T}}R & \mathbf{0} \\ * & * & * & * & -\gamma\mathbf{I} & \mathbf{0} & \mathbf{0} & \mathbf{0} \\ * & * & * & * & * & -P & \mathbf{0} & P\widehat{E} \\ * & * & * & * & * & * & -hR & hR\widehat{E} \\ * & * & * & * & * & * & * & -\varepsilon\mathbf{I} \end{bmatrix} < 0 \tag{8.31}$$

where

$$\begin{aligned} \Lambda_{11} &= -P_0 + Q + M_1 + M_1^{\mathrm{T}} + hN_{11} + \varepsilon\widehat{F}_{\mathrm{A}}^{\mathrm{T}}\widehat{F}_{\mathrm{A}} \\ \Lambda_{12} &= M_2^{\mathrm{T}} + hN_{12} + \varepsilon\widehat{F}_{\mathrm{A}}^{\mathrm{T}}\widehat{F}_{\mathrm{B}}, \\ \Lambda_{13} &= -M_1 + M_3^{\mathrm{T}} + hN_{13}, \\ \Lambda_{22} &= -P + P_0 + hN_{22} + \varepsilon\widehat{F}_{\mathrm{B}}^{\mathrm{T}}\widehat{F}_{\mathrm{B}}, \\ \Lambda_{33} &= -Q - M_3 - M_3^{\mathrm{T}} + hN_{33}. \end{aligned}$$

Otherwise, take $k = k+1$ and go back to the previous step. If the LMI conditions in (8.21), (8.22) and (8.31) could not be satisfied for a specified upper limit of iterations in practice, then increase γ and go back to **Step 1**, or terminate for no solution.

To determine the optimal $\mathcal{H}_\infty$ performance index denoted by γ^*, another algorithm is provided below based on the bisection approach, such that the computation effort could be reduced to a large extent in comparison with the previous one-dimensional searching algorithm (Najafi et al. 2014).

Algorithm II

Step 1: With a specified γ guaranteeing the above Algorithm I feasible, take $a_0 = 0$, $b_0 = \gamma$, and $i = 0$ for the initialization
Step 2: Calculate $\gamma_i = (b_i - a_i)/2$ and verify the feasibility of the above Algorithm I. If yes, proceed with the next step; Otherwise, go to **Step 4**
Step 3: Let $a_{i+1} = a_i, \quad b_{i+1} = \text{ceil}(\gamma_i), i = i + 1$. If $b_i - a_i = 1$, proceed with **Step 5**; Otherwise, go back to the above **Step 2**
Step 4: Let $a_{i+1} = \text{floor}(\gamma_i), \quad b_{i+1} = b_i, i = i + 1$. If $b_i - a_i = 1$, proceed with **Step 5**; Otherwise, go back to the above **Step 2**
Step 5: Decrease b_i by 0.1, and verify the feasibility of the above Algorithm I, till finding out the minimum of γ

When there exists no process uncertainty, it can be observed from (8.15) that the state prediction error is not related to $x_g(t,k)$ and $\delta u(t-h,k)$. Hence, the state predictor is designed to ensure the stability of the error dynamics

$$\tilde{x}_g(t+1,k) = A_1\tilde{x}_g(t,k) + (A_2 + L_2G)\tilde{x}_g(t+1,k-1) + L_1G\tilde{x}_g(t-h,k). \tag{8.32}$$

Substituting the above ILC control law in (8.16) into the closed-loop 2D system expressed by (8.12) with no uncertainty, it follows that

$$\begin{aligned} x_g(t+1,k) &= A_1x_g(t,k) + A_2x_g(t+1,k-1) + B_g\delta u(t-h,k) \\ &= \big(A_1 + B_gK_1\big)x_g(t,k) + \big(A_2 + B_gK_2\big)x_g(t+1,k-1) \\ &\quad + B_gK_1\tilde{x}_g(t,k) + B_gK_2\tilde{x}_g(t+1,k-1). \end{aligned} \tag{8.33}$$

When the state estimation error becomes zero, the ILC controllers K_1 and K_2 should stabilize the following system

$$x_g(t+1,k) = \big(A_1 + B_gK_1\big)x_g(t,k) + \big(A_2 + B_gK_2\big)x_g(t+1,k-1). \tag{8.34}$$

The control task is, therefore, focused on the feedback stabilization of the above 2D system, by using the state predictor to overcome the input delay. To this end, two corollaries are presented here to design the ILC updating law together with the above state predictor.

Corollary 8.1 For a batch process having the input delay h but with no time-varying uncertainties, the state predictor in (8.32) could be asymptotically stable if there exist matrices $P > P_0 > 0$, $R > 0$, $Q = Q^{\mathrm{T}}$, $N \geq 0$, M, L_1, L_2 such that the following matrix inequalities are satisfied

$$\begin{bmatrix} -P_0 + Q + M_1 + M_1^{\mathrm{T}} + hN_{11} & M_2^{\mathrm{T}} + hN_{12} & -M_1 + M_3^{\mathrm{T}} + hN_{13} & A_1^{\mathrm{T}} & h(A_1^{\mathrm{T}} - \mathbf{I}) \\ * & -P + P_0 + hN_{22} & -M_2 + hN_{23} & A_2^{\mathrm{T}} + G^{\mathrm{T}}L_2^{\mathrm{T}} & h(A_2^{\mathrm{T}} + G^{\mathrm{T}}L_2^{\mathrm{T}}) \\ * & * & -Q - M_3 - M_3^{\mathrm{T}} + hN_{33} & G^{\mathrm{T}}L_1^{\mathrm{T}} & hG^{\mathrm{T}}L_1^{\mathrm{T}} \\ * & * & * & -P^{-1} & \mathbf{0} \\ * & * & * & * & -dR^{-1} \end{bmatrix} < 0, \tag{8.35}$$

$$\begin{bmatrix} N_{11} & N_{12} & N_{13} & M_1 \\ * & N_{22} & N_{23} & M_2 \\ * & * & N_{33} & M_3 \\ * & * & * & R \end{bmatrix} \geq 0, \tag{8.36}$$

$$\begin{bmatrix} P+R & -R \\ * & hQ+R \end{bmatrix} > 0. \tag{8.37}$$

Proof The proof follows that of Theorem 8.1 and therefore is omitted.

Concerning the delay-free part of the above 2D system in (8.32), i.e.

$$\tilde{x}_g(t+1,k) = A_1\tilde{x}_g(t,k) + (A_2 + L_2G)\tilde{x}_g(t+1,k-1), \tag{8.38}$$

it is easy to conclude that the input delay is allowed to be arbitrarily long if the following matrix inequality could be satisfied with $P > P_0 > 0$ and $L_1 = 0$ to guarantee (8.38) to be asymptotically stable

$$\begin{bmatrix} -P_0 & \mathbf{0} & A_1^{\mathrm{T}} \\ * & -P+P_0 & A_2^{\mathrm{T}} + G^{\mathrm{T}}L_2^{\mathrm{T}} \\ * & * & -P^{-1} \end{bmatrix} < 0. \tag{8.39}$$

Correspondingly, the state feedback controllers in the above 2D system described by (8.34) can be designed according to the corollary below.

Corollary 8.2 The 2D system in (8.34) maintains asymptotical stability if there exist matrices $\overline{P} > \overline{P}_0 > 0$, F_1, F_2, such that the following LMI is satisfied

$$\begin{bmatrix} -\overline{P}_0 & \mathbf{0} & \overline{P}A_1^{\mathrm{T}} + F_1^{\mathrm{T}}B_g^{\mathrm{T}} \\ * & -\overline{P}+\overline{P}_0 & \overline{P}A_2^{\mathrm{T}} + F_2^{\mathrm{T}}B_g^{\mathrm{T}} \\ * & * & -\overline{P} \end{bmatrix} < 0 \tag{8.40}$$

and correspondingly, the 2D state feedback controllers are determined by

$$\begin{cases} K_1 = F_1\overline{P}^{-1}, \\ K_2 = F_2\overline{P}^{-1}. \end{cases} \tag{8.41}$$

Proof Consider the following Lyapunov–Krasovskii functions

$$V(x_g(t,k)) = V_h(x_g(t,k)) + V_v(x_g(t,k)),$$

$$V_h(x_g(t,k)) = x_g^{\mathrm{T}}(t,k)P_0x_g(t,k),$$

$$V_v(x_g(t,k)) = x_g^T(t,k)(P - P_0)x_g(t,k).$$

For the simplicity of analysis, here define

$$\Delta V_h(x_g(t,k)) = x_g^T(t+1,k)P_0x_g(t+1,k) - x_g^T(t,k)P_0x_g(t,k),$$

$$\begin{aligned}\Delta V_v(x_g(t,k)) &= x_g^T(t+1,k)(P - P_0)x_g(t+1,k) \\ &\quad - x_g^T(t+1,k-1)(P - P_0)x_g(t+1,k-1).\end{aligned}$$

There stands

$$\begin{aligned}\Delta V(x_g(t,k)) &= \Delta V_h(x_g(t,k)) + \Delta V_v(x_g(t,k)) \\ &= \begin{bmatrix} x_g(t,k) \\ x_g(t+1,k-1) \end{bmatrix}^T \left\{ \begin{bmatrix} -P_0 & \mathbf{0} \\ * & -P + P_0 \end{bmatrix} + \tilde{\Gamma}^T P \tilde{\Gamma} \right\} \begin{bmatrix} x_g(t,k) \\ x_g(t+1,k-1) \end{bmatrix},\end{aligned}$$

where

$$\tilde{\Gamma} = \begin{bmatrix} A_1 + B_g K_1 & A_2 + B_g K_2 \end{bmatrix}.$$

By multiplying diag$\left\{\overline{P}^{-1}, \overline{P}^{-1}, \overline{P}^{-1}\right\}$ to the both sides of (8.40) and letting $P_0 = \overline{P}^{-1}\overline{P}_0\overline{P}^{-1}$, $P = \overline{P}^{-1}$, it yields that $\Delta V(x_g(t,k)) < 0$ by using (8.41) and the Schur complement. The proof is completed. □

To ensure no steady-state output error for batch run subject to process uncertainties and/or external disturbance, an integral of the output tracking error along the time is defined as

$$\Sigma e(t,k) \triangleq \sum_{i=0}^{t} e(i,k), \tag{8.42}$$

which is also taken as an augmented state in the above closed-loop 2D system for ILC design.

It is obvious from (8.42) that

$$\Sigma e(t,k) = \Sigma e(t-1,k) + e(t,k).$$

By combining (8.42) with the 2D system description in (8.10), the augmentedin (8.10), the augmented 2D system is expressed below

$$
\tilde{\Sigma}_{2D\text{-}P_\Delta}:\begin{cases}
\begin{bmatrix}\delta x(t+1,k)\\ e(t+1,k)\\ \Sigma e(t,k)\end{bmatrix}=(\tilde{A}_1+\Delta\tilde{A}_1)\begin{bmatrix}\delta x(t,k)\\ e(t,k)\\ \Sigma e(t-1,k)\end{bmatrix}+\tilde{A}_2\begin{bmatrix}\delta x(t+1,k-1)\\ e(t+1,k-1)\\ \Sigma e(t,k-1)\end{bmatrix}\\
\qquad +(\tilde{B}_g+\Delta\tilde{B}_g)\delta u(t-h,k)+\tilde{D}_g\varpi(t,k),\\
y(t,k)=\begin{bmatrix}e(t,k-1)\\ e(t,k)\\ \Sigma e(t,k)\end{bmatrix}=\tilde{G}\begin{bmatrix}\delta x(t,k)\\ e(t,k)\\ \Sigma e(t-1,k)\end{bmatrix},\\
z(t,k)=e(t,k)=\tilde{H}\begin{bmatrix}\delta x(t,k)\\ e(t,k)\\ \Sigma e(t-1,k)\end{bmatrix},
\end{cases}
\tag{8.43}
$$

where $\tilde{H}=\begin{bmatrix}\mathbf{0} & 1 & 0\end{bmatrix}$,

$$
\tilde{G}=\begin{bmatrix}-C & 1 & 0\\ \mathbf{0} & 1 & 0\\ \mathbf{0} & 1 & 1\end{bmatrix},\ \tilde{D}_g=\begin{bmatrix}I\\ C\\ 0\end{bmatrix},
$$

$$
\tilde{A}_1=\begin{bmatrix}A & 0 & 0\\ CA & 0 & 0\\ \mathbf{0} & 1 & 1\end{bmatrix},\ \Delta\tilde{A}_1=\begin{bmatrix}\Delta A(t,k) & 0 & 0\\ C\Delta A(t,k) & 0 & 0\\ \mathbf{0} & 0 & 0\end{bmatrix}=\begin{bmatrix}E\\ CE\\ \mathbf{0}\end{bmatrix}\Delta(t,k)\begin{bmatrix}F_A & 0 & 0\end{bmatrix}\triangleq\tilde{E}\Delta(t,k)\tilde{F}_A,
$$

$$
\tilde{A}_2=\begin{bmatrix}\mathbf{0} & \mathbf{0} & \mathbf{0}\\ \mathbf{0} & 1 & 0\\ \mathbf{0} & \mathbf{0} & \mathbf{0}\end{bmatrix},\ \tilde{B}_g=\begin{bmatrix}B\\ CB\\ 0\end{bmatrix},\ \Delta\tilde{B}_g=\begin{bmatrix}\Delta B(t,k)\\ C\Delta B(t,k)\\ 0\end{bmatrix}=\begin{bmatrix}E\\ CE\\ 0\end{bmatrix}\Delta(t,k)F_B\triangleq\tilde{E}\Delta(t,k)F_B
$$

Accordingly, the state predictor and ILC controllers can be similarly designed by using the proposed method in the above section.

It should be noted that using the above ILC design based on an augmented state of the integral of output tracking error could improve the control performance for each batch, but in exchange for the cost of finding a feasible solution of the objective optimization program in the above section, along with the further extended forms of the state predictor and ILC controllers for execution.

8.4 An Illustrative Example

Consider the injection molding machine for batch run shown in Fig. 2.2 as studied in the previous Sect. 2.2.3. Herein, the batch packing process is used for verifying the above ILC design, of which a cyclic transfer model was identified as (Shi et al. 2005; Wang et al. 2012)

$$
\begin{cases}
x(t+1,k)=\left(\begin{bmatrix}1.607 & 1\\ -0.6086 & 0\end{bmatrix}+\Delta A(t,k)\right)x(t,k)+\left(\begin{bmatrix}1.239\\ -0.9282\end{bmatrix}+\Delta B(t,k)\right)u(t-h,k),\\
y(t,k)=[1,0]x(t,k),
\end{cases}
$$

where the nozzle input delay h was assumed to be zero for the simplicity of control design in the references (Shi et al., 2005; Wang et al., 2012). For illustration, the time-varying uncertainties during each cycle (batch run) studied in these references are adopted for test, i.e.

$$\Delta A(t,k) = \begin{bmatrix} 0.0804\varsigma(t,k) & 0 \\ -0.0304\varsigma(t,k) & 0 \end{bmatrix} = \begin{bmatrix} 1 & 0 \\ 0 & 1 \end{bmatrix} \begin{bmatrix} \varsigma(t,k) & 0 \\ 0 & \varsigma(t,k) \end{bmatrix} \begin{bmatrix} 0.0804 & 0 \\ -0.0304 & 0 \end{bmatrix},$$

$$\Delta B(t,k) = \begin{bmatrix} 0.062\varsigma(t,k) \\ -0.0464\varsigma(t,k) \end{bmatrix} = \begin{bmatrix} 1 & 0 \\ 0 & 1 \end{bmatrix} \begin{bmatrix} \varsigma(t,k) & 0 \\ 0 & \varsigma(t,k) \end{bmatrix} \begin{bmatrix} 0.062 \\ -0.0464 \end{bmatrix},$$

where $\varsigma(t,k)$ is a time-varying factor satisfying $|\varsigma(t,k)| \leq 1$. Note that the nonlinear behavior of the hydraulic control valve is also merged into the above process uncertainties to deal with.

For the set-point tracking, the target output profile (y_{d}) studied in the reference (Wang et al., 2012) is adopted here

$$y_{\mathrm{d}}(t) = \begin{cases} 200, & 0 \leq t \leq 100, \\ 200 + 5(t - 100), & 100 < t \leq 120, \\ 300, & 120 < t \leq N = 200. \end{cases}$$

In view of that a step change of the set-point profile (y_{d}) cannot be strictly implemented in engineering practice, a low-pass filter in the form of $G_{\mathrm{f}}(z) = (z^{-1} + z^{-2})/(3 - z^{-1})$ is taken to smooth the step change of y_{d} as studied in the reference (Liu et al., 2010).

To evaluate the tracking performance, the following index is adopted

$$\mathrm{ATE}(k) = \frac{\sum_{t=1}^{N} |e(t,k)|}{N}. \tag{8.44}$$

Three cases are considered here to validate the proposed ILC scheme.

Case 1 (Input delay)**:** Suppose that no process uncertainty occurs when applying the above two ILC designs for comparison, i.e., with or without an augmented state of the integral of output tracking error, corresponding to the closed-loop 2D system descriptions in (8.10) and (8.43), respectively. The above Algorithm I is used to determine the state predictor gain L_2, which is listed in Table 8.1 for both ILC designs, guaranteeing the convergence of state prediction error as shown in (8.38) to zero. Figure 8.1 shows the set-point tracking results under the input delay of $h = 40$(step), indicating that perfect tracking is realized under such input delay. Figure 8.2 shows that the ILC design based on the integral of tracking error could guarantee no steady-state output error for each batch run, in contrast with the ILC design without the integral of tracking error for the 2D FM system in (8.10). Note that the initial cycle results in a small amplitude far below the desired set-point profile by using the ILC design without the integral of tracking error, and at least 8 cycles are required to realize perfect tracking, as shown in Fig. 8.1. By comparison, only two or three cycles are needed to realize perfect tracking by using the enhanced ILC design for the 2D FM system in (8.43), as shown in Fig. 8.2.

It should be noted that the previous ILC methods by using a delay-free process model or state delay estimation, e.g., Shi et al. (2005) and Liu & Gao (2010), cannot guarantee perfect tracking of the desired set-point profile under the input delay, and thus are omitted for comparison.

Case 2 (Time-invariant uncertainties)**:** The process uncertainties denoted by $\Delta A(t,k)$ and $\Delta B(t,k)$ are supposed to be time-invariant from cycle to cycle, i.e., $\Delta A(t,k) = \Delta A$ and $\Delta B(t,k) = \Delta B$, taken as the upper bounds. For using the ILC design of (8.10) (with no integral of tracking error), the above Algorithm I is adopted to solve the optimization problem in (8.27) under these uncertainties and the input delay. The optimal $\mathcal{H}_\infty$ robust performance index is obtained as.

- $\gamma^* = 55.4$, when $h = 1$ and $|\varsigma(t,k)| \le 1$;
- $\gamma^* = 297.4$, when $h = 2$ and $|\varsigma(t,k)| \le 0.7$;
- $\gamma^* = 127$, when $h = 5$ and $|\varsigma(t,k)| \le 0.1$.

Table 8.2 lists the corresponding controllers when $h = 2$ and $|\varsigma(t,k)| \le 0.7$. In contrast, the ILC design of (8.43) based on an augmented state of the integral of output tracking error results in the controllers listed in Table 8.3 under the input delay of $h = 2$ and $|\varsigma(t,k)| \le 0.1$, corresponding to the optimal $\mathcal{H}_\infty$ performance index, $\gamma^* = 28$. The corresponding set-point tracking results are shown in Figs. 8.3 and 8.5, along with the output error indices of ATE plotted in Figs. 8.4 and 8.6, respectively.

Table 8.1 ILC and feedback controller gains for Case 1 with $h = 40$

System description	(8.10)	(8.43)
ILC gains	$L_1 = \begin{bmatrix} 0 & 0 \\ 0 & 0 \\ 0 & 0 \end{bmatrix}$ $L_2 = \begin{bmatrix} 0.0000 & 0.0000 \\ -0.0000 & -0.0000 \\ -0.0000 & -1.0000 \end{bmatrix}$	$L_1 = \begin{bmatrix} 0 & 0 & 0 \\ 0 & 0 & 0 \\ 0 & 0 & 0 \\ 0 & 0 & 0 \end{bmatrix}$ $L_2 = \begin{bmatrix} -0.0000 & -0.0000 & -0.0000 \\ 0.0000 & 0.0000 & 0.0000 \\ -0.0000 & -1.0000 & -0.0000 \\ 0.0000 & 0.0000 & 0.0000 \end{bmatrix}$
Feedback controller gains	$K_1 = \begin{bmatrix} -1.1896 & -0.6736 & -0.0014 \end{bmatrix}$ $K_2 = \begin{bmatrix} 0.0021 & -0.0050 & -0.4330 \end{bmatrix}$	$K_1 = \begin{bmatrix} -1.2637 & -0.7701 & -0.2986 & -0.2985 \end{bmatrix}$ $K_2 = \begin{bmatrix} 0.0027 & -0.0006 & -0.5953 & -0.0055 \end{bmatrix}$

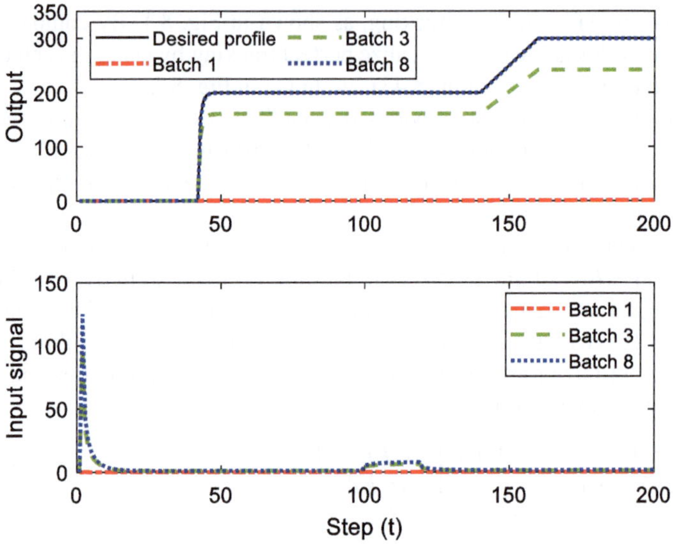

Fig. 8.1 Tracking performance for Case 1 without using the integral of tracking error as the extended learning information

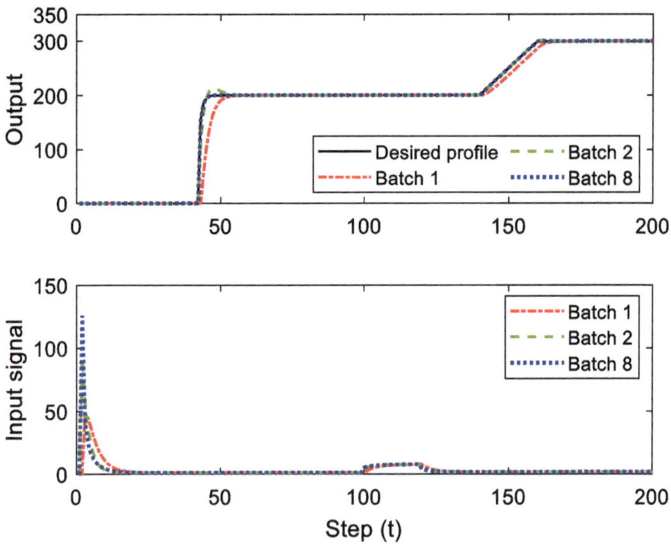

Fig. 8.2 Tracking performance for Case 1 by using the integral of tracking error as the extended learning information

It is observed from these figures that the tracking performance is gradually improved based on the cycle run and perfect tracking is realized by two ILC designs via almost 20 and 10 cycles, respectively. It should be mentioned that the robust ILC designs presented in the cited references (Shi et al. 2005; Wang et al. 2012) could not maintain batch run stability over 10 cycles, and thus are omitted for comparison.

Case 3 (Time-varying uncertainties)**:** Suppose that the process uncertainties denoted by $\Delta A(t,k)$ and $\Delta B(t,k)$ are time-varying during the cycle run. Figure 8.7 plots the ATE of the set-point tracking results on 200 cycles based on the above ILC controller design with regard to $h = 2$ and $|\varsigma(t,k)| \leq 0.7$ as shown in Table 8.2. It can be observed that the resulting ILC system holds robust stability well for batch run. Figure 8.7 also indicates that the output tracking error is further reduced when the

Table 8.2 ILC and feedback controller gains for Case 2 with $h = 2$ and $|\varsigma(t,k)| \leq 0.7$

ILC gains	$L_1 = \begin{bmatrix} 0.3280 & -0.3284 \\ -0.2260 & 0.2263 \\ 0.2185 & -0.2189 \end{bmatrix}$, $L_2 = \begin{bmatrix} 0.0005 & -0.0021 \\ -0.0013 & 0 \\ 0.4111 & -0.5485 \end{bmatrix}$
Feedback controller gains	$K_1 = \begin{bmatrix} -1.4258 & -0.8937 & -0.0149 \end{bmatrix}$, $K_2 = \begin{bmatrix} -0.0576 & 0.0716 & -0.2366 \end{bmatrix}$

Table 8.3 ILC and feedback controller gains for Case 2 with $h = 2$ and $|\varsigma(t,k)| \leq 0.1$

ILC gains	$L_1 = \begin{bmatrix} 0.3224 & -0.3376 & -0.0022 \\ -0.2254 & 0.2352 & 0.0027 \\ 0.1178 & -0.1286 & 0.0018 \\ 0.2608 & -0.2070 & -0.0359 \end{bmatrix}$, $L_2 = \begin{bmatrix} 0.0444 & -0.0420 & -0.0003 \\ -0.0373 & 0.0341 & 0.0003 \\ -0.3573 & -0.5429 & 0.0002 \\ 0.0462 & 0.1663 & -0.0048 \end{bmatrix}$
Feedback controller gains	$K_1 = \begin{bmatrix} -1.3103 & -0.8238 & -0.1119 & -0.1117 \end{bmatrix}$, $K_2 = \begin{bmatrix} -0.0966 & -0.0163 & -1.1945 & -0.0814 \end{bmatrix}$

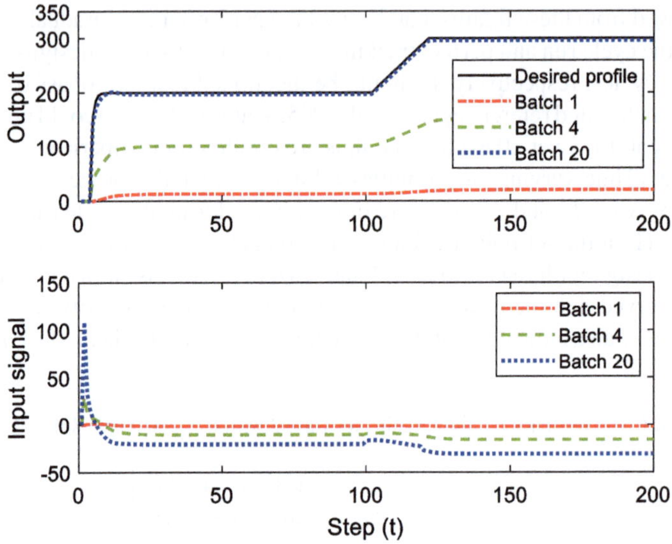

Fig. 8.3 Tracking performance for Case 2 with $h = 2$ and $|\varsigma(t, k)| \leq 0.7$

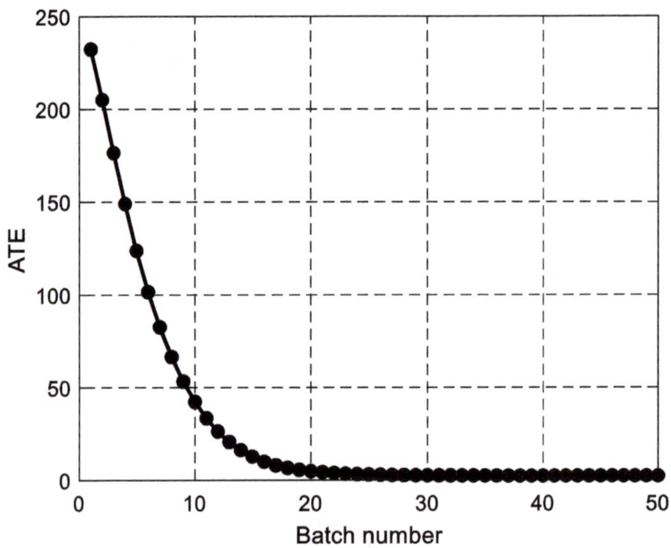

Fig. 8.4 Plot of ATE for Case 2 with $h = 2$ and $|\varsigma(t, k)| \leq 0.7$

time-varying factor is reduced to $|\varsigma(t, k)| \leq 0.1$, as indicated from the 101st to 200th cycle. This demonstrates that the proposed ILC designs can be reliably utilized for the set-point tracking and performance optimization under the time-varying process uncertainties. In contrast, the robust ILC methods given in the cited references (Shi

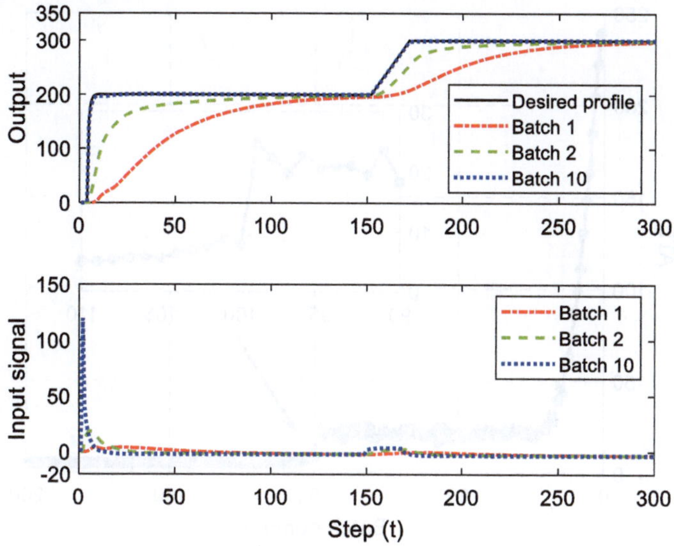

Fig. 8.5 Tracking performance for Case 2 with $h = 2$ and $|\varsigma(t, k)| \leq 0.1$

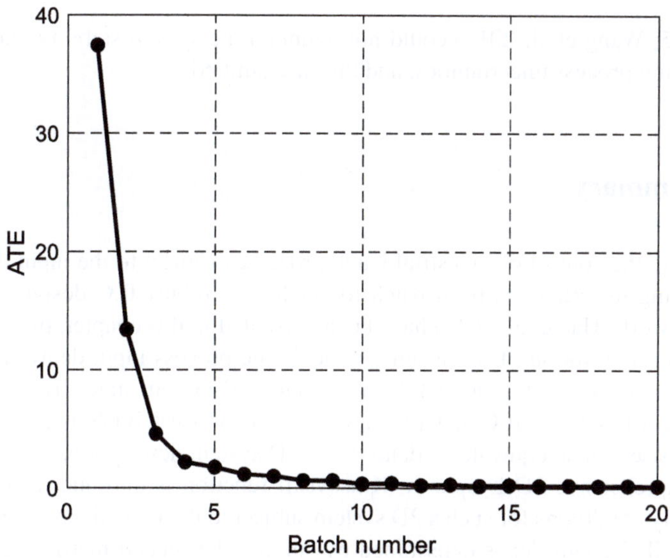

Fig. 8.6 Plot of ATE for Case 2 with $h = 2$ and $|\varsigma(t, k)| \leq 0.1$

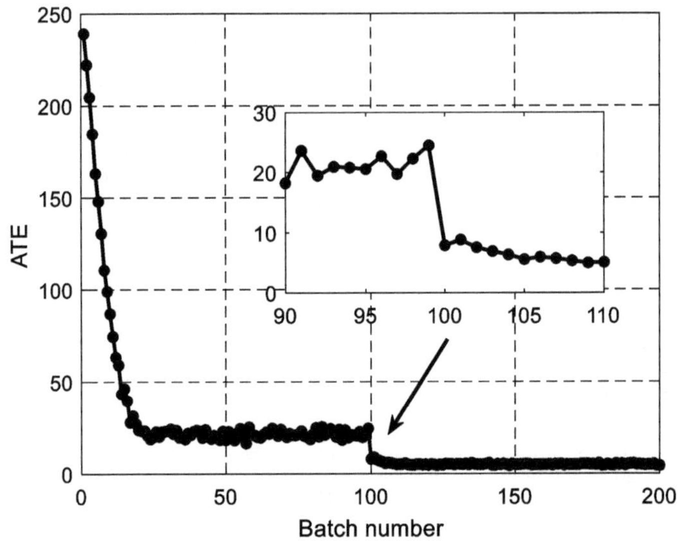

Fig. 8.7 Plot of ATE for Case 3 with $h = 2$

et al. 2005; Wang et al. 2012) could not maintain the system stability under these time-varying process uncertainties, and thus are omitted.

8.5 Summary

Concerning the control of industrial batch processes subject to the input delay and time-varying uncertainties from batch to batch, two robust ILC designs from the proposed work (Hao et al. 2017) have been presented in this chapter, by using a 2D FM system description of batch run. To tackle the process input delay, a 2D state predictor is constructed to predict the augmented 2D system states free of the time delay. Accordingly, the ILC updating law is designed by establishing a stabilization program based on an equivalent "delay-free" 2D system description.

A comprehensive 2D Lyapunov–Krasovskii function is introduced to establish the stability conditions for such a 2D system subject to the input delay. Correspondingly, the ILC controller is determined by solving the related matrix inequalities. The proposed Algorithm I is given to acquire a feasible solution of these matrix inequalities. In contrast, the proposed Algorithm II can be used to find the optimal $\mathcal{H}_\infty$ performance index, while the computation load could be significantly alleviated compared to the traditional one-dimensional searching approach (Kaczorek 1895). Moreover, it is suggested to take an augmented state of the integral of output tracking error in the closed-loop 2D system description to design the ILC updating law, such that no steady-state output error could be guaranteed from the initial batch run, but at the cost of more computation effort for solving the corresponding matrix inequalities to acquire a feasible solution of the ILC controller for implementation.

The application to batch packing process of an injection molding machine has well demonstrated that the proposed ILC designs could realize perfect set-point tracking with respect to a long input delay when there is no time-varying uncertainties. Moreover, both of the proposed ILC designs can be adopted for robust set-point tracking and batch performance optimization under time-varying process uncertainties.

References

Bonvin D, Srinivasan B, Hunkeler D (2006) Control and optimization of batch processes. IEEE Trans Control Systems Magazine 26(1):34–45

Cai Z, Freeman CT, Lewin PL, Rogers E (2008) Iterative learning control for a nonminimum phase plant based on a reference shift algorithm. Control Eng Pract 16(6):633–643

Gao F, Yang Y, Shao C (2001) Robust iterative learning control with applications to injection molding process. Chem Eng Sci 56(24):7025–7034

Ghaoui LE, Oustry F, AitRami M (1997) A cone complementarity linearization algorithm for static output-feedback and related problems. IEEE Trans Autom Control 42(8):1171–1176

Hao S, Liu T, Zhou B (2017) Predictor-based output feedback control design for sampled systems with input delay subject to disturbance. IET Control Theory & Appl, 11(18), 3329–3340

Kaczorek T (1985) Two-Dimensional Linear Systems. Springer-Verlag, Berlin, Germany

Li H, Bai J, Zou H, Yin X, Zhang R (2024) A two-dimensional model predictive iterative learning control based on the set-point learning strategy for batch processes. J Process Control 133(103133):1–22

Liu T, Dong S, Rong S, Zhong C (2017) Identification of discrete-time model with integer delay and control design for cooling processes with application to jacketed crystallizers. IEEE Trans Control Syst Technol, 25 (5), 1775–1789

Liu T, Gao F (2010) Robust two-dimensional iterative learning control for batch processes with state delay and time-varying uncertainties. Chem Eng Sci 65(23):6134–6144

Liu T, Gao F, Wang Y (2010) IMC-based iterative learning control for batch processes with uncertain time delay. J Process Control, 20(2), 173–180

Liu T, Tian H, Rong S, Zhong C (2018) Heating-up control with delay-free output prediction for industrial jacketed reactors based on step response identification. ISA Trans 83:227–238

Park KH, Bien Z, Hwang DH (1998) Design of an iterative learning control for a class of linear dynamic systems with time delay. IET Control Theory Appl 145(6):507–512

Najafi M, Sheikholeslam F, Wang Q-G, Hosseinnia S (2014) Robust H infinity control of single input-delay systems based on sequential sub-predictors. IET Control Theory Appl 8(13):1175–1184

Shao HY, Han QL (2011) New stability criteria for linear discrete-time systems with interval-like time-varying delays. IEEE Trans Autom Control 56(3):619–625

Shen D, Wang Y (2014) Survey on stochastic iterative learning control. J Process Control 24(1):64–77

Shi J, Gao F, Wu T-J (2005) Robust design of integrated feedback and iterative learning control of a batch process based on a 2D Roesser system. J Process Control 15(8):907–924

Shokri-Ghaleh H, Ganjefar S, Shahri A (2021) Robust iterative learning control for uncertain continuous-time system with input delay and random iteration-varying uncertainties. IET Control Theory Appl 15(13):1749–1761

Shokri-Ghaleh H, Ganjefar S, Shahri A (2023) Extension of iterative learning control design for batch processes with time-delay in the input subject to random cycle-varying uncertainties. J Franklin Inst 360(12):8528–8549

Su XJ, Shi P, Wu LG, Song YD (2013) A novel control design on discrete-time Takagi-Sugeno fuzzy systems with time-varying delays. IEEE Trans Fuzzy Syst 21(4):655–671

Tan KK, Zhao S, Huang S, Lee TH, Tay A (2009) A new repetitive control for LTI systems with input delay. J Process Control 19(4):711–716

Tao H, Zheng J, Wei J, Paszke W, Rogers E, Stojanovic V (2023) Repetitive process based indirect-type iterative learning control for batch processes with model uncertainty and input delay. J Process Control, 132, 103112, 1–12

Wang Y, Gao F, Doyle F (2009) Survey on iterative learning control, repetitive control, and run-to-run control. J Process Control 19(10):1589–1600

Wang Y, Liu T, Zhao Z (2012) Advanced PI control with simple learning set-point design: Application on batch processes and robust stability analysis. Chem Eng Sci 71(1):153–165

Wu LG, Yang XZ, Lam HK (2014) Dissipativity analysis and synthesis for discrete-time T-S fuzzy stochastic systems with time-varying delay. IEEE Trans Fuzzy Syst 22(2):380–394

Xu JX, Hu Q, Lee TH, Yamamoto S (2001) Iterative learning control with Smith time delay compensator for batch processes. J Process Control 11(3):321–328

Xu JX, Xu J, Lee TH (2005) Iterative learning control for systems with input deadzone. IEEE Trans Autom Control 50(9):1455–1459

Chapter 9
Predictive State Observer (PSO) Based ILC Design Under Output Delay

9.1 Introduction

In contrast to the input delay involved with industrial batch process operation, as studied in the previous chapter, there also exists the output delay in a lot of manufacturing systems, such as the permanent magnet synchronous motor (Mandra et al. 2019) and crystallizer thermocirculator (Liu et al. 2017). In fact, it has been widely practiced that the input delay is specifically regarded as output delay for the convenience of control design and stability analysis in the process industry (Kharitonov, 1999). For instance, the internal model control (IMC) scheme was adopted for designing an indirect-type ILC method to improve the set-point tracking performance of open-loop stable type batch manufacturing systems subject to the input or output delay (Liu et al. 2010). Another IMC-based ILC design (Shokri-Ghaleh et al. 2021) was presented for batch processes with output delay and iteration-varying uncertainties, which was further extended by using a unified controller and a learning weighting function (Shokri-Ghaleh et al. 2023). By comparison, a modified SP structure was used to design a direct-type ILC scheme for improving the set-point tracking performance of open-loop unstable type batch processes subject to output delay (Tian and Song, 2020). By taking frequency domain specifications for stability analysis, a robust ILC design (Tao et al. 2019) was suggested to optimize batch performance subject to multiple system state and output delays. However, only a few papers addressed the challenging issue of designing a set-point learning control scheme for industrial batch process operation subject to the delay response (Liu et al. 2024), especially under nonrepetitive disturbance and process uncertainties.

Different from the developed direct-type ILC designs that integrate a feedforward learning control strategy together with a real-time feedback control structure, a set-point learning control scheme, which is also customarily called indirect-type ILC, has received an increasing amount of attention in the past decade, by virtue of the feedback controller in such a closed-loop system and the set-point learning controller may be independently designed for practical implementation. It is therefore motivated to present a predictive state observer (PSO) based set-point learning control

T. Liu et al., *Robust Iterative Learning Control of Industrial Batch Systems*, Intelligent Control and Learning Systems 22, https://doi.org/10.1007/978-981-96-9778-6_9

method for industrial batch process operation with output delay subject to time-varying process uncertainties and nonrepetitive external disturbances, by using only the measured process output for ILC design. The proposed PSO-based closed-loop control scheme for set-point learning possesses a two-degree-of-freedom (2DOF) feature, which can not only compensate for the process delay response, but also suppress the impact from time-varying process uncertainties and disturbance for each batch run. An important merit of the proposed set-point learning control scheme is that only the set-point of the closed-loop system is regulated by an indirect-type ILC design based on a proportional-type (P-type) learning controller, thus facilitating batch-to-batch optimization in practice. Meanwhile, the convergence conditions of tracking error are established for either the delay-dependent or delay-independent case. These conditions can be easily used for batch performance evaluation in engineering practice.

9.2 PSO-Based ILC Scheme

An SISO batch manufacturing system with output delay subject to nonrepetitive uncertainties and disturbances is generally described by

$$\begin{cases} x(t+1,k) = A(t,k)x(t,k) + B(t,k)[u(t,k) + \omega(t,k)], \\ y(t,k) = Cx(t-h,k), \quad t \in \mathbb{Z}_N, \quad k \in \mathbb{Z}_+, \end{cases} \tag{9.1}$$

where t and k indicate the time and batch indices, respectively, and the total number of time steps in each batch run is denoted by N; the process input, state, and output are denoted by $u(t,k) \in \Re$, $x(t,k) \in \Re^n$ and $y(t,k) \in \Re$, respectively; Denote by $\omega(t,k) \in \Re$ the load disturbance seeping into the system from its input side, while $h \in \mathbb{Z}_+$ indicates the output response delay. Without loss of generality, the initial system state is supposed to be reset to the identical value for batch run, i.e., $x(\theta,k) = x_0, \theta = -h, \ldots, 0$. Denote by $A(t,k) = A + E\Delta(t,k)F_{\mathrm{A}}$ and $B(t,k) = B + E\Delta(t,k)F_{\mathrm{B}}$ the bounded time-varying uncertainties that are nonrepetitive during batch run, where E, F_{A} and F_{B} are known matrices reflecting the upper bounds of system uncertainties in practice, and $\Delta(t,k)$ denotes time- and batch-varying uncertainty, which is bounded by $\Delta^{\mathrm{T}}(t,k)\Delta(t,k) \leq \mathbf{I}$. Note that the output delay variation may be counted into the system uncertainties or disturbance to cope with. Moreover, the relative degree of the system model is generally supposed to be nonzero, i.e., $CB(t,k) \neq 0$.

Define by $e(t,k) = y_{\mathrm{d}}(t) - y(t,k)$ the output tracking error, where $y_{\mathrm{d}}(t)$ is the desired output trajectory that satisfies $|y_{\mathrm{d}}(t)| \leq \beta_{\mathrm{d}} < \infty$. In the presence of a delay response, the output tracking error can be written by

$$e(t,k) = y_{\mathrm{d}}(t-h) - y(t,k), \tag{9.2}$$

where the initial value of $e(t,k)$ is supposed to be zero for any $t \leq h$.

Concerning the design of ILC to track the above reference trajectory, it requires that

$$\lim_{k\to\infty} e(t,k) = 0, \forall t \in \mathbb{Z}_{N/h}. \tag{9.3}$$

However, the above goal could not be realized under time-varying system uncertainties and/or nonrepetitive disturbances. It is therefore expected to keep the output tracking error as small as possible for $t \in \mathbb{Z}_{N/h}$ and $k \in \mathbb{Z}_+$, i.e.

$$\sup_{k\geq 0}|e(t,k)| \leq \beta_e, \ \lim_{k\to\infty}\sup|e(t,k)| \leq \beta_{e_{\text{sup}}}, \tag{9.4}$$

where $\beta_e > \beta_{e_{\text{sup}}} \geq 0$ are specified bounds in practice. In the meantime, the system input and state in (9.1) should also be bounded, i.e.

$$\begin{aligned}&\sup_{k\geq 0}\max_{t\in\mathbb{Z}_{N-1}} |u(t,k)| \leq \beta_u < \infty,\\ &\sup_{k\geq 0}\max_{t\in\mathbb{Z}_{N}} \|x(t,k)\| \leq \beta_x < \infty,\end{aligned} \tag{9.5}$$

where $\beta_u \geq 0$ and $\beta_x \geq 0$ are also specified upper bounds in practice.

For the design of ILC, the following assumption is made without loss of generality, together with two lemmas for stability analysis.

Assumption 9.1 (Meng and Moore 2017) Suppose that $\omega(t,k)$, $A(t,k)$ and $B(t,k)$ are bounded for any $t \in \mathbb{Z}_N$ and $k \in \mathbb{Z}_+$ during batch run, i.e.

$$\begin{aligned}&\sup_{k\geq 0}\max_{t\in\mathbb{Z}_N}|\omega(t,k)| \leq \beta_\omega,\\ &\sup_{k\geq 0}\max_{t\in\mathbb{Z}_N}\|A(t,k)\| \leq \beta_{\text{A}},\\ &\sup_{k\geq 0}\max_{t\in\mathbb{Z}_N}\|B(t,k)\| \leq \beta_{\text{B}},\end{aligned}$$

where $\beta_\omega \geq 0$, $\beta_{\text{A}} \geq 0$, $\beta_{\text{B}} \geq 0$.

Lemma 9.1 Concerning a linear batch system with $t \in \mathbb{Z}_N$ and $k \in \mathbb{Z}_+$

$$x(t,k+1) = [1 - CB(t,k)\gamma(t,k)]x(t,k) + u(t,k), \tag{9.6}$$

where the system input and state are denoted by $u(t,k)$ and $x(t,k)$, respectively, $\gamma(t,k)$ is a tuning parameter, and the initial condition $x(t,0)$ and system input $u(t,k)$ are bounded, i.e., $\beta_{x_0} = \max_{t\in\mathbb{Z}_N}|x(t,0)| < \infty$ and $\beta_u(t) = \sup_{k\in\mathbb{Z}_+}|u(t,k)| < \infty$, there exists a finite bound $\beta_x(t) \geq 0$ satisfying $\sup_{k\in\mathbb{Z}_+}|x(t,k)| \leq \beta_x(t)$ for this system if the condition $|1 - CB(t,k)\gamma(t,k)| < 1$ stands true for any $k \in \mathbb{Z}_+$ and $t \in \mathbb{Z}_N$.

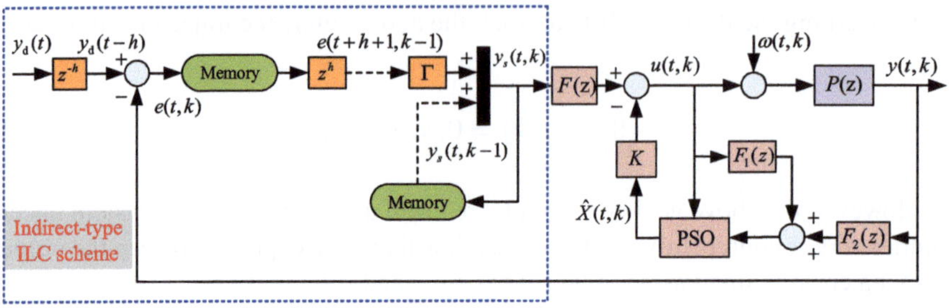

Fig. 9.1 Diagram of the PSO-based indirect-type ILC scheme

Proof It can be easily verified by using Lemma 9 in the reference (Meng and Moore, 2017) and taking the iteration sequence $\{\xi_s(t) : s \in \mathbb{Z}_+\}$ to be $\xi_s(t) = s$.

Figure 9.1 shows the proposed PSO based indirect-type ILC scheme, where the plant is denoted by $P(z)$, while the control input, process output, and disturbance in each batch run are denoted by $u(t, k)$, $y(t, k)$, and $\omega(t, k)$, respectively; the reference output trajectory is denoted by $y_d(t)$, and there follows $y_d(t-h) = z^{-h}y_d(t)$, where z^{-h} denotes the discrete-time delay also terms as h steps; $F_1(z)$ and $F_2(z)$ are two stable filters used for predicting the delay-free system output, while $F(z)$ is the set-point filter used for smoothing the set-point command and eliminating the steady-state output tracking error, K is the closed-loop feedback controller set for disturbance rejection, Γ is a P-type set-point learning controller, "Memory" is a storage taken for recording the output tracking error and set-point command ($y_s(t, k)$) in each batch run, and offering the tracking error ($e(t, k-1)$) and set-point command ($y_s(t, k-1)$) in the previous batch run. It should be mentioned that only the PSO-based feedback control structure with no memory is executed in the initial batch run.

9.2.1 *Feedback Control Structure with PSO*

In the nominal case, i.e., $\omega(t) = 0$, of the batch process described by (9.1), the discrete-time domain transfer function may be written by

$$P(z) = \frac{b_1 z^{n-1} + \cdots + b_{n-1}z + b_n}{z^n + a_1 z^{n-1} + \cdots + a_{n-1}z + a_n} z^{-d} \triangleq G_n(z)z^{-h}. \tag{9.7}$$

For the convenience of designing PSO, the delay-free part in (9.1) are rewritten in the observable canonical state-space form

$$A = \begin{bmatrix} -a_1 & 1\,0 \cdots 0 \\ -a_2 & 0\,1 \cdots 0 \\ \vdots & \vdots\,\vdots \ddots \vdots \\ -a_{n-1} & 0\,0 \cdots 1 \\ -a_n & 0\,0 \cdots 0 \end{bmatrix}, B = \begin{bmatrix} b_1 \\ b_2 \\ \vdots \\ b_{n-1} \\ b_n \end{bmatrix}, C = \begin{bmatrix} 1 \\ 0 \\ \vdots \\ 0 \\ 0 \end{bmatrix}^{\mathrm{T}}.$$

Meanwhile, the process uncertainties, external disturbances, and unsteady initial process state are lumped together and denoted by $f(t)$, which is therefore termed as the total disturbance. Accordingly, the batch process described by (9.1) is expressed by

$$\begin{cases} x(t+1) = Ax(t) + B[u(t) + f(t)], \\ y(t) = Cx(t-h). \end{cases} \tag{9.8}$$

Herein, the batch index k is omitted for brevity, in view of its irrelevance to the control system design in the time domain. For estimating the process state and $f(t)$ to design the feedback control scheme, the following augmented system description is established by assigning $f(t)$ as an extended state

$$\begin{cases} X(t+1) = \hat{A}X(t) + \hat{B}u(t) + G\Delta f(t+1), \\ y(t) = \hat{C}X(t-h), \end{cases} \tag{9.9}$$

where $\hat{C} = [\,C\ 0\,]$, $\Delta f(t+1) = f(t+1) - f(t)$ and

$$\hat{A} = \begin{bmatrix} A & B \\ 0 & 1 \end{bmatrix}, \hat{B} = \begin{bmatrix} B \\ 0 \end{bmatrix}, G = \begin{bmatrix} 0 \\ 1 \end{bmatrix}, X(t) = \begin{bmatrix} x(t) \\ f(t) \end{bmatrix}.$$

Taking advantage of delay-free output prediction, a PSO is designed to simultaneously estimate the process state and total disturbance

$$\hat{X}(t+1) = \hat{A}\hat{X}(t) + \hat{B}u(t) + L[y_{\mathrm{p}}(t) - \hat{C}\hat{X}(t)], \tag{9.10}$$

where $\hat{X}(t) = [\hat{x}^{\mathrm{T}}(t)\,\hat{f}(t)\,]^{\mathrm{T}}$, $\hat{x}(t)$ and $\hat{f}(t)$ are the estimations on $x(t)$ and $f(t)$, respectively; denoted by L the PSO gain, which is used to set all eigenvalues of $\hat{A} - L\hat{C}$ at α_0 and $\alpha_0 \in (0, 1)$ is an adjustable parameter that may be initially taken in a range of [0.9, 0.99] for implementation; the predicted delay-free output response is denoted by $y_{\mathrm{p}}(t)$, which is obtained by

$$y_{\mathrm{p}}(t) = F_1(z)u(z) + F_2(z)y(z), \tag{9.11}$$

where $F_1(z)$ and $F_2(z)$ are designed by using the approach given in the reference (Liu et al. (2018) as below.

Decompose the process description in (9.7) as

$$G_n(z) = \frac{N(z)}{D(z)} = \tilde{G}(z)\Lambda(z), \tag{9.12}$$

where

$$\begin{aligned}\Lambda(z) &= \frac{N(z)}{(z-\lambda)^m} H(z,\lambda),\\ \tilde{G}(z) &= \frac{(z-\lambda)^m}{D(z)} H^{-1}(z,\lambda) = C_{\rm p}(z\mathbf{I} - A_{\rm p})^{-1} B_{\rm p},\end{aligned}$$

and $\lambda \in (0, 1)$ is a tuning parameter, m is the number of zeros in $G_n(z)$, $H(z, \lambda)$ is a filter to be designed, $(A_{\rm p}, B_{\rm p}, C_{\rm p})$ is a minimum-order state space representation of $\tilde{G}(z)$.

To suppress the measurement noise typically encountered in practical applications, $H(z, \lambda)$ is configured as

$$H(z,\lambda) = \frac{z-1}{z-\lambda} \cdot \frac{(1-\lambda)^{n_k} z^{n_k}}{(z-\lambda)^{n_k}}, \tag{9.13}$$

which takes a unity static gain and $(1-\lambda)^{n_k} z^{n_k}/(z-\lambda)^{n_k}$ is an all-pass filter for reducing the sensitivity to output measurement noise, where n_k is a user-specified order that may be initially taken as zero or one for practical implementation.

By defining the following new transfer function

$$\tilde{G}^*(z) = C_{\rm p}(z\mathbf{I} - A_{\rm p})^{-1} A_{\rm p}^d B_{\rm p} = \frac{\tilde{N}^*(z)}{\tilde{D}(z)}, \tag{9.14}$$

the above two filters for predicting delay-free output response are constructed by

$$F_1(z) = C_{\rm p} \sum_{i=1}^{d} A_{\rm p}^{i-1} B_{\rm p} z^{-i} \Lambda(z), \tag{9.15}$$

$$F_2(z) = \frac{\tilde{N}^*(z)}{(z-\lambda)^{m+n_k+1}}. \tag{9.16}$$

Hence, the delay-free output $y_{\rm p}(t)$ reconstructed by (9.11) can be used to estimate the delay-free system state and total disturbance, which could alleviate the influence of the delay response to the closed-loop control performance.

Based on the above design of PSO, the closed-loop control law is given by

$$u(t) = \hat{y}_s(t) - K\hat{X}(t), \tag{9.17}$$

where $K = [K_0 \ 1]$ indicates a gain vector of the closed-loop controller, and K_0 is selected to set all eigenvalues of $A - BK_0$ at α_c in order to maintain the closed-loop stability, while $\alpha_c \in (0, 1)$ is another adjustable parameter that may be initially taken in a range of [0.9, 0.95] for implementation. Note that $\hat{y}_s(t)$ is a modified set-point command, which is defined by

$$\hat{y}_s(z)=F(z)y_s(z), \tag{9.18}$$

where $\hat{y}_s(z)$ and $y_s(z)$ indicate the discrete-time domain transforms of $\hat{y}_s(t)$ and $y_s(t)$, respectively; The set-point filter is designed by

$$F(z)=\frac{K_s(1-\lambda_s)^{m_s}z^{m_s}}{(z-\lambda_s)^{m_s}}, \tag{9.19}$$

where K_s denotes the filter gain to be determined by the set-point tracking performance specification; λ_s is a tuning parameter that may be initially taken in a range of [0.95, 0.99] for implementation; m_s is a user-specified filter order to comply with the control constraint in practice.

Ideally, $\hat{X}(t)$ should offer an accurate h-step ahead prediction of $X(t)$ to compensate for the delay response, so that the transfer function from the set-point command to the system output could be acquired as

$$y(z)=C(\mathbf{I}-A+BK_0)^{-1}Bz^{-h}F(z)y_s(z).$$

It is therefore determined to take $K_s = 1/[C(\mathbf{I}-A+BK_0)^{-1}B]$, such that $\lim_{z\to 1} C(\mathbf{I}-A+BK_0)^{-1}Bz^{-h}F(z) = 1$ could be realized for obtaining no steady-state output tracking error during each batch run (Liu et al. 2019).

9.2.2 Set-Point Learning

From Fig. 9.1, it can be seen that the dash-line box outlines the proposed set-point learning law of P-type, that is, the set-point command is updated by

$$y_s(t,k) = y_s(t,k-1) + \Gamma e(t+h+1, k-1), \tag{9.20}$$

where Γ denotes the learning gain.

For the convenience of designing the set-point learning law, the state-space expressions of the above system and delay-free output predictor are written below

$$P(z): \begin{cases} x(t+1) = A(t)x(t) + B(t)[u(t) + \omega(t)], \\ y(t) = Cx(t-h), \end{cases} \tag{9.21}$$

$$F_1(z): \begin{cases} x_1(t+1) = A_1 x_1(t) + B_1 u(t), \\ y_1(t) = C_1 x_1(t), \end{cases} \tag{9.22}$$

$$F_2(z): \begin{cases} x_2(t+1) = A_2 x_2(t) + B_2 y(t), \\ y_2(t) = C_2 x_2(t) + D_2 y(t). \end{cases} \tag{9.23}$$

It is obvious that $y_p(t)$ given by the PSO in (9.10) is expressed by $y_p(t) = y_1(t) + y_2(t)$. Note that the initial states of $F_1(z)$, $F_2(z)$, and PSO are reset to zero for batch implementation.

Hence, the overall system is reformulated by

$$\begin{cases} \mathcal{X}(t+1,k) = \mathcal{A}(t,k)\mathcal{X}(t,k) + \mathcal{A}_h \mathcal{X}(t-h,k) + \mathcal{B}(t,k) y_s(t,k) + \mathcal{D}(t,k)\omega(t,k), \\ y(t,k) = \mathcal{C}\mathcal{X}(t-h,k), \end{cases} \tag{9.24}$$

where $\mathcal{A}(t,k) = \mathcal{A} + \mathcal{E}\Delta(t,k)\mathcal{F}_A$, $\mathcal{B}(t,k) = \mathcal{B} + \mathcal{E}\Delta(t,k)\mathcal{F}_B$, $\mathcal{F}_B = F_B F$, $\mathcal{C} = \begin{bmatrix} C & 0 & 0 & 0 \end{bmatrix}$, $\mathcal{F}_A = \begin{bmatrix} F_A & -F_B K & 0 & 0 \end{bmatrix}$, and

$$\mathcal{A} = \begin{bmatrix} A & -BK & 0 & 0 \\ 0 & \hat{A} - L\hat{C} - \hat{B}K & LC_1 & LC_2 \\ 0 & -B_1 K & A_1 & 0 \\ 0 & 0 & 0 & A_2 \end{bmatrix}, \mathcal{A}_h = \begin{bmatrix} 0 & 0\,0\,0 \\ LD_2 C & 0\,0\,0 \\ 0 & 0\,0\,0 \\ B_2 C & 0\,0\,0 \end{bmatrix},$$

$$\mathcal{B} = \begin{bmatrix} BF \\ \hat{B}F \\ B_1 F \\ 0 \end{bmatrix}, \quad \mathcal{D}(t,k) = \begin{bmatrix} B(t,k) \\ 0 \\ 0 \\ 0 \end{bmatrix}, \mathcal{E} = \begin{bmatrix} E \\ 0 \\ 0 \\ 0 \end{bmatrix}, \mathcal{X}(t,k) = \begin{bmatrix} x(t,k) \\ \hat{X}(t,k) \\ x_1(t,k) \\ x_2(t,k) \end{bmatrix}.$$

Along the batch-direction, an incremental form of the above system can be written as

$$\begin{aligned} \delta\mathcal{X}(t+1,k) = \mathcal{A}(t,k)\delta\mathcal{X}(t,k) + \mathcal{A}_h \delta\mathcal{X}(t-h,k) \\ + \mathcal{B}(t,k)\Gamma e(t+h+1,k-1) + \varpi(t,k), \end{aligned} \tag{9.25}$$

where $\varpi(t,k) = \delta\mathcal{A}(t,k)\mathcal{X}(t,k-1) + \delta\mathcal{B}(t,k)y_s(t,k-1) + \mathcal{D}(t,k)\omega(t,k) - \mathcal{D}(t,k-1)\omega(t,k-1)$. Accordingly, the batchwise error dynamics can be derived as

$$\begin{aligned} e(t+h+1,k) &= e(t+h+1,k-1) - \mathcal{C}\delta\mathcal{X}(t+1,k) \\ &= [1 - \mathcal{C}\mathcal{B}(t,k)\Gamma]e(t+h+1,k-1) - \mathcal{C}\mathcal{A}(t,k)\delta\mathcal{X}(t,k) \\ &\quad - \mathcal{C}\mathcal{A}_h\delta\mathcal{X}(t-h,k) - \mathcal{C}\varpi(t,k). \end{aligned} \tag{9.26}$$

The guideline for choosing the learning gain of Γ will be provided based on the convergence analysis to be presented in the next section.

9.3 Convergence Analysis on ILC

Theorem 9.1 Concerning a batch process described in (9.1), the tracking objective in (9.4) and the operational bound in (9.5) could be satisfied by using the set-point learning law in (9.20) under Assumption 9.1, if there exists a proper scalar gain Γ that could guarantee

$$|1 - CB(t,k)F\Gamma| < \rho \tag{9.27}$$

with $\rho \in (0, 1]$.

Proof It follows from rewriting (9.20) that

$$\begin{aligned} y_s(t,k) &= y_s(t,k-1) + \Gamma e(t+h+1,k-1) \\ &= [1 - \Gamma C\mathcal{B}(t,k)]y_s(t,k-1) - \Gamma C\mathcal{A}(t,k-1)\mathcal{X}(t,k-1) \\ &- \Gamma C\mathcal{A}_{\text{h}}\mathcal{X}(t-h,k-1) + \phi_s(t,k), \end{aligned} \tag{9.28}$$

where $\phi_s(t,k) = \Gamma y_{\text{d}}(t+1) - \Gamma C\mathcal{D}(t,k-1)\omega(t,k-1)$. It is obvious that $|\phi_s(t,k)| \leq |\Gamma|(\beta_{\text{d}} + \|C\|\beta_{\text{B}}\beta_{\omega})$ for any $k \in \mathbb{Z}_+$ and $t \in \mathbb{Z}_N$.

Based on Assumption 9.1, it can be easily verified that both $\mathcal{A}(t,k)$ and $\mathcal{B}(t,k)$ are bounded for any $k \in \mathbb{Z}_+$ and $t \in \mathbb{Z}_N$ when setting the PSO and feedback controller K with finite gains denoted by $\beta_{\mathcal{A}}$ and $\beta_{\mathcal{B}}$, respectively.

The boundedness described by (9.5) for any $t \in \mathbb{Z}_N$ and $k \in \mathbb{Z}_+$ can be clarified by using the inductive method. When $t = 0$, there stands $|-\Gamma C\mathcal{A}(0,k-1)\mathcal{X}(0,k-1) - \Gamma C\mathcal{A}_{\text{h}}\mathcal{X}(-h,k-1)| \leq |\Gamma|\|C\|\beta_{\mathcal{X}}(0)(\beta_{\mathcal{A}} + \|\mathcal{A}_{\text{h}}\|)$ for any $k \in \mathbb{Z}_+$, where $\beta_{\mathcal{X}}(0) = \|x_0\|$. According to Lemma 9.1, $y_s(0,k)$ is bounded for any $k \in \mathbb{Z}_+$ and satisfies $\sup_{k\in\mathbb{Z}_+}|y_s(0,k)| \leq \beta_s(0) < \infty$ when the condition in (9.27) stands true. Considering that there are $1 - \Gamma C\mathcal{B}(t,k) = 1 - \Gamma CB(t,k)F$ and $1 - \Gamma CB(t,k)F = 1 - CB(t,k)F\Gamma$ for an SISO batch process, it yields from (9.24) and (9.28) that

$$\begin{aligned} \|\mathcal{X}(1,k)\| &\leq \|\mathcal{A}(0,k)\mathcal{X}(0,k)\| + \|\mathcal{A}_{\text{h}}\mathcal{X}(-h,k)\| + \|\mathcal{B}(0,k)y_S(0,k)\| + \|\mathcal{D}(0,k)\omega(0,k)\| \\ &\leq \beta_{\mathcal{A}}\beta_{\mathcal{X}}(0) + \|\mathcal{A}_{\text{h}}\|\beta_{\mathcal{X}}(0) + \beta_{\mathcal{B}}\beta_s(0) + \beta_{\mathcal{B}}\beta_{\omega} \triangleq \beta_{\mathcal{X}}(1) < \infty. \end{aligned}$$

Using Lemma 9.1 to (9.28) together with bounded $y_s(0,k)$, one can ensure that $y_s(1,k)$ is bounded, which is herein denoted by $\beta_s(1)$.

Similarly, when $t = 1$, there follows

$$\begin{aligned}\|\mathcal{X}(2,k)\| &\leq \|\mathcal{A}(1,k)\mathcal{X}(1,k)\| + \|\mathcal{A}_{\rm h}\mathcal{X}(1-h,k)\| + \|\mathcal{B}(1,k)y_s(1,k)\| + \|\mathcal{D}(1,k)\omega(1,k)\| \\ &\leq \beta_{\mathcal{A}}\beta_{\mathcal{X}}(1) + \|\mathcal{A}_{\rm h}\|\beta_{\mathcal{X}}(0) + \beta_{\mathcal{B}}\beta_s(1) + \beta_{\mathcal{B}}\beta_\omega \triangleq \beta_{\mathcal{X}}(2) < \infty.\end{aligned}$$

Accordingly, $y_s(2,k)$ is also bounded for all $k \in \mathbb{Z}_+$ under the condition in (9.27), which is indicated by $\beta_s(2)$.

When $t = N-1$, there follows

$$\begin{aligned}\|\mathcal{X}(N,k)\| &\leq \|\mathcal{A}(N-1,k)\mathcal{X}(N-1,k)\| + \|\mathcal{A}_{\rm h}\mathcal{X}(N-1-h,k)\| \\ &\quad + \|\mathcal{B}(N-1,k)y_s(N-1,k)\| + \|\mathcal{D}(N-1,k)\omega(N-1,k)\| \\ &\leq \beta_{\mathcal{A}}\beta_{\mathcal{X}}(N-1) + \|\mathcal{A}_{\rm h}\|\beta_{\mathcal{X}}(N-1-h) + \beta_{\mathcal{B}}\beta_s(N-1) + \beta_{\mathcal{B}}\beta_\omega \\ &\triangleq \beta_{\mathcal{X}}(N) < \infty.\end{aligned}$$

Therefore, it is concluded that

$$\sup_{k\in\mathbb{Z}_+}\max_{t\in\mathbb{Z}_N}\|\mathcal{X}(t,k)\| \leq \beta_{\mathcal{X}}, \quad \sup_{k\in\mathbb{Z}_+}\max_{t\in\mathbb{Z}_{N-1}}|y_s(t,k)| \leq \beta_s,$$

where $\beta_{\mathcal{X}} = \max_{t\in\mathbb{Z}_N}\beta_{\mathcal{X}}(t)$ and $\beta_s = \max_{t\in\mathbb{Z}_{N-1}}\beta_s(t)$. By using (9.17), (9.24) and (9.27) together, it is sure that the objective in (9.5) could be satisfied.

Based on the boundedness of $\mathcal{X}(t,k)$ and $y_s(t,k)$, it yields

$$|\Theta(t,k)| \leq 2\|C\|\beta_{\mathcal{A}}\beta_{\mathcal{X}} + 2\|C\|\|\mathcal{A}_{\rm h}\|\beta_{\mathcal{X}} + 2\|C\|(\beta_{\mathcal{A}}\beta_{\mathcal{X}} + \beta_{\mathcal{B}}\beta_s + \beta_{\mathcal{B}}\beta_\omega) < \infty,$$

where $\Theta(t,k) = -\mathcal{C}\mathcal{A}(t,k)\delta\mathcal{X}(t,k) - \mathcal{C}\mathcal{A}_{\rm h}\delta\mathcal{X}(t-h,k) - \mathcal{C}\varpi(t,k)$.

Then by using Lemma 9.1 and the condition in (9.27) to the error dynamics described by (9.26), it is concluded that the tracking objective in (9.4) could be realized. This completes the proof. □

It should be pointed out that the sufficient condition in (9.27) in Theorem 9.1 is independent of the time delay, which could be used for industrial batch processes with a long output delay. Nevertheless, the convergence of tracking error in the time domain may not be guaranteed under the delay variation. To address this issue, a delay-dependent condition is offered in the next theorem, where the error dynamics described by (9.25) and (9.26) is rewritten into a 2D state delay FM model to facilitate the analysis

$$\begin{cases}\chi(t+1,k) = \mathbb{A}_1(t,k)\chi(t,k) + \mathbb{A}_2(t,k)\chi(t+1,k-1) \\ \qquad\qquad\qquad + \mathbb{A}_{\rm h}\chi(t-h,k) + \mathbb{D}\varpi(t,k), \\ z(t,k) = \mathbb{C}\chi(t,k),\end{cases} \tag{9.29}$$

where $\mathbb{A}_1(t,k) = \mathbb{A}_1 + \mathbb{E}\Delta(t,k)\mathbb{F}_{\rm A}$, $\mathbb{A}_2(t,k) = \mathbb{A}_2 + \mathbb{E}\Delta(t,k)\mathbb{F}_{\rm B}$, and

$$\mathbb{A}_1 = \begin{bmatrix}\mathcal{A} & 0\\ -\mathcal{C}\mathcal{A} & 0\end{bmatrix}, \quad \mathbb{A}_2 = \begin{bmatrix}0 & \mathcal{B}\Gamma\\ 0 & 1-\mathcal{C}\mathcal{B}\Gamma\end{bmatrix}, \quad \mathbb{A}_{\rm h} = \begin{bmatrix}\mathcal{A}_{\rm h} & 0\\ -\mathcal{C}\mathcal{A}_{\rm h} & 0\end{bmatrix},$$

$$\mathbb{D} = \begin{bmatrix} \mathbf{I} \\ -\mathcal{C} \end{bmatrix}, \quad \mathbb{C} = \begin{bmatrix} 0 \\ 1 \end{bmatrix}^{\mathrm{T}}, \mathbb{E} = \begin{bmatrix} \mathcal{E} \\ -\mathcal{C}\mathcal{E} \end{bmatrix}, \quad \chi(t,k) = \begin{bmatrix} \delta\mathcal{X}(t,k) \\ e(t+h,k) \end{bmatrix},$$
$$\mathbb{F}_{\mathrm{A}} = \begin{bmatrix} \mathcal{F}_{\mathrm{A}} \; 0 \end{bmatrix}, \quad \mathbb{F}_{\mathrm{B}} = \begin{bmatrix} 0 \; \mathcal{F}_{\mathrm{B}}\Gamma \end{bmatrix}.$$

Concerning a 2D FM model in (9.29), the boundary conditions are denoted by

$$\begin{cases} \chi(t,0) = 0, \;\; t \in \mathbb{Z}_N, \\ \chi(t,k) = 0, \;\; t = -h, -h+1, \ldots, 0, \; k \in \mathbb{Z}_+, \end{cases} \tag{9.30}$$

based on an assumption that the system disturbance satisfies

$$\sum_{t=0}^{N} \sum_{k=0}^{\infty} \varpi^{\mathrm{T}}(t,k)\varpi(t,k) < \infty. \tag{9.31}$$

Theorem 9.2 The 2D system in (9.29) with a user-specified learning gain Γ could maintain asymptotical stability with a prescribed 2D $\mathcal{H}_\infty$ performance γ, if there exist matrices $P > P_0 > 0$, $Q > 0$, $R > 0$, N_1, N_2, N_3, M_{11}, M_{12}, M_{13},M_{22}, M_{23},M_{33} and a scalar $\varepsilon > 0$, such that the following LMIs are satisfied

$$\begin{bmatrix} \overline{\Upsilon}_{11} & \Upsilon_{12} & \overline{\Upsilon}_{13} & 0 & \mathbb{A}_1^{\mathrm{T}}P & \overline{\Upsilon}_{16} & \mathbb{C}^{\mathrm{T}} & 0 \\ * & \Upsilon_{22} & \Upsilon_{23} & 0 & \mathbb{A}_{\mathrm{h}}^{\mathrm{T}}P & \sqrt{h}\mathbb{A}_{\mathrm{h}}^{\mathrm{T}}R & 0 & 0 \\ * & * & \overline{\Upsilon}_{33} & 0 & \mathbb{A}_2^{\mathrm{T}}P & \sqrt{h}\mathbb{A}_2^{\mathrm{T}}R & 0 & 0 \\ * & * & * & -\gamma\mathbf{I} & \mathbb{D}^{\mathrm{T}}P & \sqrt{h}\mathbb{D}^{\mathrm{T}}R & 0 & 0 \\ * & * & * & * & -P & 0 & 0 & P\mathbb{E} \\ * & * & * & * & * & -R & 0 & \sqrt{h}R\mathbb{E} \\ * & * & * & * & * & * & -\gamma\mathbf{I} & 0 \\ * & * & * & * & * & * & * & -\varepsilon\mathbf{I} \end{bmatrix} < 0, \tag{9.32}$$

$$\begin{bmatrix} M_{11} & M_{12} & M_{13} & N_1 \\ * & M_{22} & M_{23} & N_2 \\ * & * & M_{33} & N_3 \\ * & * & * & R \end{bmatrix} \geq 0, \tag{9.33}$$

where $\overline{\Upsilon}_{11} = \Upsilon_{11} + \varepsilon\mathbb{F}_{\mathrm{A}}^{\mathrm{T}}\mathbb{F}_{\mathrm{A}}, \overline{\Upsilon}_{13} = \Upsilon_{13} + \varepsilon\mathbb{F}_{\mathrm{A}}^{\mathrm{T}}\mathbb{F}_{\mathrm{B}}, \overline{\Upsilon}_{33} = \Upsilon_{33} + \varepsilon\mathbb{F}_{\mathrm{B}}^{\mathrm{T}}\mathbb{F}_{\mathrm{B}}, \overline{\Upsilon}_{16} = \sqrt{h}(\mathbb{A}_1^{\mathrm{T}} - \mathbf{I})R$, and

$$\Upsilon_{11} = -P_0 + Q + \mathrm{sym}\{N_1\} + hM_{11}, \; \Upsilon_{33} = -P + P_0 + hM_{33},$$
$$\Upsilon_{12} = -N_1 + N_2^{\mathrm{T}} + hM_{12}, \; \Upsilon_{22} = -Q - \mathrm{sym}\{N_2\} + hM_{22},$$
$$\Upsilon_{13} = N_3^{\mathrm{T}} + hM_{13}, \; \Upsilon_{23} = -N_3^{\mathrm{T}} + hM_{23}.$$

Proof The following Lyapunov–Krasovskii function is considered for a 2D system in (9.29) under $\varpi(t,k)=0$

$$V(t,k)=V_h(t,k)+V_v(t,k),$$

where $\sigma(i,k)=\chi(i+1,k)-\chi(i,k)$, $V_v(t,k)=\chi^{\mathrm{T}}(t,k)(P-P_0)\chi(t,k)$ and

$$V_h(t,k)=\sum_{i=1}^{3}V_i(t,k),\ V_1(t,k)=\chi^{\mathrm{T}}(t,k)P_0\chi(t,k),$$
$$V_2(t,k)=\sum_{i=t-h}^{t-1}\chi^{\mathrm{T}}(i,k)Q\chi(i,k),\ V_3(t,k)=\sum_{j=-h}^{-1}\sum_{i=t+j}^{t-1}\sigma^{\mathrm{T}}(i,k)R\sigma(i,k).$$

An increment form of the above Lyapunov–Krasovskii function is written as

$$\Delta V(t,k)=\sum_{i=1}^{3}\Delta V_i(t,k)+\Delta V_v(t,k),$$

where $\Delta V_i(t,k)=V_i(t+1,k)-V_i(t,k)$, $i=1,2,3$ and $\Delta V_v(t,k)=V_v(t+1,k)-V_v(t+1,k-1)$.

Then it follows that

$$\begin{aligned}\Delta V(t,k)&=\chi^T(t+1,k)P\chi(t+1,k)-\chi^T(t,k)P_0\chi(t,k)+\chi^T(t,k)Q\chi(t,k)\\&-\chi^T(t-h,k)Q\chi(t-h,k)+h\sigma^T(t,k)R\sigma(t,k)\\&-\sum_{i=t-h}^{t-1}\sigma^T(i,k)R\sigma(i,k)-\chi^T(t+1,k-1)(P-P_0)\chi(t+1,k-1)\\&+2\Big[\chi^T(t,k)N_1+\chi^T(t-h,k)N_2+\chi^T(t+1,k-1)N_3\Big]\times\\&\left[\chi(t,k)-\chi(t-h,k)-\sum_{i=t-h}^{t-1}\sigma(i,k)\right]+h\phi^{\mathrm{T}}(t,k)M\phi(t,k)-\sum_{i=t-h}^{t-1}\phi^{\mathrm{T}}(t,k)M\phi(t,k)\\&=\phi^T(t,k)(\Upsilon+\Gamma_1^{\mathrm{T}}P\Gamma_1+h\Gamma_2^{\mathrm{T}}R\Gamma_2)\phi(t,k)-\sum_{i=t-h}^{t-1}\begin{bmatrix}\phi(t,k)\\\sigma(i,k)\end{bmatrix}^{\mathrm{T}}\begin{bmatrix}M&N\\ *&R\end{bmatrix}\begin{bmatrix}\phi(t,k)\\\sigma(i,k)\end{bmatrix},\end{aligned}$$

where $\phi(t,k)=\begin{bmatrix}\chi^{\mathrm{T}}(t,k)\ \chi^{\mathrm{T}}(t-h,k)\ \chi^{\mathrm{T}}(t+1,k-1)\end{bmatrix}^{\mathrm{T}}$ and

$$\Upsilon=\begin{bmatrix}\Upsilon_{11}&\Upsilon_{12}&\Upsilon_{13}\\ *&\Upsilon_{22}&\Upsilon_{23}\\ *&*&\Upsilon_{33}\end{bmatrix},M=\begin{bmatrix}M_{11}&M_{12}&M_{13}\\ *&M_{22}&M_{23}\\ *&*&M_{33}\end{bmatrix},N=\begin{bmatrix}N_1\\N_2\\N_3\end{bmatrix},$$
$$\Gamma_1=\begin{bmatrix}\mathbb{A}_1(t,k)\ \mathbb{A}_h\ \mathbb{A}_2(t,k)\end{bmatrix},\Gamma_2=\begin{bmatrix}\mathbb{A}_1(t,k)-\mathbf{I}\ \mathbb{A}_h\ \mathbb{A}_2(t,k)\end{bmatrix}.$$

Applying the Schur complement and Lemma 2.1 to the LMI in (9.32) and (9.33), it follows that $\Delta V(t,k) < 0$. Thus, there stands

$$V_h(t+1,k) + V_v(t+1,k) < V_h(t,k) + V_v(t+1,k-1).$$

Owing to $V_h(0,r) = V_v(r,0) = 0$ for any $r > 0$, it is verified that

$$\begin{aligned}\sum_{t+k=r} V(t,k) &= V_h(0,r) + V_h(1,r-1) + \cdots + V_h(r-1,1) + V_h(r,0)\\ &\quad + V_v(0,r) + V_v(1,r-1) + \cdots + V_v(r-1,1) + V_v(r,0)\\ &< V_h(0,r-1) + V_h(1,r-2) \cdots + V_h(r-2,1) + V_h(r-1,0)\\ &\quad + V_v(0,r-1) + V_v(1,r-2) \cdots + V_v(r-1,0)\\ &= \sum_{t+k=r-1} V(t,k).\end{aligned}$$

According to the 2D control theory (Kaczorek 1985), there stands $\lim_{t+k\to\infty} \|\chi(t,k)\| = 0$, which means that the closed-loop 2D system in (9.29) under $\varpi(t,k) = 0$ could guarantee asymptotical stability.

For analyzing the robust control performance against external disturbance, define the following cost function

$$J = \sum_{t=0}^{N}\sum_{k=0}^{\infty}\big[\gamma^{-1}z^{\mathrm{T}}(t,k)z(t,k) - \gamma\varpi^{\mathrm{T}}(t,k)\varpi(t,k)\big].$$

Based on resetting zero initial conditions for batch run, there follows

$$\begin{aligned}J &< \sum_{t=0}^{N}\sum_{k=0}^{\infty}\big[\gamma^{-1}z^{\mathrm{T}}(t,k)z(t,k) - \gamma\varpi^{\mathrm{T}}(t,k)\varpi(t,k) + \Delta V(t,k)\big]\\ &= \sum_{t=0}^{N}\sum_{k=0}^{\infty}\Big\{\hat{\phi}^{\mathrm{T}}(t,k)(\hat{\Upsilon} + \hat{\Gamma}_1^{\mathrm{T}}P\hat{\Gamma}_1 + h\hat{\Gamma}_2^{\mathrm{T}}R\hat{\Gamma}_2)\hat{\phi}(t,k)\\ &\quad - \sum_{i=t-h}^{t-1}\begin{bmatrix}\phi(t,k)\\ \sigma(i,k)\end{bmatrix}^{\mathrm{T}}\begin{bmatrix}M & N\\ * & R\end{bmatrix}\begin{bmatrix}\phi(t,k)\\ \sigma(i,k)\end{bmatrix}\Big\},\end{aligned}$$

where $\hat{\phi}(t,k) = \big[\phi^{\mathrm{T}}(t,k)\ \varpi^{\mathrm{T}}(t,k)\big]^{\mathrm{T}}$, $\hat{\Gamma}_1 = \big[\Gamma_1\ \mathbb{D}\big]$, $\hat{\Gamma}_2 = \big[\Gamma_2\ \mathbb{D}\big]$ and

$$\hat{\Upsilon} = \begin{bmatrix}\Upsilon_{11}+\gamma^{-1}\mathbb{C}^{\mathrm{T}}\mathbb{C} & \Upsilon_{12} & \Upsilon_{13}\\ * & \Upsilon_{22} & \Upsilon_{23}\\ * & * & \Upsilon_{33}\end{bmatrix}.$$

Following a similar analysis as above, it can be easily verified that $J < 0$, which means that the robust $\mathcal{H}_\infty$ performance could be guaranteed for the above 2D system. This completes the proof. □

9.4 An Illustrative Example

Consider a benchmark example of batch process with output delay studied in Liu et al. (2010)

$$G(s) = \frac{1}{s+1}e^{-s}.$$

By taking a sampling period $T_s = 0.1$ (sec) for implementation, a discrete-time process model is obtained as

$$G(z) = \frac{0.09516}{z - 0.9048}z^{-10}.$$

For illustration, a desired output trajectory is taken as

$$y_{\mathrm{d}}(t) = \begin{cases} 0, & t \leq 1, \\ 2(t-1), & 1 < t \leq 7, \\ 12, & 7 < t \leq 10. \end{cases}$$

Suppose that a step change of external disturbance with a negative magnitude of 4 is added to the process input at $t = 2$ (sec).

By setting $\alpha_0 = 0.7$ and $\alpha_c = 0.8$ in the proposed control scheme, the PSO gain and the feedback controller are determined, respectively, as

$$L = \begin{bmatrix} 0.5048 \; 0.9457 \end{bmatrix}^{\mathrm{T}}, \; K = \begin{bmatrix} 1.1017, \; 1 \end{bmatrix}.$$

By taking $\lambda_s = 0.2$ and $m_s = 1$, the set-point filter is determined as $F = 1.681z/(z-0.2)$. Meanwhile, another smaller parameter setting of $\alpha_0 = 0.4$ and $\alpha_c = 0.4$ is also chosen for comparison, leading to $L = \begin{bmatrix} 1.1048 \; 3.7830 \end{bmatrix}^{\mathrm{T}}$, $K = \begin{bmatrix} 5.3050 \; 1 \end{bmatrix}$ and $F = 5.044z/(z-0.2)$. Identically, the set-point learning gain is taken as $\Gamma = 0.8$, which satisfies the condition in (9.27) in Theorem 9.1. The delay-free output predictor is configured by taking $\lambda = 0.6$ and $n_k = 0$

$$F_1(z) = \Phi(z)\Lambda(z), \; F_2(z) = \frac{3.025z - 2.625}{z - 0.6},$$

where $\Phi(z) = C_{\mathrm{p}} \sum_{i=1}^{10} A_{\mathrm{p}}^{i-1} B_{\mathrm{p}} z^{-i}$, $\Lambda(z) = \frac{0.09516(z-1)}{z-0.6}$ and

$$A_\mathrm{p} = \begin{bmatrix} 1.9048 & -0.9048 \\ 1 & 0 \end{bmatrix}, \; B_\mathrm{p} = \begin{bmatrix} 1 \\ 0 \end{bmatrix}, \; C_\mathrm{p} = \begin{bmatrix} 1 \\ -0.6 \end{bmatrix}^\mathrm{T}.$$

To make comparison with the existing methods, a classical open-loop ILC with a P-type learning gain is performed by assuming that the output delay is known in advance and can be compensated for implementation. The corresponding control law is taken as $u(t,k) = u(t,k-1) + \Gamma_0 e(t+h+1,k-1)$, where $\Gamma_0 = 1.6813$ is chosen to satisfy the condition $|1 - CB(t,k)\Gamma_0| < 1$ in Lemma 9.1.

In addition, the IMC-based ILC given in Liu et al. (2010) and another weighted IMC-based ILC given in Shokri-Ghaleh et al. (2021) are also performed for comparison, where a uniform controller in these ILC methods is taken as $C(s) = (s+1)/(0.55s+1)$ based on the controller design formula given therein, so as to procure a similar initial set-point tracking speed with that of the proposed method by taking $\alpha_0 = 0.7, \alpha_c = 0.8$ and $\lambda_s = 0.2$ for implementation. It should be mentioned that the learning function in Shokri-Ghaleh et al. (2021) is taken as $W_b(s) = 10/(0.5s+10)$ in order to satisfy the system stability condition discussed therein. Figure 9.2 shows the output tracking results for batch runs. Table 9.1 lists the corresponding MSE indices by different ILC methods, where the proposed (a) corresponds to the parameter setting $\alpha_0 = 0.4$, $\alpha_c = 0.4$, $\lambda_s = 0.2$, while the proposed (b) corresponds to another parameter setting $\alpha_0 = 0.7$, $\alpha_c = 0.8$, $\lambda_s = 0.2$. It is obvious that perfect tracking of the desired output trajectory is almost realized after 10 batch runs by the proposed ILC scheme, and the convergence rate is evidently faster than the cited ILC methods. Moreover, it is seen that the smaller parameter setting ($\alpha_0 = 0.4, \alpha_c = 0.4$, and $\lambda_s = 0.2$) procures better tracking performance leading to a smaller MSE in the initial batch run, in contrast with the larger parameter setting ($\alpha_0 = 0.7$, $\alpha_c = 0.8$ and $\lambda_s = 0.2$). It should be noted that the cited open-loop ILC method results in a notable steady-state tracking error in the initial batch run, due to no feedback control mechanism is included for the disturbance rejection.

Then suppose that the process model parameters like gain and time constant vary in a range of [0.8, 1.2] during a batch run. Figure 9.3 shows the resulting MSE indices under nonrepetitive process gain and/or time constant, by using different ILC methods for comparison. It is observed that the proposed ILC method obtains apparently smaller tracking errors along either the time- or batch-direction. In the meantime, the bounded tracking objectives in (9.4) and (9.5) are satisfied, therefore omitted for brevity. It should be noted that the larger parameter setting ($\alpha_0 = 0.7$,$\alpha_c = 0.8$) results in better robustness for batch run, exactly verifying the controller tuning guideline given in the previous Sect. 9.2.

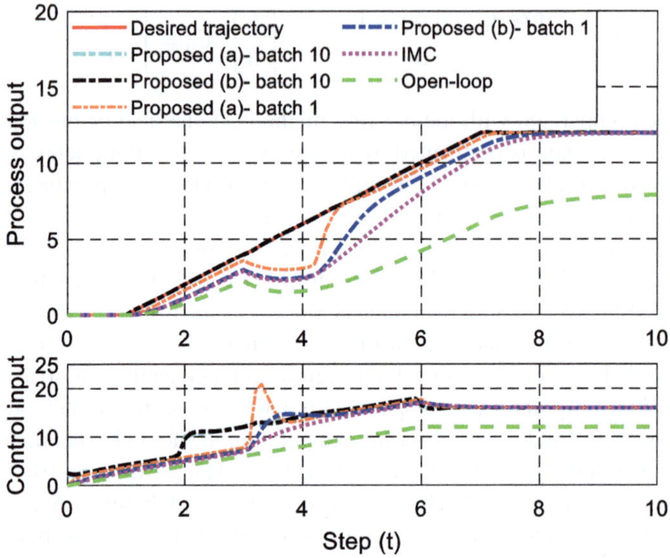

Fig. 9.2 Tracking results for the nominal case

Table 9.1 MSE indices of different methods for the nominal case

Batch	Proposed (a)	Proposed (b)	Liu et al. (2010)	Shokri-Ghaleh et al. (2021)	Open-loop ILC
1	**0.7524**	2.0550	3.0726	3.0726	3.7791
2	1.1018	0.8629	0.8205	0.9446	0.6201
4	0.0040	0.0357	0.0638	0.1417	0.0992
6	2.6×10^{-4}	0.0084	0.0069	0.0512	0.0157
8	5.2×10^{-5}	0.0030	0.0016	0.0352	0.0074
10	1.3×10^{-5}	0.0014	7.5×10^{-4}	0.0313	0.0046

9.5 Experiments on the Batch Temperature Regulation of a 4-Litre Crystallizer

Figure 9.4 shows a temperature regulation device and a 4-L crystallizer containing 2-L distilled water for real experiment, where the temperature regulation device consists of a thermostatic circulator filled with 6-L medium solution composed of purified water and ethylene glycol in a proportion of 3:2, an electric heater with a rated power of 2000(W), a cooling compressor with a rated power of 1000(W), a zero-crossing solid state relay module with the pulse-width modulator (PWM) for altering the heating power, a PT100 thermometer, and a programmable logic controller (PLC) plus voltage-to-frequency converter for supervising the PWM and cooling compressor.

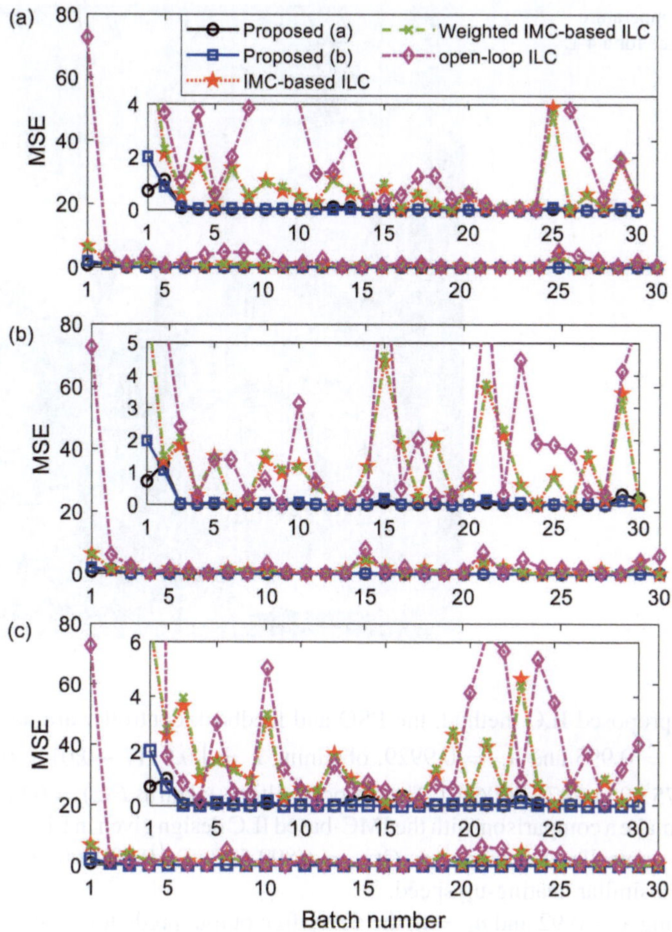

Fig. 9.3 MSE indices under different processes uncertainties: (a) nonrepetitive process gain; (b) nonrepetitive time constant; (c) nonrepetitive process gain and time constant

For the control test, the main task is specified as heating up the crystallizer solution should be from the room temperature (25 °C) to 60 °C as quickly as possible, and then maintaining the operating temperature of 60 °C against external disturbance (e.g., feeding the crystallization solution of distilled water at the room temperature).

To identify a dynamic model of the temperature response for control design, an open-loop step response test was conducted by fully turning on the electric heater to heating up the crystallizer solution from 25 °C to 60 °C. By taking the sampling period of $T_s = 3$(s) for control execution, the identification method in the reference (Liu and Gao 2012) is adopted to acquire a transfer function model

$$G(z) = \frac{1.8290 \times 10^{-6}(z + 0.9989)}{(z - 1)(z - 0.9967)} z^{-38}.$$

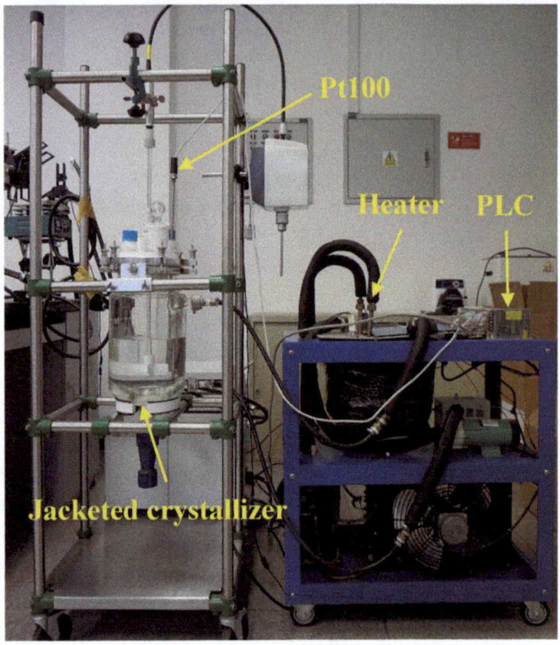

Fig. 9.4 Temperature control device for a 4-L crystallizer

In the proposed ILC method, the PSO and feedback controller are designed by taking $\alpha_0 = 0.995$ and $\alpha_c = 0.9929$, obtaining $L = \begin{bmatrix} 0.0117 & -0.0116 & 0.0342 \end{bmatrix}^T$, $K = \begin{bmatrix} 2975.2954 & 2971.4006 & 1 \end{bmatrix}$. The set-point filter is taken as $F(z) = 0.0689z/(z-0.995)$ to make a comparison with the IMC-based ILC design given in Liu et al. (2010) where the controller is designed as $C(s) = s(899.5s+1)/0.000366(800s+1)^3$ for obtaining a similar heating-up speed.

By taking $\lambda = 0.92$ and $n_k = 0$, the delay-free output predictor is designed as

$$F_1(z) = \Phi(z)\Lambda(z),\ F_2(z) = \frac{10.9255z^2 - 21.4479z + 10.5287}{z^2 - 1.8400z + 0.8464},$$

where $C_p = \begin{bmatrix} 1 & -1.8400 & 0.8464 \end{bmatrix}$, $\Phi(z) = C_p \sum_{i=1}^{38} A_p^{i-1} B_p z^{-i}$

$$\Lambda(z) = \frac{1.8290 \times 10^{-6} z^2 - 2.0322 \times 10^{-9} z - 1.8270 \times 10^{-6}}{z^2 - 1.8400z + 0.8464},$$

$$A_p = \begin{bmatrix} 2.9967 & -2.9933 & 0.9967 \\ 1 & 0 & 0 \\ 0 & 1 & 0 \end{bmatrix},\ B_p = \begin{bmatrix} 1 \\ 0 \\ 0 \end{bmatrix}.$$

During the heat-up process of the crystallizer solution, a load disturbance is mimicked by adding 300mL solution of distilled water at room temperature into the crystallizer at 12000s for each batch.

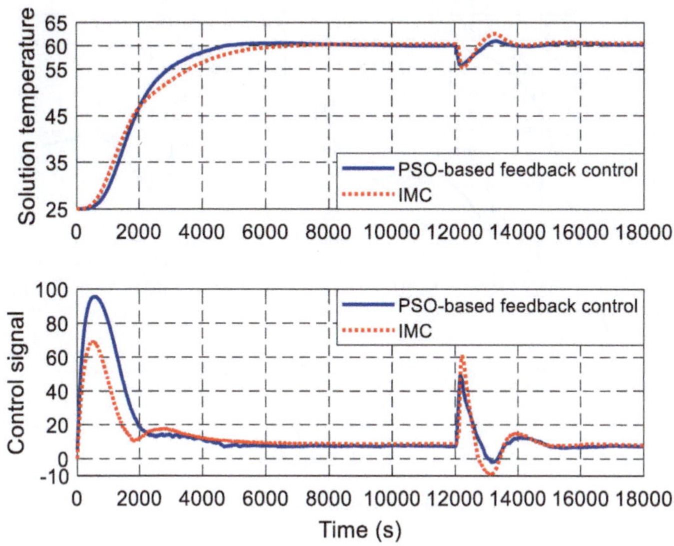

Fig. 9.5 Experiment results by two control methods for the initial batch

In the first batch, only the feedback control structure with PSO was implemented in the proposed ILC method, and in contrast, only the IMC structure was used in the indirect-type ILC (Liu et al. 2010). Figure 9.5 shows the experimental results. It is observed that the proposed PSO-based closed-loop control system results in obviously improved set-point tracking performance and disturbance rejection, in comparison with that of the IMC structure presented in Liu et al. (2010), owing to the use of PSO for estimating process uncertainties and disturbance to make counteraction. It should be noted that the classical open-loop ILC scheme could not maintain the temperature control system stability in each batch owing to lacking the feedback control mechanism, and thus is omitted for comparison.

Then, from the second batch and on, a set-point learning gain $\Gamma = 1$ is taken to execute the proposed ILC scheme, which satisfies the convergence condition in (9.27). Figure 9.6 shows the batch run results. It is observed that the heat-up process from 25 °C to 60 °C only needs about 3200 s after 15 batch runs, compared to almost 4700 s spent by the initial batch. The results demonstrate that over 1500 s could be saved for the heating-up process by the proposed set-point learning control scheme. It should be noted that the overshoot in the set-point tracking arises from the upper limit of the control input as shown in Fig. 9.6. However, the overshoot is gradually alleviated by taking the proposed set-point updating law. It is a trade-off that has to be made between a small overshoot and a fast heating-up speed, due to the fact that there is a control saturation constraint on the heating power ($0 \leq u(t) \leq 100$). Since a longer saturation in the control signal may provoke system instability, such saturation should be avoided in practice. Regarding the disturbance rejection, almost 5000 s are needed in the initial batch to recover the solution temperature to the desired

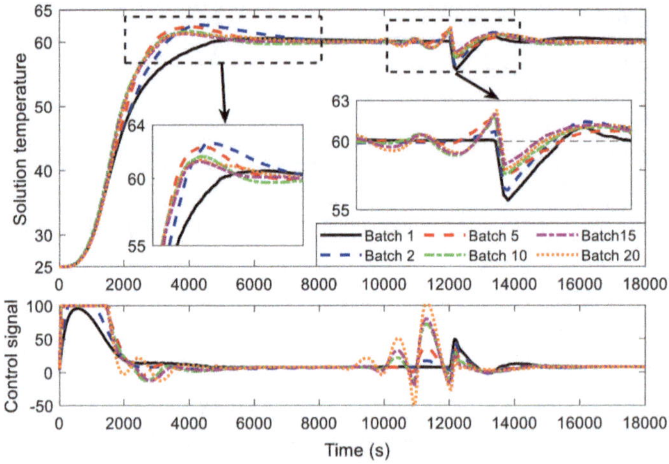

Fig. 9.6 Experimental results under the proposed ILC scheme

operating temperature zone of $(60 \pm 0.2)^{\circ}C$ under the load disturbance of feeding distilled water. Based on 15 batch runs, over 30 min could be saved for recovering the solution temperature to the aforementioned temperature zone, while the peak of disturbance response could be significantly reduced. Note that it may not further improve the disturbance rejection performance by taking more batch runs, due to the presence of time- and batch-varying process uncertainties, which may be verified by the control result of the 20th batch as shown in Fig. 9.6. However, the control system stability maintains well under such time-varying uncertainty, owing to the PSO-based feedback control structure in the proposed ILC scheme.

Figure 9.7 shows a comparison of the optimal batch run results by using the proposed ILC method and the IMC-based ILC in Liu et al. (2010), respectively. It is observed that over 1000 s could be saved for heating up the crystallization solution by using the proposed ILC method, and moreover, the peak of temperature drop in the disturbance rejection is obviously smaller, compared to that of the IMC-based ILC presented in Liu et al. (2010). It should be mentioned that if the MSE criterion is adopted for evaluating the tracking performance, the 10th batch result should be regarded as the optimal, which is also shown in Fig. 9.7. In contrast, any further batch run by the ILC given in Liu et al. (2010) could not obtain better performance due to the occurrence of batch-varying process uncertainties, and thus are omitted for comparison.

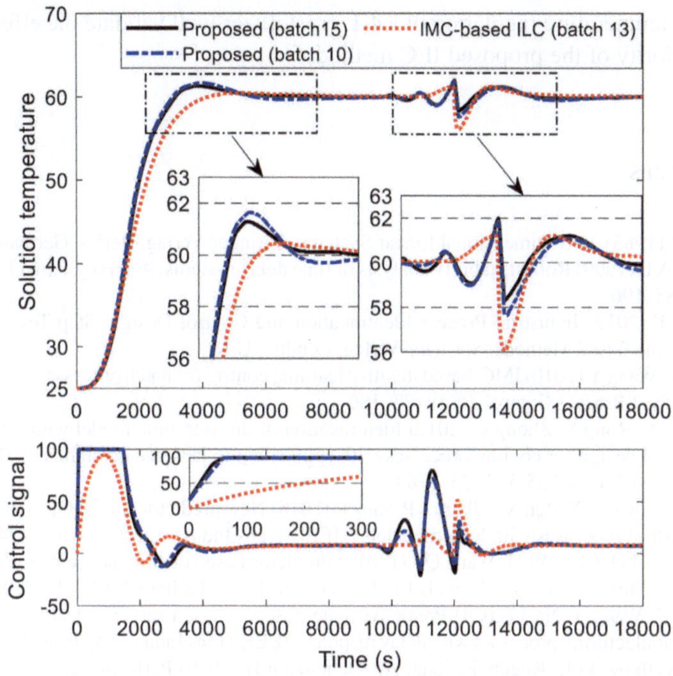

Fig. 9.7 Comparison of optimal tracking results by two ILC methods

9.6 Summary

For industrial batch manufacturing systems with delay response subject to nonrepetitive disturbances and process uncertainties, an indirect-type ILC scheme based on the design of PSO (Liu et al. 2024) is presented in this chapter. The feedback control loop with PSO is essentially a 2DOF control structure, which could separately adjust the output tracking performance and disturbance rejection. In the meantime, the influence from nonrepetitive disturbances and process uncertainties could be effectively alleviated for batch run. For the convenience of engineering application, a simple P-type learning controller is designed to only update the system set-point for batch performance optimization.

The convergence of output tracking error is clarified with rigorous analysis. The sufficient conditions for both delay-independent and delay-dependent cases are established to guarantee robust convergence of the tracking error, together with the boundedness of system input and state.

The control tests on a benchmark batch process example from the existing literature well demonstrate the merits of the proposed ILC scheme, in contrast with the previously developed direct-type ILC methods, where the control performance in each batch is coupled with that of batch run, or open-loop ILC methods that overlook the control performance in the time direction of each batch. The experiment results on temperature regulation of a 4-L crystallizer well validate the effectiveness and superiority of the proposed ILC method.

References

Kaczorek T (1985) Two-Dimensional Linear Systems. Springer-Verlag, Berlin, Germany

Kharitonov VL (1999) Robust stability analysis of time delay systems: A survey. Annu Rev Control 23(2):185–196

Liu T, Gao F (2012) Industrial Process Identification and Control Design: Step-Test and Relay-Experiment-Based Methods. Springer-Verlag, London, U.K.

Liu T, Gao F, Wang Y (2010) IMC-based iterative learning control for batch processes with uncertain time delay. J Process Control 20(2):173–180

Liu T, Dong S, Rong S, Zhong C (2017) Identification of discrete-time model with integer delay and control design for cooling processes with application to jacketed crystallizers. IEEE Trans Control Syst Technol 25(5):1775–1789

Liu T, García P, Chen Y, Ren X, Albertos P, Sanz R (2018) New predictor and 2DOF control scheme for industrial processes with long time delay. IEEE Trans Industr Electron 65(5):4247–4256

Liu T, Hao S, Li D, Chen W-H, Wang Q-G (2019) Predictor-based disturbance rejection control for sampled systems with input delay. IEEE Trans Control Syst Technol 27(2):772–780

Liu T, Hao S, Wang Y, Na J (2024) Predictive state observer based set-point learning control for batch manufacturing processes with delay response. IEEE Trans Industr Electron 71(1):788–797

Mandra S, Gałkowski L, Rogers E, Rauh A, Aschemann H (2019) Performance-enhanced robust iterative learning control with experimental application to PMSM position tracking. IEEE Trans Control Syst Technol 27(4):1813–1819

Meng D, Moore K (2017) Convergence of iterative learning control for SISO nonrepetitive systems subject to iteration-dependent uncertainties. Autom 79:167–177

Shokri-Ghaleh H, Ganjefar S, Shahri A (2021) Robust iterative learning control for uncertain continuous-time system with input delay and random iteration-varying uncertainties. IET Control Theory Appl 15(13):1749–1761

Shokri-Ghaleh H, Ganjefar S, Shahri A (2023) Extension of iterative learning control design for batch processes with time-delay in the input subject to random cycle-varying uncertainties. J Franklin Inst 360(12):8528–8549

Tao H, Paszke W, Yang H, Gałkowski K (2019) Finite frequency range robust iterative learning control of linear discrete system with multiple time-delays. J Franklin Inst 356(5):2690–2708

Tian D, Song X (2020) Iterative learning control for a type of modified Smith predictor. J Dyn Syst, Meas, Control 142(5):041009

Chapter 10
Robust ILC Design Under Process State Delay

10.1 Introduction

The state delay is usually associated with batch processes with a time delay in response to the setpoint change, when these processes are described by state-space models for control system design and batch run optimization. In practical applications, e.g., polymer injection molding and robotic manipulators, time delayed state variables that reflect important process response characteristics may be required under supervision along with the process output(s). Also note that time delay in control input or output measurement may be transformed to state delay in a state-space model for analysis (Chen and Gu 2000; Seborg et al. 2016). For the purpose of realizing no steady tracking error in both the time direction within a batch and the batchwise direction from cycle to cycle, ILC design for a batch process or repetitive system may be transformed to a control scheme for a class of 2D systems with two temporal directions (Owens et al. 2000). Based on the 2D repetitive system theory, a few robust ILC methods were developed for time-invariant linear repetitive systems with state delay and uncertainties (Li et al. 2005; Wang et al. 2013a, 2013b). For batch processes with time-varying state delays and uncertainties, a static output feedback based PD-type robust ILC was presented by using a heuristic approach to iteratively calculate the feasible ILC gains (Wei et al. 2022). In contrast, an adaptive ILC scheme by integrating a P-type feedback structure was suggested (Wu et al. 2023) for nonlinear multiple-input multiple-output systems subject to unknown state delays and non-repetitive uncertainties.

To address model-plant mismatches and time-varying process uncertainties from batch to batch, a number of ILC methods have been developed in the past decades (Lee et al. 2000; Shi et al. 2005a, b, c; Zhang 2008). By comparison, Chen and Kong (2009) presented an on-line adaptive ILC strategy to optimize tracking performance, based on estimating the minimum variance bounds of tracking error. Furthermore, Liu et al. (2010) gave an internal model control (IMC) based ILC method to cope with uncertain input delay. Robust convergence constraints for ILC design and stability

T. Liu et al., *Robust Iterative Learning Control of Industrial Batch Systems*, Intelligent Control and Learning Systems 22, https://doi.org/10.1007/978-981-96-9778-6_10

analysis can be seen in the references (Huang et al. 2002; Harte et al. 2005; Ahn et al. 2007; Tayebi 2007; Dabkowski et al. 2009; Wijdeven et al. 2009; Zhuang et al. 2023; Zhou et al. 2023).

In this Chapter, a robust closed-loop ILC method is presented for batch processes with state delay and time-varying uncertainties, in order to realize robust tracking and on-line optimization. To reflect time-varying uncertainties of the process dynamic response from batch to batch, norm-bounded process uncertainties are prescribed in terms of a state-space model, together with estimated lower and upper bounds of state delay variation. To facilitate closed-loop ILC design, the state-space process description is equivalently transformed into a 2D system description. For the convenience of implementation, only measured output errors of current and previous cycles are used to design an ILC scheme composed of dynamic output feedback plus feedforward control. By introducing a comprehensive 2D difference Lyapunov function to keep monotonical decrease of state energy along both the time and batch directions, sufficient conditions for holding robust stability of the closed-loop ILC system are established in terms of LMI constraints. By solving these LMI constraints, the 2D ILC controller is correspondingly formulated, together with an adjustable robust $\mathcal{H}_\infty$ performance level. A practical example of injection molding is given to demonstrate the effectiveness of the proposed ILC method.

10.2 Robust 2D ILC Design

Consider a batch process with state delay and time-varying uncertainties from batch to batch described by

$$\begin{cases} x(t+1,k) = \left[A_m + \Delta\tilde{A}(t,k)\right]x(t,k) + \left[A_{hm} + \Delta\tilde{A}_h(t,k)\right]x(t-h(t),k) \\ \quad + Bu(t,k), \\ y(t,k) = Cx(t,k), \quad 0 \le t \le N, \\ x(0,k) = x(0), \quad k = 0, 1, \ldots, \end{cases} \tag{10.1}$$

where $\Delta\tilde{A}(t,k)$ and $\Delta\tilde{A}_h(t,k)$ denotes time-varying uncertainties from batch to batch. $h(t) = \hat{h} + \Delta h(t)$ indicates a state delay that is also uncertain in a range estimated as $[h_m, h_M]$, and $\hat{h}$ is a nominal value modeled for controller design. Note that $h(t)$ is hereafter abbreviated as h for simplicity when no confusion will be caused. N is the time period of each batch run; $x(0)$ denotes the initial resetting condition of each batch. Practically, the process uncertainty bounds may be specified as $\Delta\tilde{A}(t,k) = \overline{F}_{1,1}\tilde{\Theta}_1(t)\overline{F}_{1,2}$, $\Delta\tilde{A}_h(t,k) = \overline{F}_{3,1}\tilde{\Theta}_3(t)\overline{F}_{3,2}$, where $F_{i,1}$ and $F_{i,2}$ are constant matrices, and $\tilde{\Theta}_i^{\mathrm{T}}(t)\tilde{\Theta}_i(t) < \mathbf{I}$, $i = 1, 3$ for $0 \le t \le N$. Note that other process uncertainties, such as from the input actuator and output measurement, may also be lumped into $\Delta\tilde{A}(t,k)$ and $\Delta\tilde{A}_h(t,k)$ for analysis.

It should be noted that nonlinear process characteristics, e.g., high-frequency dynamics, which cannot be represented by a finite-dimensional linear space model as above, are hereby considered into the norm-bounded uncertainties or load disturbance to treat with, as will be illustrated in the later example in Sect. 10.4. For those batch processes with high nonlinearity that cannot be effectively described in this manner, a piecewise modeling strategy may be adopted for application of the robust ILC control to be developed, or alternatively, based on a nonlinear model likely obtained from the first-principle approach, nonlinear model predictive control in conjunction with ILC may be developed for robust control and perfection tracking, as introduced in the literature (Nagy et al. 2007; Wang et al. 2009).

For the above batch process, an ILC updating law generally has the following form

$$\Sigma_{\text{ILC}}:\begin{cases} u(t,k)=u(t,k-1)+\delta u(t,k), \\ u(0,0)=u(0), \quad 0\le t\le N, \quad k=0,1,\ldots, \end{cases} \tag{10.2}$$

where $\delta u(t, k)$ is the updating law to be determined in the current batch (k), and $u(0)$ denotes the initial value of iteration that is usually reset to zero for implementation.

To facilitate closed-loop ILC design to accommodate for time-varying uncertainties from batch to batch, the output tracking error in the current batch is defined as

$$e(t,k)\triangleq y(t,k)-y_{\text{d}}(t), \tag{10.3}$$

where $y_{\text{d}}(t)$denotes the desired reference trajectory for each batch.

It follows from (10.1–10.3) that

$$e(t,k)=e(t,k-1)+C\delta x(t,k), \tag{10.4}$$

$$\begin{aligned}\delta x(t+1,k)=&\left[A_m+\Delta\tilde{A}(t,k)\right]\delta x(t,k)+\left[A_{hm}+\Delta\tilde{A}_h(t,k)\right]\delta x(t-h,k)\\ &+B\delta u(t,k)+\omega(t,k),\end{aligned} \tag{10.5}$$

where

$$\begin{aligned}\omega(t,k)=&[\Delta\tilde{A}(t,k)-\Delta\tilde{A}(t,k-1)]x(t,k-1)\\ &+\Delta\tilde{A}_h(t,k)x(t-h(t,k),k-1)\\ &-\Delta\tilde{A}_h(t,k-1)x(t-h(t,k-1),k-1).\end{aligned} \tag{10.6}$$

It is obvious that $\omega(t,k)\neq 0$ in the presence of nonrepetitive model uncertainties, and therefore, is viewed as a nonrepeatable load disturbance to treat with. Meanwhile, it is also used to denote external disturbance, for the analysis of control system stability and performance.

To meet with a practical requirement of no steady-state tracking error for each batch, the information of tracking error integral is introduced as an additional state variable for ILC design, defined by

$$\Sigma e(t,k) \triangleq \sum_{i=0}^{t} e(i,k), \quad 1 \le t \le N. \tag{10.7}$$

It follows that

$$\Sigma e(t,k) = \Sigma e(t-1,k) + e(t,k). \tag{10.8}$$

Thereby, an equivalent 2D system description of the above batch process, together with two augmented state variables of tracking error information, can be expressed as

$$\Sigma P_\Delta : \begin{cases} \begin{bmatrix} \delta x(t+1,k) \\ e(t+1,k) \\ \Sigma e(t,k) \end{bmatrix} = (A_1 + \Delta A_1) \begin{bmatrix} \delta x(t,k) \\ e(t,k) \\ \Sigma e(t-1,k) \end{bmatrix} + A_2 \begin{bmatrix} \delta x(t+1,k-1) \\ e(t+1,k-1) \\ \Sigma e(t,k-1) \end{bmatrix} \\ +(A_3 + \Delta A_3) \begin{bmatrix} \delta x(t-h,k) \\ e(t-h,k) \\ \Sigma e(t-1-h,k) \end{bmatrix} + B_g \delta u(t,k) + D_g \omega(t,k), \\ y_e(t,k) = \begin{bmatrix} e(t,k-1) \\ e(t,k) \\ \Sigma e(t,k) \end{bmatrix} = G \begin{bmatrix} \delta x(t,k) \\ e(t,k) \\ \Sigma e(t-1,k) \end{bmatrix}, 1 \le t \le N, \\ z(t,k) = e(t,k) = L \begin{bmatrix} \delta x(t,k) \\ e(t,k) \\ \Sigma e(t-1,k) \end{bmatrix}, \\ \delta x(0,k) = 0, k = 0, 1, \ldots . \end{cases} \tag{10.9}$$

where $L = [\mathbf{0}\ 1\ 0]$, and

$$A_1 = \begin{bmatrix} A_m & 0\ 0 \\ CA_m & 0\ 0 \\ \mathbf{0} & 1\ 1 \end{bmatrix}, A_2 = \begin{bmatrix} \mathbf{0}\ 0\ 0 \\ \mathbf{0}\ 1\ 0 \\ \mathbf{0}\ 0\ 0 \end{bmatrix}, A_3 = \begin{bmatrix} A_{hm} & 0\ 0 \\ CA_{hm} & 0\ 0 \\ \mathbf{0} & 0\ 0 \end{bmatrix},$$

$$B_g = \begin{bmatrix} B \\ CB \\ 0 \end{bmatrix}, \ D_g = \begin{bmatrix} \mathbf{I} \\ C \\ \mathbf{0} \end{bmatrix}, \ G = \begin{bmatrix} -C\ 1\ 0 \\ \mathbf{0}\ \ 1\ 0 \\ \mathbf{0}\ \ 1\ 1 \end{bmatrix}, \ \Delta A_i = F_{i,1}\Theta_i(t)F_{i,2},$$

$$F_{i,1} = \begin{bmatrix} \overline{F}_{i,1} & 0\ 0 \\ C\overline{F}_{i,1} & 0\ 0 \\ 0 & 0\ 0 \end{bmatrix}, \ F_{i,2} = \mathbf{I}, \ \Theta_i^{\mathrm{T}}(t)\Theta_i(t) \le \mathbf{I}, \quad i = 1, 3.$$

From a view of control design, all the state variables in (10.9), denoted by $x_g(t,k) = \left[\delta x^T(t,k),\ e(t,k),\ \sum e(t-1,k)\right]^T$, should be stabilized to final steady-state point of the origin (or a constant value for the tracking error integral) in the state space. For batch process operation, individual measurements of all the process state variables are usually inaccurate or even unavailable in practice. Only measured output tracking errors in current and previous cycles are therefore used for ILC design to facilitate practical implementation. Note that the last cycle error, $e(t, k-1)$ in $y_e(t,k)$, is used for the feedforward control design in the batch direction, while $e(t,k)$ and $\sum e(t,k)$ are used for the feedback control design in the time direction (during the current batch). $z(t,k)$ indicates measured tracking error in the current batch to be minimized against load disturbance, i.e.

$$\|z(t,k)\|_2 \le \gamma \|\omega(t,k)\|_2, \tag{10.10}$$

where γ is a prescribed level of disturbance attenuation, i.e., a robust $\mathcal{H}_\infty$ control performance level.

Based on the above 2D system description, a closed-loop ILC scheme is proposed as shown in Fig. 10.1, where a 2D ILC controller is constructed in the form of

$$\Sigma_{\text{2D - ILC}}: \begin{cases} x_c(t+1,k) = A_{c1}x_c(t,k) + A_{c2}x_c(t+1,k-1) + A_{c3}x_c(t-\hat{h},k) \\ \qquad + B_{c1}y_e(t,k) + B_{c2}y_e(t+1,k-1) + B_{c3}y_e(t-\hat{h},k), \\ \delta u(t,k) = C_{c1}x_c(t,k) + C_{c2}x_c(t+1,k-1) + C_{c3}x_c(t-\hat{h},k) \\ \qquad + D_{c1}y_e(t,k) + D_{c2}y_e(t+1,k-1) + D_{c3}y_e(t-\hat{h},k). \end{cases} \tag{10.11}$$

Note that the controller output, $\delta u(t,k)$,is the ILC updating law which corresponds to the control input of the above 2D system description of a batch process.

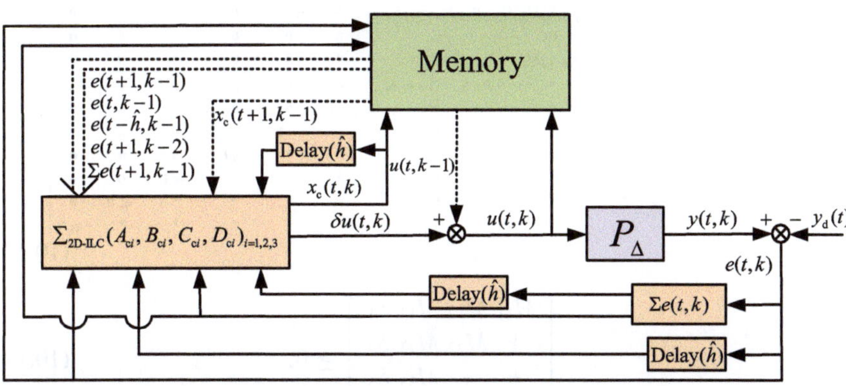

Fig. 10.1 Schematic of the robust 2D ILC scheme

By substituting (10.11) into (10.9) and letting $\hat{x}(t,k) = [x_{\mathrm{g}}^{\mathrm{T}}(t,k), x_{\mathrm{c}}^{\mathrm{T}}(t,k)]^{\mathrm{T}}$, the closed-loop 2D control system based on using $h(t) = \hat{h}$ (by lumping the influence from batch-to-batch state delay mismatch into $\omega(t,k)$ as shown in (10.6) for consideration) can be written as

$$\Sigma_{\mathrm{ILC\text{-}BP}}: \begin{cases} \hat{x}(t+1,k) = (\hat{A}_1 + \Delta\hat{A}_1)\hat{x}(t,k) + \hat{A}_2\hat{x}(t+1,k-1) \\ \qquad\qquad\qquad + (\hat{A}_3 + \Delta\hat{A}_3)\hat{x}(t-\hat{h},k) + \hat{D}\omega(t,k), \\ y(t,k) = \hat{G}\hat{x}(t,k), \\ \hat{z}(t,k) = \hat{L}\hat{x}(t,k), \end{cases} \tag{10.12}$$

where $\hat{G} = GH, \hat{L} = LH, \hat{D} = H^{\mathrm{T}}D_{\mathrm{g}}, H = \begin{bmatrix}\mathbf{I}\ \mathbf{0}\end{bmatrix}$, and

$$\hat{A}_i = \begin{bmatrix} A_i + B_{\mathrm{g}}D_{\mathrm{c}i}G & B_{\mathrm{g}}C_{\mathrm{c}i} \\ B_{\mathrm{c}i}G & A_{\mathrm{c}i} \end{bmatrix}, \quad i = 1,2,3, \quad \Delta\hat{A}_i = \hat{F}_{i,1}\Theta_i\hat{F}_{i,2},$$

$$\hat{F}_{i,1} = \begin{bmatrix} F_{i,1}^{\mathrm{T}}\ \mathbf{0}\end{bmatrix}^{\mathrm{T}}, \quad \hat{F}_{i,2} = \begin{bmatrix} F_{i,2}\ \mathbf{0}\end{bmatrix}, \quad i = 1,3.$$

To derive a robust ILC updating law for the above closed-loop system, the following theorem is given in the first place which will play a key role in the derivation:

Theorem 10.1 The closed-loop 2D system in (10.12) is asymptotically stable for $h(t) \in [h_m, h_M]$ and has robust $\mathcal{H}_\infty$ performance level in (10.10), if there exist $P > Q > 0, W > 0, R > 0$, matrices, M_{ij} and S_i $(i,j = 1,2,3)$ with appropriate dimensions, and positive scalars, ε_1 and ε_3, such that the following two LMI constraints are satisfied

$$\begin{bmatrix} -P & \mathbf{0} & P\hat{A}_1 & P\hat{A}_2 & P\hat{A}_3 & \mathbf{0} & P\hat{D} & \mathbf{0} & P\hat{F}_{1,1} & P\hat{F}_{3,1} & \mathbf{0} & \mathbf{0} \\ * & -h_M W & h_M W(\hat{A}_1 - \mathbf{I}) & h_M W\hat{A}_2 & h_M W\hat{A}_3 & \mathbf{0} & h_M W\hat{D} & \mathbf{0} & h_M W\hat{F}_{1,1} & h_M W\hat{F}_{3,1} & \mathbf{0} & \mathbf{0} \\ * & * & \Gamma_{11} & \mathbf{0} & S_2^{\mathrm{T}} + h_M M_{12} & -S_1 + S_3^{\mathrm{T}} + h_M M_{13} & \mathbf{0} & H^{\mathrm{T}}L^{\mathrm{T}} & \mathbf{0} & \mathbf{0} & \hat{F}_{1,2}^{\mathrm{T}} & \mathbf{0} \\ * & * & * & -P+Q & \mathbf{0} & \mathbf{0} & \mathbf{0} & \mathbf{0} & \mathbf{0} & \mathbf{0} & \mathbf{0} & \mathbf{0} \\ * & * & * & * & -R + h_M M_{22} & -S_2^{\mathrm{T}} + h_M M_{23} & \mathbf{0} & \mathbf{0} & \mathbf{0} & \mathbf{0} & \mathbf{0} & \mathbf{0} \\ * & * & * & * & * & -R - S_3 - S_3^{\mathrm{T}} + h_M M_{33} & \mathbf{0} & \mathbf{0} & \mathbf{0} & \mathbf{0} & \mathbf{0} & \hat{F}_{3,2}^{\mathrm{T}} \\ * & * & * & * & * & * & -\gamma\mathbf{I} & \mathbf{0} & \mathbf{0} & \mathbf{0} & \mathbf{0} & \mathbf{0} \\ * & * & * & * & * & * & * & -\gamma\mathbf{I} & \mathbf{0} & \mathbf{0} & \mathbf{0} & \mathbf{0} \\ * & * & * & * & * & * & * & * & -\varepsilon_1\mathbf{I} & \mathbf{0} & \mathbf{0} & \mathbf{0} \\ * & * & * & * & * & * & * & * & * & -\varepsilon_3\mathbf{I} & \mathbf{0} & \mathbf{0} \\ * & * & * & * & * & * & * & * & * & * & -\varepsilon_1^{-1}\mathbf{I} & \mathbf{0} \\ * & * & * & * & * & * & * & * & * & * & * & -\varepsilon_3^{-1}\mathbf{I} \end{bmatrix} < 0, \tag{10.13}$$

$$\begin{bmatrix} M_{11} & M_{12} & M_{13} & S_1 \\ * & M_{22} & M_{23} & S_2 \\ * & * & M_{33} & S_3 \\ * & * & * & W \end{bmatrix} \geq 0, \tag{10.14}$$

where $\Gamma_{11} = -Q + (h_M - h_m + 2)R + S_1 + S_1^{\mathrm{T}} + h_M M_{11}$.

Proof Firstly, the asymptotic stability of the unforced closed-loop 2D system (i.e., external $\omega(t,k) = 0$, $h(t,0) = \hat{h}$, $\Delta\hat{A}_1 = \Delta\hat{A}_3 = 0$) but with $h(t) \in [h_m, h_M]$ from batch to batch should be clarified. Denote three shorthands, $V_{i,j} = V(t+i, k+j)$, $\hat{x}_{i,j} = \hat{x}(t+i, k+j)$, $\eta_{l,0} = \hat{x}_{l+1,0} - \hat{x}_{l,0}$, for simplicity. Inspired by the Lyapunov–Krasovskii theorem for analyzing asymptotic stability of 1D systems (Gu et al. 2003), the following state energy functions are introduced

$$V_{1,0} = \hat{x}_{1,0}^{\mathrm{T}} P \hat{x}_{1,0}, \tag{10.15}$$

$$V_{0,0} = \hat{x}_{0,0}^{\mathrm{T}} Q \hat{x}_{0,0}, \tag{10.16}$$

$$V_{1,-1} = \hat{x}_{1,-1}^{\mathrm{T}} (P - Q) \hat{x}_{1,-1}, \tag{10.17}$$

$$V_{-h+1,0} = \sum_{l=-h}^{-1} \hat{x}_{l+1,0}^{\mathrm{T}} R \hat{x}_{l+1,0} + \sum_{j=-h_M}^{-h_m} \sum_{l=j}^{-1} \hat{x}_{l+1,0}^{\mathrm{T}} R \hat{x}_{l+1,0} + \sum_{j=-h_M}^{-1} \sum_{l=j}^{-1} \eta_{l+1,0}^{\mathrm{T}} W \eta_{l+1,0}, \tag{10.18}$$

$$V_{-h,0} = \sum_{I=-h}^{-1} \hat{x}_{l,0}^{\mathrm{T}} R \hat{x}_{l,0} + \sum_{j=-h_M}^{-h_m} \sum_{l=j}^{-1} \hat{x}_{l,0}^{\mathrm{T}} R \hat{x}_{l,0} + \sum_{j=-h_M}^{-1} \sum_{l=j}^{1} \eta_{l,0}^{\mathrm{T}} W \eta_{l,0}. \tag{10.19}$$

Correspondingly, it is desired that state energy increment along both the time and batch directions is negative, i.e.

$$\Delta V = \Delta V_1 + \Delta V_2 + \Delta V_3 + \Delta V_4 < 0, \tag{10.20}$$

where

$$\Delta V_1 = \hat{x}_{1,0}^{\mathrm{T}} P \hat{x}_{1,0} - \hat{x}_{0,0}^{\mathrm{T}} Q \hat{x}_{0,0} - \hat{x}_{1,-1}^{\mathrm{T}} (P - Q) \hat{x}_{1,-1}, \tag{10.21}$$

$$\Delta V_2 = \sum_{l=-h(t+1)}^{-1} \hat{x}_{l+1,0}^{\mathrm{T}} R \hat{x}_{l+1,0} - \sum_{l=-h(t)}^{-1} \hat{x}_{l,0}^{\mathrm{T}} R \hat{x}_{l,0}, \tag{10.22}$$

$$\Delta V_3 = \sum_{j=-h_M}^{-h_m} \sum_{l=j}^{-1} \hat{x}_{l+1,0}^{\mathrm{T}} R \hat{x}_{l+1,0} - \sum_{j=-h_M}^{-h_m} \sum_{l=j}^{-1} \hat{x}_{l,0}^{\mathrm{T}} R \hat{x}_{l,0}, \tag{10.23}$$

$$\Delta V_4 = \sum_{j=-h_M}^{-1} \sum_{l=j}^{-1} \eta_{l+1,0}^{\mathrm{T}} W \eta_{l+1,0} - \sum_{j=-h_M}^{-1} \sum_{l=j}^{-1} \eta_{l,0}^{\mathrm{T}} W \eta_{l,0}. \tag{10.24}$$

Note that $\Delta V_1 = V_{1,0} - V_{0,0} - V_{1,-1}$ reflects state energy increment from a 2D view of energy transfer, that is, state energy transferred to the state point, $i + j =$

$N+1$ (where N is any nonnegative integer), comes from two state points that satisfy $i+j=N$ along both the time and batch directions (Shi et al. 2005c). In addition, $\Delta V_2+\Delta V_3+\Delta V_4=V_{-h+1,0}-V_{-h,0}$ reflects the state energy increment related to state delay in the time direction during a batch.

It follows from (10.22)–(10.24) that

$$\begin{aligned}\Delta V_2 &= \hat{x}_{0,0}^{\mathrm{T}}R\hat{x}_{0,0}-\hat{x}_{-h(t),0}^{\mathrm{T}}R\hat{x}_{-h(t),0}+\sum_{l=-h(t+1)+1}^{-1}\hat{x}_{l,0}^{\mathrm{T}}R\hat{x}_{l,0}-\sum_{l=-h(t)+1}^{-1}\hat{x}_{l,0}^{\mathrm{T}}R\hat{x}_{l,0}\\ &= \hat{x}_{0,0}^{\mathrm{T}}R\hat{x}_{0,0}-\hat{x}_{-h,0}^{\mathrm{T}}R\hat{x}_{-h,0}+\sum_{l=-h(t+1)+1}^{-h_m}\hat{x}_{l,0}^{\mathrm{T}}R\hat{x}_{l,0}+\sum_{l=-h_m+1}^{-1}\hat{x}_{l,0}^{\mathrm{T}}R\hat{x}_{l,0}\\ &\quad -\sum_{l=-h(t)+1}^{-1}\hat{x}_{l,0}^{\mathrm{T}}R\hat{x}_{l,0}\le \hat{x}_{0,0}^{\mathrm{T}}R\hat{x}_{0,0}-\hat{x}_{-h,0}^{\mathrm{T}}R\hat{x}_{-h,0}+\sum_{l=-h_M+1}^{-h_m}\hat{x}_{l,0}^{\mathrm{T}}R\hat{x}_{l,0}\\ &\quad +\sum_{l=-h(t)+1}^{-1}\hat{x}_{l,0}^{\mathrm{T}}R\hat{x}_{l,0}-\sum_{l=-h(t)+1}^{-1}\hat{x}_{l,0}^{\mathrm{T}}R\hat{x}_{l,0}\\ &= \hat{x}_{0,0}^{\mathrm{T}}R\hat{x}_{0,0}-\hat{x}_{-h,0}^{\mathrm{T}}R\hat{x}_{-h,0}+\sum_{l=-h_M+1}^{-h_m}\hat{x}_{l,0}^{\mathrm{T}}R\hat{x}_{l,0},\end{aligned}\tag{10.25}$$

$$\Delta V_3=\sum_{j=-h_M}^{-h_m}\left(\hat{x}_{0,0}^{\mathrm{T}}R\hat{x}_{0,0}-\hat{x}_{j,0}^{\mathrm{T}}R\hat{x}_{j,0}\right)=(h_M-h_m+1)\hat{x}_{0,0}^{\mathrm{T}}R\hat{x}_{0,0}-\sum_{j=-h_M}^{-h_m}\hat{x}_{j,0}^{\mathrm{T}}R\hat{x}_{j,0},\tag{10.26}$$

$$\Delta V_4=\sum_{j=-h_M}^{-1}\left(\eta_{0,0}^{\mathrm{T}}W\eta_{0,0}-\eta_{j,0}^{\mathrm{T}}W\eta_{j,0}\right)=h_M\eta_{0,0}^{\mathrm{T}}W\eta_{0,0}-\sum_{j=-h_M}^{-1}\eta_{j,0}^{\mathrm{T}}W\eta_{j,0}.\tag{10.27}$$

Summing up (10.21), and (10.25)–(10.27) along the closed-loop 2D system expression in (10.12) and using the following identical equation for any matrices, S_1, S_2, and S_3 with appropriate dimensions

$$\chi=2\left(\hat{x}_{0,0}^{\mathrm{T}}S_1+\hat{x}_{-h,0}^{\mathrm{T}}S_2+\hat{x}_{-h_M,0}^{\mathrm{T}}S_3\right)\left(\hat{x}_{0,0}-\hat{x}_{-h_M,0}-\sum_{j=-h_M}^{-1}\eta_{j,0}\right)=0,\tag{10.28}$$

it yields

$$\Delta V=\Delta V+\chi<\xi^{\mathrm{T}}\hat{\Omega}_1\xi-\sum_{j=-h_M}^{-1}\zeta^{\mathrm{T}}\Omega_2\zeta,\tag{10.29}$$

where $\xi(t,k) = \left[\hat{x}_{0,0}^{\mathrm{T}}, \hat{x}_{1,-1}^{\mathrm{T}}, \hat{x}_{-h,0}^{\mathrm{T}}, \hat{x}_{-h_M,0}^{\mathrm{T}}\right]^{\mathrm{T}}$, $\zeta(t,k) = \left[\hat{x}_{0,0}^{\mathrm{T}}, \hat{x}_{-h,0}^{\mathrm{T}}, \hat{x}_{-h_M,0}^{\mathrm{T}}, \eta_{j,0}^{\mathrm{T}}\right]^{\mathrm{T}}$, Ω_2 is the matrix at the left-hand side of (10.14), and

$$\hat{\Omega}_1 = \begin{bmatrix} \hat{\Gamma}_{11} & \hat{\Gamma}_{12} & \hat{\Gamma}_{13} & -S_1 + S_3^{\mathrm{T}} + h_M M_{13} \\ * & \hat{\Gamma}_{22} & \hat{\Gamma}_{23} & \mathbf{0} \\ * & * & \hat{\Gamma}_{33} & -S_2 + h_M M_{23} \\ * & * & * & -R - S_3 - S_3^{\mathrm{T}} + h_M M_{33} \end{bmatrix},$$

$$\begin{aligned}
\hat{\Gamma}_{11} &= \hat{A}_1^{\mathrm{T}} P \hat{A}_1 - Q + (h_M - h_m + 2)R + h_M \left(\hat{A}_1^{\mathrm{T}} - \mathbf{I}\right) W (\hat{A}_1 - \mathbf{I}) + S_1 + S_1^{\mathrm{T}} + h_M M_{11}, \\
\hat{\Gamma}_{12} &= \hat{A}_1^{\mathrm{T}} P \hat{A}_2 + h_M \left(\hat{A}_1^{\mathrm{T}} - \mathbf{I}\right) W \hat{A}_2, \\
\hat{\Gamma}_{13} &= \hat{A}_1^{\mathrm{T}} P \hat{A}_3 + h_M \left(\hat{A}_1^{\mathrm{T}} - \mathbf{I}\right) W \hat{A}_3 + S_2^{\mathrm{T}} + h_M M_{12}, \\
\hat{\Gamma}_{22} &= \hat{A}_2^{\mathrm{T}} P \hat{A}_2 - P + Q + h_M \hat{A}_2^{\mathrm{T}} W \hat{A}_2, \\
\hat{\Gamma}_{23} &= \hat{A}_2^{\mathrm{T}} P \hat{A}_3 + h_M \hat{A}_2^{\mathrm{T}} W \hat{A}_3, \\
\hat{\Gamma}_{33} &= \hat{A}_3^{\mathrm{T}} P \hat{A}_3 + h_M \hat{A}_3^{\mathrm{T}} W \hat{A}_3 - R + h_M M_{22}.
\end{aligned} \tag{10.30}$$

By the Schur complement, one can transform (10.30) into a leading principal minor of the matrix at the left-hand side in (10.13), i.e.

$$\Omega_1 = \begin{bmatrix} -P & \mathbf{0} & P\hat{A}_1 & P\hat{A}_2 & P\hat{A}_3 & \mathbf{0} \\ * & -h_M W & h_M W(\hat{A}_1 - \mathbf{I}) & h_M W \hat{A}_2 & h_M W \hat{A}_3 & \mathbf{0} \\ * & * & \Gamma_{11} & \mathbf{0} & S_2^{\mathrm{T}} + h_M M_{12} & -S_1 + S_3^{\mathrm{T}} + h_M M_{13} \\ * & * & * & -P + Q & \mathbf{0} & \mathbf{0} \\ * & * & * & * & -R + h_M M_{22} & -S_2 + h_M M_{23} \\ * & * & * & * & * & -R - S_3 - S_3^{\mathrm{T}} + h_M M_{33} \end{bmatrix}. \tag{10.31}$$

Therefore, $\Delta V < 0$ is ensured if the LMI constraints in (10.13) and (10.14) hold, which guarantees asymptotic stability of the unforced closed-loop 2D system with state delay uncertainty.

Next, it should be proved that the robust $\mathcal{H}_\infty$ performance level in (10.10) is satisfied for $\omega(t,k) \neq 0$ together with other process uncertainties ($\Delta\hat{A}_1 \neq 0$ and $\Delta\hat{A}_3 \neq 0$). To this end, define the control objective function as

$$J_{\mathrm{BP}} = \sum_{t=0}^{N_1=N} \sum_{k=0}^{N_2\to\infty} \left[\gamma^{-1} \hat{z}^{\mathrm{T}}(t,k)\hat{z}(t,k) - \gamma \omega^{\mathrm{T}}(t,k)\omega(t,k)\right], \tag{10.32}$$

which can be equivalently rewritten as

$$\begin{aligned} J_{\mathrm{BP}} = &\sum_{t=0}^{N_1=N} \sum_{k=0}^{N_2\to\infty} \left[\gamma^{-1} \hat{z}^{\mathrm{T}}(t,k)\hat{z}(t,k) - \gamma \omega^{\mathrm{T}}(t,k)\omega(t,k) + \Delta V\right] \\ &- \sum_{t=0}^{N_1=N} \sum_{k=0}^{N_2\to\infty} \Delta V(t,k). \end{aligned} \tag{10.33}$$

Owing to the initial resetting condition of batch process operation as shown in (10.1), there stands $\hat{x}(t,k) = 0$ for $t \le 0$ and $\forall k$, and for $0 \le t \le N$ and $k \le 0$. It follows that

$$\begin{aligned}\sum_{t=0}^{N}\sum_{k=0}^{N_2}\Delta V_1(t,k) &= \sum_{t=0}^{N}\sum_{k=0}^{N_2}\big[\hat{x}_{1,0}^{\mathrm{T}}(P-Q)\hat{x}_{1,0} + \hat{x}_{1,0}^{\mathrm{T}}Q\hat{x}_{1,0} - \hat{x}_{0,0}^{\mathrm{T}}Q\hat{x}_{0,0} \\ &\quad -\hat{x}_{1,-1}^{\mathrm{T}}(P-Q)\hat{x}_{1,-1}\big] \\ &= \sum_{t=0}^{N}\hat{x}^{\mathrm{T}}(t+1,N_2)(P-Q)\hat{x}(t+1,N_2) \\ &\quad + \sum_{k=0}^{N_2}\hat{x}^{\mathrm{T}}(N,k)Q\hat{x}(N,k) > 0,\end{aligned}$$

$$\begin{aligned}\sum_{t=0}^{N}\sum_{k=0}^{N_2}\Delta V_2(t,k) &= \sum_{t=0}^{N}\sum_{k=0}^{N_2}\left(\sum_{l=-h}^{-1}\hat{x}_{l+1,0}^{T}R\hat{x}_{l+1,0} - \sum_{l=-h}^{-1}\hat{x}_{l,0}^{T}R\hat{x}_{l,0}\right) \\ &= \sum_{k=0}^{N_2}\sum_{l=-h}^{-1}\big[\hat{x}^{T}(N+l+1,k)R\hat{x}(N+l+1,k) \\ &\quad -\hat{x}^{T}(l,k)R\hat{x}(l,k)\big] \\ &= \sum_{k=0}^{N_2}\sum_{l=-h}^{-1}\hat{x}^{T}(N+l+1,k)R\hat{x}(N+l+1,k) > 0,\end{aligned}$$

$$\begin{aligned}\sum_{t=0}^{N}\sum_{k=0}^{N_2}\Delta V_3(t,k) &= \sum_{t=0}^{N}\sum_{k=0}^{N_2}\left(\sum_{j=-h_M}^{-h_m}\sum_{l=j}^{-1}\hat{x}_{l+1,0}^{T}R\hat{x}_{l+1,0} - \sum_{j=-h_M}^{-h_m}\sum_{l=j}^{-1}\hat{x}_{l,0}^{T}R\hat{x}_{l,0}\right) \\ &= \sum_{k=0}^{N_2}\sum_{j=-h_M}^{-h_m}\sum_{l=j}^{-1}\big[\hat{x}^{T}(N+l+1,k)R\hat{x}(N+l+1,k) \\ &\quad -\hat{x}^{T}(l,k)R\hat{x}(l,k)\big] \\ &= \sum_{k=0}^{N_2}\sum_{j=-h_M}^{-h_m}\sum_{l=j}^{-1}\hat{x}^{T}(N+l+1,k)R\hat{x}(N+l+1,k) > 0,\end{aligned}$$

$$\begin{aligned}\sum_{t=0}^{N}\sum_{k=0}^{N_2}\Delta V_4(t,k) &= \sum_{t=0}^{N}\sum_{k=0}^{N_2}\left(\sum_{j=-h_M}^{-1}\sum_{l=j}^{-1}\eta_{l+1,0}^{T}W\eta_{l+1,0} - \sum_{j=-h_M}^{-1}\sum_{l=j}^{-1}\eta_{l,0}^{T}W\eta_{l,0}\right) \\ &= \sum_{k=0}^{N_2}\sum_{j=-h_M}^{-h_m}\sum_{l=j}^{-1}\big[\eta_{N+l+1,0}^{T}W\eta_{N+l+1,0} - \eta_{l,0}^{T}W\eta_{l,0}\big]\end{aligned}$$

$$= \sum_{k=0}^{N_2} \sum_{j=-h_M}^{-h_m} \sum_{l=j}^{-1} \eta_{N+l+1,0}^{T} W \eta_{N+l+1,0} > 0.$$

Accordingly

$$\sum_{t=0}^{N_1=N} \sum_{k=0}^{N_2\to\infty} \Delta V(t,k) = \sum_{t=0}^{N_1=N} \sum_{k=0}^{N_2\to\infty} [\Delta V_1(t,k) + \Delta V_2(t,k) + \Delta V_3(t,k) + \Delta V_4(t,k)] > 0, \tag{10.34}$$

and therefore

$$J_{\mathrm{BP}} < \sum_{t=0}^{N} \sum_{k=0}^{N_2\to\infty} \left[\gamma^{-1}\hat{z}^{\mathrm{T}}(t,k)\hat{z}(t,k) - \gamma\omega^{\mathrm{T}}(t,k)\omega(t,k) + \Delta V\right]. \tag{10.35}$$

By substituting (10.29) into (10.35) along with the above process uncertainties, it yields

$$J_{\mathrm{BP}} < \tilde{\xi}^{\mathrm{T}}\tilde{\Omega}_1\tilde{\xi} - \sum_{j=-h_M}^{-1} \zeta^{\mathrm{T}}\Omega_2\zeta, \tag{10.36}$$

where $\tilde{\xi}(t,k) = \left[\hat{x}_{0,0}^{\mathrm{T}}, \hat{x}_{1,-1}^{\mathrm{T}}, \hat{x}_{-h,0}^{\mathrm{T}}, \omega^{\mathrm{T}}, \hat{x}_{-h_M,0}^{\mathrm{T}}\right]^{\mathrm{T}}$, ζ and Ω_2 are the same in (10.29), and

$$\tilde{\Omega}_1 = \begin{bmatrix} \tilde{\Gamma}_{11} & \tilde{\Gamma}_{12} & \tilde{\Gamma}_{13} & \tilde{\Gamma}_{14} & -S_1 + S_3^{\mathrm{T}} + h_M M_{13} \\ * & \tilde{\Gamma}_{22} & \tilde{\Gamma}_{23} & \tilde{\Gamma}_{24} & \mathbf{0} \\ * & * & \tilde{\Gamma}_{33} & \tilde{\Gamma}_{34} & -S_2 + h_M M_{23} \\ * & * & * & \tilde{\Gamma}_{44} & \mathbf{0} \\ * & * & * & * & -R - S_3 - S_3^{\mathrm{T}} + h_M M_{33} \end{bmatrix}. \tag{10.37}$$

$$
\begin{aligned}
&\tilde{\Gamma}_{11} = (\hat{A}_1 + \Delta\hat{A}_1)^{\mathrm{T}} P(\hat{A}_1 + \Delta\hat{A}_1) - Q + (h_M - h_m + 2)R \\
&+h_M(\hat{A}_1^{\mathrm{T}} + \Delta\hat{A}_1^{\mathrm{T}} - \mathbf{I})W(\hat{A}_1 + \Delta\hat{A}_1 - \mathbf{I}) + S_1 \\
&+S_1^{\mathrm{T}} + h_M M_{11} + \gamma^{-1} H^{\mathrm{T}} L^{\mathrm{T}} L H, \\
&\tilde{\Gamma}_{12} = (\hat{A}_1 + \Delta\hat{A}_1)^{\mathrm{T}} P\hat{A}_2 + h_M(\hat{A}_1^{\mathrm{T}} + \Delta\hat{A}_1^{\mathrm{T}} - \mathbf{I})W\hat{A}_2, \\
&\tilde{\Gamma}_{13} = (\hat{A}_1 + \Delta\hat{A}_1)^{\mathrm{T}} P(\hat{A}_3 + \Delta\hat{A}_3) + h_M(\hat{A}_1^{\mathrm{T}} + \Delta\hat{A}_1^{\mathrm{T}} - \mathbf{I}) \\
&W(\hat{A}_3 + \Delta\hat{A}_3) + S_2^{\mathrm{T}} + h_M M_{12}, \\
&\tilde{\Gamma}_{14} = (\hat{A}_1 + \Delta\hat{A}_1)^{\mathrm{T}} P\hat{D} + h_M(\hat{A}_1^{\mathrm{T}} + \Delta\hat{A}_1^{\mathrm{T}} - \mathbf{I})W\hat{D}, \\
&\tilde{\Gamma}_{22} = \hat{\Gamma}_{22}, \\
&\tilde{\Gamma}_{23} = \hat{A}_2^{\mathrm{T}} P(\hat{A}_3 + \Delta\hat{A}_3) + h_M\hat{A}_2^{\mathrm{T}} W(\hat{A}_3 + \Delta\hat{A}_3), \\
&\tilde{\Gamma}_{24} = \hat{A}_2^{\mathrm{T}} P\hat{D} + h_M\hat{A}_2^{\mathrm{T}} W\hat{D}, \\
&\tilde{\Gamma}_{33} = (\hat{A}_3 + \Delta\hat{A}_3)^{\mathrm{T}} P(\hat{A}_3 + \Delta\hat{A}_3) + h_M(\hat{A}_3 + \Delta\hat{A}_3)^{\mathrm{T}} \\
&W(\hat{A}_3 + \Delta\hat{A}_3) - R + h_M M_{22}, \\
&\tilde{\Gamma}_{34} = (\hat{A}_3 + \Delta\hat{A}_3)^{\mathrm{T}} P\hat{D} + h_M(\hat{A}_3 + \Delta\hat{A}_3)^{\mathrm{T}} W\hat{D}, \\
&\tilde{\Gamma}_{44} = \hat{D}^{\mathrm{T}} P\hat{D} + h_M\hat{D}^{\mathrm{T}} W\hat{D} - \gamma\mathbf{I}.
\end{aligned}
$$

By the Schur complement, one can transform $\Omega_1 < 0$ into

$$
\begin{bmatrix} \Omega_1 & \begin{bmatrix} P\hat{D} \\ h_M W\hat{D} \\ \mathbf{0} \\ \mathbf{0} \\ \mathbf{0} \\ \mathbf{0} \end{bmatrix} \\ * & -\gamma\mathbf{I} \end{bmatrix} + \begin{bmatrix} \mathbf{0} & \mathbf{0} & P\Delta\hat{A}_1 & \mathbf{0} & P\Delta\hat{A}_3 & \mathbf{0} & \mathbf{0} \\ * & \mathbf{0} & h_M W\Delta\hat{A}_1 & \mathbf{0} & h_M W\Delta\hat{A}_3 & \mathbf{0} & \mathbf{0} \\ * & * & \gamma^{-1} H^{\mathrm{T}} L^{\mathrm{T}} L H & \mathbf{0} & \mathbf{0} & \mathbf{0} & \mathbf{0} \\ * & * & * & \mathbf{0} & \mathbf{0} & \mathbf{0} & \mathbf{0} \\ * & * & * & * & \mathbf{0} & \mathbf{0} & \mathbf{0} \\ * & * & * & * & * & \mathbf{0} & \mathbf{0} \\ * & * & * & * & * & * & \mathbf{0} \end{bmatrix} < 0. \tag{10.38}
$$

Given $\Delta\hat{A}_i = \hat{F}_{i,1}\Theta_i\hat{F}_{i,2}$, $i = 1, 3$ in (10.12), it follows from Lemma 2.1 in Chap. 2 that

$$
\begin{bmatrix} P\hat{F}_{i,1} \\ h_M W\hat{F}_{i,1} \\ \mathbf{0} \\ \mathbf{0} \\ \mathbf{0} \\ \mathbf{0} \\ \mathbf{0} \end{bmatrix} \hat{\Theta}_i \begin{bmatrix} \mathbf{0} \\ \mathbf{0} \\ \hat{F}_{i,2}^{\mathrm{T}} \\ \mathbf{0} \\ \mathbf{0} \\ \mathbf{0} \\ \mathbf{0} \end{bmatrix}^{\mathrm{T}} + \begin{bmatrix} \mathbf{0} \\ \mathbf{0} \\ \hat{F}_{i,2}^{\mathrm{T}} \\ \mathbf{0} \\ \mathbf{0} \\ \mathbf{0} \\ \mathbf{0} \end{bmatrix} \hat{\Theta}_i^{\mathrm{T}} \begin{bmatrix} P\hat{F}_{i,1} \\ h_M W\hat{F}_{i,1} \\ \mathbf{0} \\ \mathbf{0} \\ \mathbf{0} \\ \mathbf{0} \\ \mathbf{0} \end{bmatrix}^{\mathrm{T}}
$$

$$\leq \varepsilon_i^{-1}\begin{bmatrix} P\hat{F}_{i,1} \\ h_M W\hat{F}_{i,1} \\ \mathbf{0} \\ \mathbf{0} \\ \mathbf{0} \\ \mathbf{0} \\ \mathbf{0} \end{bmatrix}\begin{bmatrix} P\hat{F}_{i,1} \\ h_M W\hat{F}_{i,1} \\ \mathbf{0} \\ \mathbf{0} \\ \mathbf{0} \\ \mathbf{0} \\ \mathbf{0} \end{bmatrix} + \varepsilon_i \begin{bmatrix} \mathbf{0} \\ \mathbf{0} \\ \hat{F}_{i,2}^{\mathrm{T}} \\ \mathbf{0} \\ \mathbf{0} \\ \mathbf{0} \\ \mathbf{0} \end{bmatrix}\begin{bmatrix} \mathbf{0} \\ \mathbf{0} \\ \hat{F}_{i,2}^{\mathrm{T}} \\ \mathbf{0} \\ \mathbf{0} \\ \mathbf{0} \\ \mathbf{0} \end{bmatrix}^{\mathrm{T}}, i = 1, 3. \tag{10.39}$$

Substituting (10.39) into (10.38) and using the Schur complement, one can obtain the LMI constraint shown in (10.13). Hence, $J_{\mathrm{BP}} < 0$ is ensured with the LMI constraints in (10.13) and (10.14), which guarantees the robust $\mathcal{H}_\infty$ control performance level shown in (10.10). This completes the proof. □

Now it comes in a position to give a robust 2D ILC algorithm in the following theorem.

Theorem 10.2 A sufficient condition for constructing a 2D ILC controller shown in (10.11) that can guarantee the closed-loop 2D control system in (10.12) robustly stable for $h(t) \in [h_m, h_M]$, together with the $\mathcal{H}_\infty$ control performance level γ, is that there exist $X > 0$, $Y > 0$, $Q > 0$, $R > 0$, matrices, $\{A_{ci}, B_{ci}, C_{ci}, D_{ci}\}_{i=1,2,3}$, $\hat{M}_{ij}$ and $\hat{S}_i (i, j = 1, 2, 3)$ with appropriate dimension and positive scalars λ, ε_1 and ε_3, such that the following matrix inequalities are satisfied

$$\begin{bmatrix} \Xi_{11} & \Xi_{12} \\ * & \Xi_{22} \end{bmatrix} < 0, \tag{10.40}$$

$$\begin{bmatrix} \hat{M}_{11} & \hat{M}_{12} & \hat{M}_{13} & \hat{S}_1 \\ * & \hat{M}_{22} & \hat{M}_{23} & \hat{S}_2 \\ * & * & \hat{M}_{33} & \hat{S}_3 \\ * & * & * & \begin{bmatrix} \lambda Y & \lambda \mathbf{I} \\ * & \lambda X \end{bmatrix} \end{bmatrix} \geq 0, \tag{10.41}$$

where

$$\Xi_{11} = \begin{bmatrix} -Y & -\mathbf{I} & \mathbf{0} & \mathbf{0} & \Pi_{1,1} & \Pi_{2,1} & \Pi_{1,2} & \Pi_{2,2} & \Pi_{1,3} & \Pi_{2,3} \\ * & -X & \mathbf{0} & \mathbf{0} & \Pi_{3,1} & \Pi_{4,1} & \Pi_{3,2} & \Pi_{4,2} & \Pi_{3,3} & \Pi_{4,3} \\ * & * & -h_M\lambda Y & -h_M\lambda \mathbf{I} & h_M\lambda\Pi_{1,1} - h_M\lambda Y & h_M\lambda\Pi_{2,1} - h_M\lambda \mathbf{I} & h_M\lambda\Pi_{1,2} & h_M\lambda\Pi_{2,2} & h_M\lambda\Pi_{1,3} & h_M\lambda\Pi_{2,3} \\ * & * & * & -h_M\lambda X & h_M\lambda\Pi_{3,1} - h_M\lambda \mathbf{I} & h_M\lambda\Pi_{4,1} - h_M\lambda X & h_M\lambda\Pi_{3,2} & h_M\lambda\Pi_{4,2} & h_M\lambda\Pi_{3,3} & h_M\lambda\Pi_{4,3} \\ * & * & * & * & \hat{\Gamma}_{11}(1,1) & \hat{\Gamma}_{11}(1,2) & \mathbf{0} & \mathbf{0} & \hat{S}_2(1,1) + h_M\hat{M}_{12}(1,1) & \hat{S}_2(2,1) + h_M\hat{M}_{12}(1,2) \\ * & * & * & * & * & \hat{\Gamma}_{11}(2,2) & \mathbf{0} & \mathbf{0} & \hat{S}_2^T(1,2) + h_M\hat{M}_{12}(2,1) & \hat{S}_2(2,2) + h_M\hat{M}_{12}(2,2) \\ * & * & * & * & * & * & -Y + \hat{Q}_{11} & -\mathbf{I} + \hat{Q}_{12} & \mathbf{0} & \mathbf{0} \\ * & * & * & * & * & * & * & -X + \hat{Q}_{22} & \mathbf{0} & \mathbf{0} \\ * & * & * & * & * & * & * & * & -\hat{R}_{11} + h_M\hat{M}_{22}(1,1) & -\hat{R}_{12} + h_M\hat{M}_{22}(1,2) \\ * & * & * & * & * & * & * & * & * & -\hat{R}_{22} + h_M\hat{M}_{22}(2,2) \end{bmatrix},$$

$$\Xi_{12}=\begin{bmatrix} \mathbf{0} & \mathbf{0} & D_g & \mathbf{0} & F_{11} & F_{21} & \mathbf{0} & \mathbf{0} \\ \mathbf{0} & \mathbf{0} & XD_g & \mathbf{0} & XF_{11} & XF_{21} & \mathbf{0} & \mathbf{0} \\ \mathbf{0} & \mathbf{0} & h_M\lambda D_g & \mathbf{0} & h_M\lambda F_{1,1} & h_M\lambda F_{3,1} & \mathbf{0} & \mathbf{0} \\ \mathbf{0} & \mathbf{0} & h_M\lambda XD_g & \mathbf{0} & h_M\lambda XF_{1,1} & h_M\lambda XF_{3,1} & \mathbf{0} & \mathbf{0} \\ -\widehat{S}_1(1,1)+\widehat{S}_3^{\mathrm{T}}(1,1)+h_M\widehat{M}_{13}(1,1) & -\widehat{S}_1(1,2)+\widehat{S}_3^{\mathrm{T}}(2,1)+h_M\widehat{M}_{13}(1,2) & \mathbf{0} & YL^{\mathrm{T}} & \mathbf{0} & \mathbf{0} & YF_{1,2}^{\mathrm{T}} & \mathbf{0} \\ -\widehat{S}_1(2,1)+\widehat{S}_3^{\mathrm{T}}(1,2)+h_M\widehat{M}_{13}(2,1) & -\widehat{S}_1(2,2)+\widehat{S}_3^{\mathrm{T}}(2,2)+h_M\widehat{M}_{13}(2,2) & \mathbf{0} & L^{\mathrm{T}} & \mathbf{0} & \mathbf{0} & F_{1,2}^{\mathrm{T}} & \mathbf{0} \\ \mathbf{0} & \mathbf{0} & \mathbf{0} & \mathbf{0} & \mathbf{0} & \mathbf{0} & \mathbf{0} & \mathbf{0} \\ \mathbf{0} & \mathbf{0} & \mathbf{0} & \mathbf{0} & \mathbf{0} & \mathbf{0} & \mathbf{0} & \mathbf{0} \\ -\widehat{S}_2(1,1)+h_M\widehat{M}_{23}(1,1) & -\widehat{S}_2(1,2)+h_M\widehat{M}_{23}(1,2) & \mathbf{0} & \mathbf{0} & \mathbf{0} & \mathbf{0} & \mathbf{0} & YF_{3,2}^{\mathrm{T}} \\ -\widehat{S}_2(2,1)+h_M\widehat{M}_{23}(2,1) & -\widehat{S}_2(2,2)+h_M\widehat{M}_{23}(2,2) & \mathbf{0} & \mathbf{0} & \mathbf{0} & \mathbf{0} & \mathbf{0} & F_{3,2}^{\mathrm{T}} \end{bmatrix},$$

$$\Xi_{22}=\begin{bmatrix} -\widehat{R}_{11}-\widehat{S}_3(1,1)-\widehat{S}_3^{\mathrm{T}}(1,1)+h_M\widehat{M}_{33}(1,1) & -\widehat{R}_{12}-\widehat{S}_3(1,2)-\widehat{S}_3^{\mathrm{T}}(2,1)+h_M\widehat{M}_{33}(1,2) & \mathbf{0} & \mathbf{0} & \mathbf{0} & \mathbf{0} & \mathbf{0} & \mathbf{0} \\ * & -\widehat{R}_{22}-\widehat{S}_3(2,2)-\widehat{S}_3^{\mathrm{T}}(2,2)+h_M\widehat{M}_{33}(2,2) & \mathbf{0} & \mathbf{0} & \mathbf{0} & \mathbf{0} & \mathbf{0} & \mathbf{0} \\ * & * & -\gamma\mathbf{I} & \mathbf{0} & \mathbf{0} & \mathbf{0} & \mathbf{0} & \mathbf{0} \\ * & * & * & -\gamma\mathbf{I} & \mathbf{0} & \mathbf{0} & \mathbf{0} & \mathbf{0} \\ * & * & * & * & -\varepsilon_1\mathbf{I} & \mathbf{0} & \mathbf{0} & \mathbf{0} \\ * & * & * & * & * & -\varepsilon_3\mathbf{I} & \mathbf{0} & \mathbf{0} \\ * & * & * & * & * & * & -\varepsilon_1^{-1}\mathbf{I} & \mathbf{0} \\ * & * & * & * & * & * & * & -\varepsilon_3^{-1}\mathbf{I} \end{bmatrix},$$

$$\hat{Q}=\begin{bmatrix} \widehat{Q}_{11} & \widehat{Q}_{12} \\ \widehat{Q}_{21} & \widehat{Q}_{22} \end{bmatrix},\ \hat{R}=\begin{bmatrix} \widehat{R}_{11} & \widehat{R}_{12} \\ \widehat{R}_{21} & \widehat{R}_{22} \end{bmatrix},$$

$$\hat{S}_i=\begin{bmatrix} \widehat{S}_i(1,1) & \widehat{S}_i(1,2) \\ \widehat{S}_i(2,1) & \widehat{S}_i(2,2) \end{bmatrix},\ \hat{M}_{ij}=\begin{bmatrix} \widehat{M}_{ij}(1,1) & \widehat{M}_{ij}(1,2) \\ \widehat{M}_{ij}(2,1) & \widehat{M}_{ij}(2,2) \end{bmatrix},\ i,j=1,2,3,$$

$$\Pi_{1,i}=A_iY+B_g\tilde{C}_{ci},\ \Pi_{2,i}=A_i+B_g\tilde{D}_{ci}G,\ \Pi_{3,i}=\tilde{A}_{ci},$$

$$\Pi_{4,i}=XA_i+\tilde{B}_{ci}G,\quad i=1,2,3,$$

$$\hat{\Gamma}_{11}(1,1)=-\widehat{Q}_{11}+(h_M-h_m+2)\widehat{R}_{11}+\widehat{S}_1(1,1)+\widehat{S}_1^{\mathrm{T}}(1,1)+h_M\widehat{M}_{11}(1,1),$$
$$\hat{\Gamma}_{11}(1,2)=-\widehat{Q}_{12}+(h_M-h_m+2)\widehat{R}_{12}+\widehat{S}_1(1,2)+\widehat{S}_1^{\mathrm{T}}(2,1)+h_M\widehat{M}_{11}(1,2),$$

$$\hat{\Gamma}_{11}(2,2)=-\widehat{Q}_{22}+(h_M-h_m+2)\widehat{R}_{22}+\widehat{S}_1(2,2)+\widehat{S}_1^{\mathrm{T}}(2,2)+h_M\widehat{M}_{11}(2,2),$$

and correspondingly, the 2D ILC controller is given by

$$\begin{cases} D_{ci}=\tilde{D}_{ci}, \\ B_{ci}=\Phi^{-1}\left(\tilde{B}_{ci}-XB_g\tilde{D}_{ci}\right), \\ C_{ci}=\left(\tilde{C}_{ci}-\tilde{D}_{ci}GY\right)\Psi^{-\mathrm{T}}, \\ A_{ci}=\Phi^{-1}\left[\tilde{A}_{ci}-XA_1Y-\tilde{B}_{ci}GY-XB_g\left(\tilde{C}_{ci}-\tilde{D}_{ci}GY\right)\right]\Psi^{-\mathrm{T}}, \end{cases},\ i=1,2,3 \tag{10.42}$$

where Φ and Ψ are nonsingular matrices that satisfy $\Phi\Psi^{\mathrm{T}}=\mathbf{I}-XY$.

Proof It follows from (10.40) that $\Xi_{11} < 0$, which implies that

$$\begin{bmatrix} -Y & -\mathbf{I} \\ * & -X \end{bmatrix} < 0. \tag{10.43}$$

By the Schur complement, it is easy to see $\mathbf{I} - XY$ is guaranteed nonsingular. Therefore, the following nonsingular matrices are introduced

$$\Lambda_1 = \begin{bmatrix} Y & \mathbf{I} \\ \Psi^{\mathrm{T}} & \mathbf{0} \end{bmatrix}, \quad \Lambda_2 = \begin{bmatrix} \mathbf{I} & X \\ \mathbf{0} & \Phi^{\mathrm{T}} \end{bmatrix}. \tag{10.44}$$

By letting $P = \Lambda_2 \Lambda_1^{-1}$, it follows that

$$\Lambda_1^{\mathrm{T}} P \Lambda_1 = \begin{bmatrix} Y & \mathbf{I} \\ * & X \end{bmatrix} > 0. \tag{10.45}$$

Next, let $W = \lambda P$ in (10.13), where λ is an adjustable parameter to keep the state energy in (10.21) commensurate with those of (10.22)–(10.24), so as to solve the LMI conditions in (10.13) and (10.14). Besides, define

$$Q = \Lambda_1^{-\mathrm{T}} \widehat{Q} \Lambda_1^{-1},\ R = \Lambda_1^{-\mathrm{T}} \widehat{R} \Lambda_1^{-1},$$
$$S_i = \Lambda_1^{-\mathrm{T}} \widehat{S}_i \Lambda_1^{-1},\ M_i = \Lambda_1^{-\mathrm{T}} \widehat{M}_i \Lambda_1^{-1}, \quad i, j = 1, 2, 3.$$

By performing a congruence transformation to (10.13) (or (10.14)) in terms of a pre-multiplication with $\tilde{\Lambda}_1^{-1} = \mathrm{diag}\big(\Lambda_1^{\mathrm{T}}, \Lambda_1^{\mathrm{T}}, \Lambda_1^{\mathrm{T}}, \Lambda_1^{\mathrm{T}}, \Lambda_1^{\mathrm{T}}, \Lambda_1^{\mathrm{T}}, \mathrm{I}, \mathrm{I}, \mathrm{I}, \mathrm{I}, \mathrm{I}, \mathrm{I},\big)$ (or $\widehat{\Lambda}_1^{\mathrm{T}} = \mathrm{diag}(\Lambda_1^{\mathrm{T}}, \Lambda_1^{\mathrm{T}}, \Lambda_1^{\mathrm{T}}, \Lambda_1^{\mathrm{T}})$) and a post- multiplication with $\tilde{\Lambda}_1$ (or $\widehat{\Lambda}_1$), and using the Schur complement, one can obtain the matrix inequalities shown in (10.40) and (10.41), where the following matrix variables are defined for computation,

$$\begin{cases} \tilde{D}_{\mathrm{ci}} = D_{\mathrm{ci}}, \\ \tilde{B}_{\mathrm{ci}} = XB_{\mathrm{g}}\tilde{D}_{\mathrm{ci}} + \Phi B_{ci}, \\ \tilde{C}_{ci} = \tilde{D}_{\mathrm{ci}} GY + C_{ci}\Psi^{\mathrm{T}}, \\ \tilde{A}_{ci} = XA_i Y + \tilde{B}_{\mathrm{ci}} GY + XB_{\mathrm{g}}(\tilde{C}_{ci} - \tilde{D}_{\mathrm{ci}} GY) + \Phi A_{ci}\Psi^{\mathrm{T}}, \end{cases}, \quad i = 1, 2, 3. \tag{10.46}$$

Therefore, by solving (10.40) and (10.41) to determine feasible solutions of the matrix variables in (10.46), the 2D ILC controller can be inversely retrieved by the formulae given in (10.42). This completes the proof. □

It should be noted that Φ and Ψ may be practically computed from the singular value decomposition of matrix $\mathbf{I} - XY$, to determine the ILC controller parameters from (10.42).

10.3 Performance Optimization

For practical implementation, it is desired to optimize the $\mathcal{H}_\infty$ performance level γ, to accommodate process uncertainties and nonrepeatable load disturbance as much as possible, including unmodeled nonlinear dynamic characteristics that are estimated by norm-bounded uncertainties and repetitive load disturbance.

Given uncertainty bounds of $\left\|\Delta\tilde{A}(t,k)\right\| \leq \left\|\overline{F}_{1,1}\right\|\left\|\overline{F}_{1,2}\right\|$, $\left\|\Delta\tilde{A}_d(t,k)\right\| \leq \left\|\overline{F}_{3,1}\right\|\left\|\overline{F}_{3,2}\right\|$ and $h \in [h_m,\ h_M]$, as usually prescribed in engineering practice, one can view γ as an objective function to determine its minimum and the corresponding ILC controller from the following optimization program:

$$\text{Minimize} \quad \gamma \tag{10.47}$$

(subject to $X > 0, Y > 0, \widehat{Q} > 0, \widehat{R} > 0, \lambda > 0, \varepsilon_1 > 0, \varepsilon_3 > 0$ and (10.40) and (10.41)).

Remark 10.1 Note that the sufficient condition in (10.40) is not a strict linear matrix inequality due to that ε_1 and ε_3 enter Ξ_{22} in a nonlinear manner. An easy way is to fix ε_1 and $\varepsilon_3$in using an LMI toolbox like MATLAB to obtain a feasible solution, but at the cost of conservativeness. Generally, ε_1 may be initially taken much smaller than the unity, e.g., 0.1, for the case of $\left\|F_{1,1}\right\| < \left\|F_{1,2}\right\|$. Further reduced conservativeness can be obtained by using an iterative algorithm in combination with the LMI convex optimization toolbox as introduced by Shi et al., (2005c). ◊

Remark 10.2 It is evident that λ has an effect on the solvability of matrix inequalities (10.40) and (10.41). Owing to that λ linearly enters both (10.40) and (10.41), the optimization program of (10.47) can be conveniently solved by monotonically decreasing or increasing λ to reduce conservativeness.

Moreover, one can prescribe a specified robust $\mathcal{H}_\infty$ control performance level γ^*, which may be taken with reference to the minimal solution of (10.47), to evaluate an admissible upper (or lower) bound of state delay variation from the following optimization program:

$$\text{Maximize} \quad h_M\,(\text{or } h_m) \tag{10.48}$$

(subject to γ^*, $X > 0, Y > 0, \widehat{Q} > 0, \widehat{R} > 0, \lambda > 0, \varepsilon_1 > 0, \varepsilon_3 > 0$ and (10.40) and (10.41)).

Similarly, an optimization program can be formulated to evaluate an admissible upper norm bound of $\Delta\tilde{A}(t,k)$ or $\Delta\tilde{A}_h(t,k)$, for reliable implementation of the above ILC algorithm against process uncertainties from batch to batch. ◊

10.4 An Illustrative Example

Consider the batch packing process of an injection molding machine as shown in Fig. 2.2, where the nozzle pressure is a key variable that should be controlled to follow a desired profile to ensure product quality and consistency from batch to batch. A state-space model reflecting the packing process dynamics could be identified as (Gao et al. 2001)

$$P_{\Delta}:\begin{cases} x(t+1)=\left(\begin{bmatrix}1.607 & -0.6086\\ 1 & 0\end{bmatrix}+\Delta\tilde{A}(t)\right)x(t)\\ \qquad +\left(\begin{bmatrix}-0.202 & 0.196\\ 0 & 0\end{bmatrix}+\Delta\tilde{A}_h(t)\right)x(t-h)+\begin{bmatrix}1\\ 0\end{bmatrix}u(t),\\ y(t)=[1.239,\ -0.9282]x(t).\end{cases}$$

Note that the process state delay h was assumed to be zero or fixed for ILC design in the referred references (Gao et al. 2001; Shi et al. 2006; Wang et al. 2013a; Liu et al. 2010). In fact, the state delay randomly varies from batch to batch, which may be estimated by a range of [2, 5] with regard to the sampling time of 0.5 s, and the nominal value of $\hat{h}=3$ is used herein for ILC design. Batch-to-batch time-varying uncertainties of the state transfer matrices are estimated as

$$\Delta\tilde{A}(t)=\begin{bmatrix}0.08\delta(t) & 0.05\delta(t)\\ 0 & 0\end{bmatrix}=\begin{bmatrix}0.08 & 0.05\\ 0 & 0\end{bmatrix}\begin{bmatrix}\delta(t) & 0\\ 0 & \delta(t)\end{bmatrix}\begin{bmatrix}1 & 0\\ 0 & 1\end{bmatrix},$$

$$\Delta\tilde{A}_h(t)=\begin{bmatrix}0.05\delta(t) & 0.04\delta(t)\\ 0 & 0\end{bmatrix}=\begin{bmatrix}0.05 & 0.04\\ 0 & 0\end{bmatrix}\begin{bmatrix}\delta(t) & 0\\ 0 & \delta(t)\end{bmatrix}\begin{bmatrix}1 & 0\\ 0 & 1\end{bmatrix},$$

where $|\delta(t)|\leq 1$. Note that the influence from nonlinear dynamics of the hydraulic control valve is also considered in these norm-bounded uncertainties, and therefore, may be further enlarged by simply increasing the magnitude of $|\delta(t)|$, to estimate worse nonlinear dynamics in practice.

By performing the optimization program in (10.47) in terms of taking $\lambda=0.01$, $\varepsilon_1=\varepsilon_2=0.1$, one can obtain the minimal $\mathcal{H}_\infty$robust performance level $\gamma^*=99.8$, which corresponds to the ILC controller parameters listed in Table 10.1.

The desired output trajectory (y_d) for tracking is shown in Fig. 10.2a (dashed line). In view of the fact that a step change is not a continuous signal, the initial part of y_d, corresponding to a step change of the setpoint, is smoothed by a user-specified prefilter of $G_f(z)=(z^{-1}+z^{-2})/(3-z^{-1})$ for tracking, so as to avoid surpassing the upper limit of the hydraulic control valve. A repetitive load disturbance is imitated by passing a step change with a magnitude of 180 through a slow transfer function $G_d(z)=(z^{-1}+z^{-2})/(11-4z^{-1})$, which is added to the process output at $t=30$ (step), and another one with a magnitude of 150 and the same dynamics is added at $t=50$ (step), where each step corresponds to the sampling time of 0.5 s. For illustration, the following cases of process uncertainties are tested.

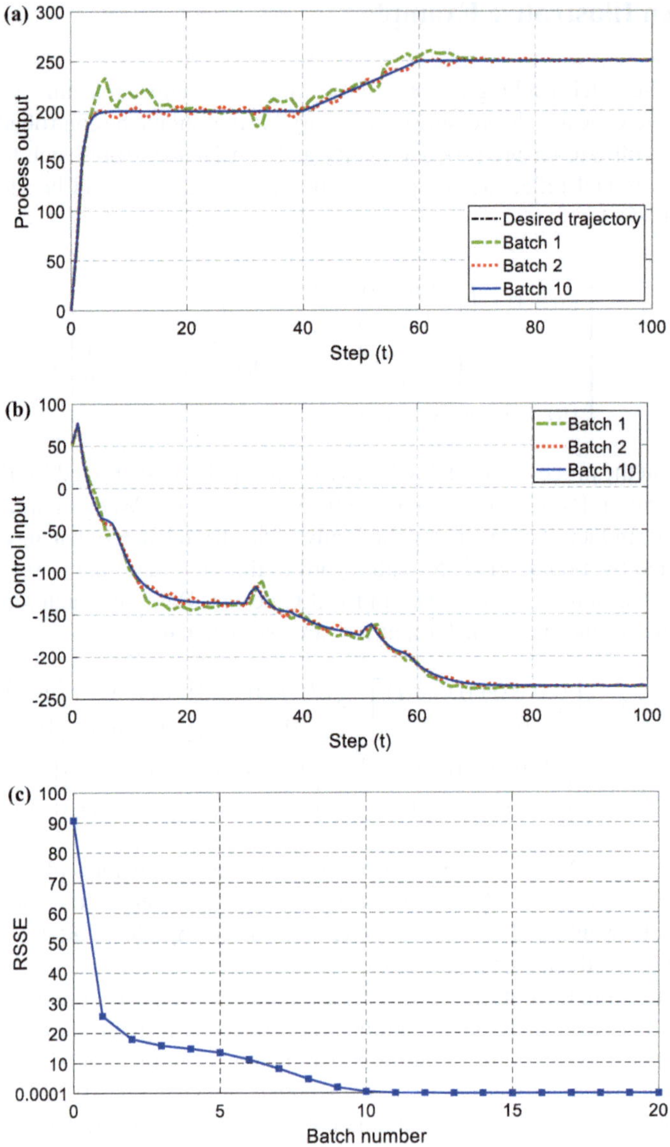

Fig. 10.2 Tracking results for Case 1 in the presence of the fixed uncertainties and repetitive disturbance

Table 10.1 ILC controller gains for the robust control performance level $\gamma^* = 99.8$

	A_{ci}	B_{ci}	C_{ci}	D_{ci}
$i = 1$	$\begin{bmatrix} 0.7492 & 0.0004 & 0.0003 & -0.0002 \\ 0.0014 & 0.0606 & 0.0000 & 0.0000 \\ 0.0013 & 0.0000 & 0.0464 & -0.0001 \\ 0.0002 & 0.0000 & -0.0018 & 0.0477 \end{bmatrix}$	$10^4 \times \begin{bmatrix} 0.9797 & -0.2431 & -0.0016 \\ 0.0020 & 0.2566 & 4.0268 \\ 0.0018 & -0.0005 & 0.0000 \\ 0.0003 & -0.0001 & -0.0005 \end{bmatrix}$	$10^{-4} \times \begin{bmatrix} 0.2089 \\ 0.0000 \\ -0.2236 \\ -0.0007 \end{bmatrix}^{\mathrm{T}}$	$\begin{bmatrix} 0.9289 \\ -0.7142 \\ -0.5321 \end{bmatrix}^{\mathrm{T}}$
$i = 2$	$10^{-3} \times \begin{bmatrix} -0.0007 & 0.3308 & 0.0000 & -0.0001 \\ 0.0000 & 0.0067 & 0.0000 & 0.0002 \\ -0.0001 & -0.0080 & -0.0380 & 0.4744 \\ -0.0002 & -0.0017 & -0.0358 & 0.6453 \end{bmatrix}$	$\begin{bmatrix} 0.0000 & -9782.684 & -14.1696 \\ 0.0000 & -19.6274 & 0.2890 \\ 0.0011 & 6.6979 & 0.3403 \\ 0.0018 & 1.3453 & 0.0711 \end{bmatrix}$	$10^{-5} \times \begin{bmatrix} -0.0002 \\ 0.0039 \\ -0.0634 \\ 0.8335 \end{bmatrix}^{\mathrm{T}}$	$\begin{bmatrix} 0.0000 \\ -0.7595 \\ -0.0017 \end{bmatrix}^{\mathrm{T}}$
$i = 3$	$10^{-5} \times \begin{bmatrix} -0.0003 & 0.1285 & 0.0000 & 0.0000 \\ 0.0000 & 0.0003 & 0.0000 & 0.0000 \\ -0.0001 & -0.0012 & 0.0861 & 0.0827 \\ -0.0002 & -0.0003 & 0.1787 & 0.1056 \end{bmatrix}$	$\begin{bmatrix} 0.0000 & 0.0550 & -0.0550 \\ 0.0000 & 0.0001 & -0.0001 \\ 0.0000 & -0.0005 & 0.0005 \\ 0.0000 & -0.0001 & 0.0001 \end{bmatrix}$	$10^{-4} \times \begin{bmatrix} 0.1467 \\ -0.0001 \\ 0.0003 \\ 0.0001 \end{bmatrix}^{\mathrm{T}}$	$\begin{bmatrix} -0.0193 \\ 0.1629 \\ 0.0001 \end{bmatrix}^{\mathrm{T}}$

Case 1. (Time-invariant process uncertainties) Assume that all elements of state transfer matrices, $\Delta\tilde{A}(t)$ and $\Delta\tilde{A}_h(t)$, and state delay $h(t)$ are fixed as their upper bounds, respectively. The tracking results are shown in Fig. 10.2a, b, while the output tracking error in terms of the root-sum-squared-error (RSSE) criterion is plotted in Fig. 10.2c. It is seen that perfect tracking is realized almost through 10 cycles by the proposed ILC method. Moreover, there exists no steady-state output tracking error from the first batch, owing to the use of the integral error information for ILC design. Note that, for a larger fixed upper bounds of $\Delta\tilde{A}(t)$ and $\Delta\tilde{A}_h(t)$, e.g., letting $|\delta(t)| = 2$ to estimate worse nonlinear dynamics of the hydraulic control valve, similar tracking performance can also be obtained by the proposed ILC method, almost through 15 cycles for perfect tracking, and thus is omitted.

Case 2. (Time-varying state delay) Assume that the state delay varies randomly in a range as above estimated, while the state transfer matrices remain fixed as in Case 1, the resulting RSSEs for 100 cycles are plotted in Fig. 10.3a, and correspondingly, Fig. 10.3b shows the tracking results of representative batches of 1, 50 and 100. It is seen that the proposed method holds good robust stability against state delay variation from batch and batch.

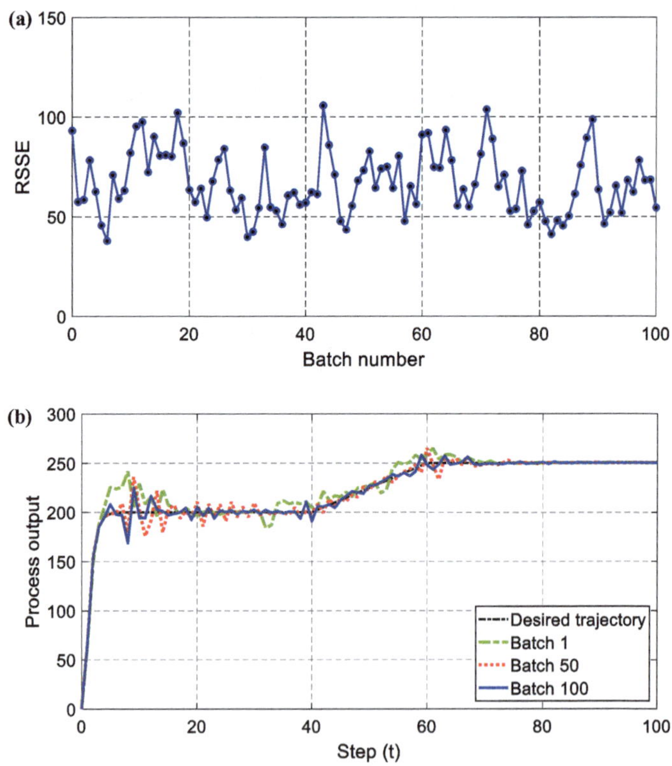

Fig. 10.3 RSSE and process output for case 2 in the presence of state delay variation

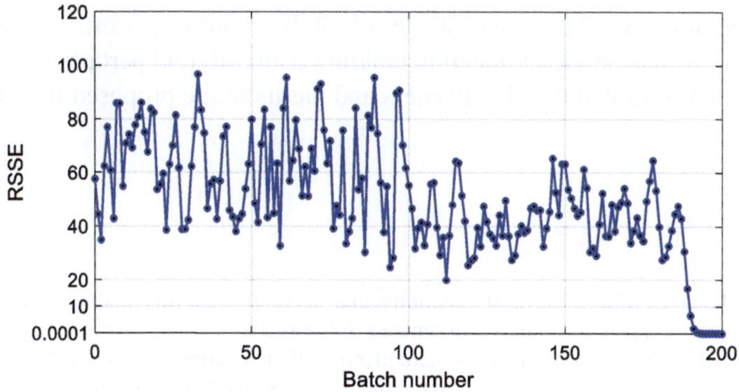

Fig. 10.4 RSSE for case 3 in the presence of time-varying uncertainties

Case 3. (Time-varying uncertainties of both state transfer matrices and state delay). Assume that time-varying uncertainties occur as estimated in the process modeling. The RSSE plot for 100 batches is shown in Fig. 10.4, which demonstrates that the closed-loop 2D ILC system maintains robust stability well in both time and batch directions. Note that if the state transfer matrix uncertainties are then reduced to $\Delta\tilde{A}(t) = \Delta\tilde{A}_h(t) = [0.04\delta(t), 0.03\delta(t); 0, 0]$, as illustrated from the 101st to 180th batches in Fig. 10.4, the output tracking error is reduced accordingly. If the process uncertainties further become time-invariant, which is illustrated from the 181st to 200th batches in Fig. 10.4, it is evidently seen that the corresponding RSSE goes to zero, indicating that perfect tracking is realized thereafter. It is thus demonstrated that the proposed ILC method can be well used for robust tracking and on-line optimization against batch-to-batch time-varying uncertainties.

10.5 Summary

A robust closed-loop ILC method has been presented for industrial batch processes with state delay and time-varying uncertainties (Liu et al. 2010). Based on an equivalent 2D system description of such a batch process, a 2D ILC scheme is provided for robust tracking of the desired output trajectory, which integrates the feedback control with feedforward control to guarantee robust convergence along both the time and batch directions, while maintaining no steady-state output tracking error from the first batch, compared to open-or closed-loop ILC methods developed in terms of 1D system description that can only guarantee robust stability in the time direction. Given upper bounds of state delay variation and norm-bounded parameter uncertainties of such a batch process, as usually specified in practice, a stability theorem is proposed for assessing robust stability of the closed-loop system, by introducing a comprehensive 2D difference Lyapunov function for asymptotic stability analysis. Through solving the established LMI conditions, the ILC controller is explicitly formulated, together with an adjustable robust $\mathcal{H}_\infty$ control performance level. The application to an example of injection molding with different perturbation cases has evidently demonstrated the effectiveness and merits of the proposed ILC method.

References

Ahn H-S, Chen YQ, Moore KL (2007) Stability analysis of discrete-time iterative learning control systems with interval uncertainty. Automatica 43(5):892–902

Chen J, Gu G (2000) Control-oriented system identification: an approach. Wiley, New York

Chen J, Kong C-K (2009) Performance assessment for iterative learning control of batch units. J Process Control 19(6):1043–1053

Dabkowski P, Galkowski K, Rogers E, Kummert A (2009) Strong practical stability and stabilization of discrete linear repetitive processes. Multi Syst Signal Process 20(4):311–331

Gao F, Yang Y, Shao C (2001) Robust iterative learning control with applications to injection molding process. Chem Eng Sci 56(24):7025–7034

Gu K, Kharitonov VL, Chen J (2003) Stability of time-delay systems. Birkhäuser Springer, Boston

Harte TJ, Hätönen J, Owens DH (2005) Discrete-time inverse model-based iterative learning control stability, monotonicity and robustness. Int J Control 78(8):577–586

Huang SN, Tan KK, Lee TH (2002) Necessary and sufficient condition for convergence of iterative learning algorithm. Automatica 38(7):1257–1260

Lee JH, Lee KS, Kim WC (2000) Model-based iterative learning control with a quadratic criterion for time-varying linear systems. Automatica 36(5):641–657

Li X-D, Chow TWS, Ho JKL (2005) 2-D system theory based iterative learning control for linear continuous systems with time delays. IEEE Trans Circuits Syst-i: Regul Pap 52(7):1421–1430

Liu T, Gao F (2010) Robust two-dimensional iterative learning control for batch processes with state delay and time-varying uncertainties. Chem Eng Sci 65(23):6134–6144

Liu T, Gao F, Wang YQ (2010) IMC-based iterative learning control for batch processes with uncertain time delay. J Process Control 20(2):173–180

Nagy ZK, Mahn B, Franke R, Allgöwer F (2007) Efficient output feedback nonlinear model predictive control for temperature control of industrial batch reactors. Control Eng Pract 15(7):839–850

Owens DH, Amann N, Rogers E, French M (2000) Analysis of linear iterative learning control schemes-a 2D systems/repetitive processes approach. Multi Dimension Syst Signal Process 11(2):125–177

Seborg DE, Edgar TF, Mellichamp DA (2016) Process dynamic and control, 4th edn. Wiley, New Jersey

Shi J, Gao F, Wu T-J (2005a) Robust design of integrated feedback and iterative learning control of a batch process based on a 2D Roesser system. J Process Control 15(8):907–924

Shi J, Gao F, Wu T-J (2005a) From two-dimensional linear quadratic optimal control to iterative learning control. Paper 1. Two-dimensional linear quadratic optimal controls and system analysis. Ind Eng Chem Res 45(13):4603–4616

Shi J, Gao F, Wu T-J (2005b) From two-dimensional linear quadratic optimal control to iterative learning control. Paper 2. Iterative learning controls for batch processes. Ind Eng Chem Res 45(13):4617–4628

Shi J, Gao F, Wu T-J (2006) Robust iterative learning control design for batch processes with uncertain perturbations and initialization. AIChE J 52(6):2171–2187

Tayebi A (2007) Analysis of two particular iterative learning control schemes in frequency and time domains. Automatica 43(9):1565–1572

Wang L, Mo S, Zhou D, Gao F, Chen X (2013a) Delay-range-dependent robust 2D iterative learning control for batch processes with state delay and uncertainties. J Process Control 23(5):715–730

Wang L, Mo S, Qu H, Zhou D, Gao F (2013b) Design of 2D controller for batch processes with uncertainties and interval time-varying delays. Control Eng Pract 21(10):1321–1333

Wang YQ, Gao F, Doyle FJ III (2009) Survey on iterative learning control, repetitive control, and run-to-run control. J Process Control 19(10):1589–1600

Wei J, Tao H, Hao S, Paszke W, Galkowskii K (2022) Output feedback based robust iterative learning control via a heuristic approach for batch processes with time-varying state delays and uncertainties. J Process Control 116:159–171

Wijdeven J, Donkers T, Bosgra O (2009) Iterative learning control for uncertain systems: robust monotonic convergence analysis. Automatica 45(10):2383–2391

Wu X, Park JH, Shen M (2023) Adaptive iterative learning control for continuous systems with non-repetitive uncertainties and unknown state delays. Int J Robust Nonlinear Control 33(4):2796–2810

Zhang J (2008) Batch-to-batch optimal control of a batch polymerisation process based on stacked neural network models. Chem Eng Sci 63(5):1273–1281

Zhuang Z, Tao H, Chen Y, Stojanovic V, Paszke W (2023) An optimal iterative learning control approach for linear systems with nonuniform trial lengths under input constraints. IEEE Trans Syst Mans Cybern: Syst 53(6):3461–3473

Zhou Y, Gao K, Tang X, Hu H, Li D, Gao F (2023) Conic input mapping design of constrained optimal iterative learning controller for uncertain systems. IEEE Trans Cybern 53(3):1843–1855

Chapter 11
Robust Data-Driven ILC Design for Unknown System Dynamics

11.1 Introduction

To overcome the difficulties of modelling complex manufacturing systems for designing ILC to realize batch run optimization, data-driven control (DDC), based on available system input/output (I/O) data, has received considerable attention, see, e.g., the monographs (Hou and Jin 2013; Chi et al. 2022) and survey papers (Hou et al. 2017; Benosman 2018). To enumerate a few, Meng and Zhang (2022) put forth an optimization-based data-driven learning control (DDLC) design for performance optimization of a nonlinear system with unknown dynamics. By estimating the pseudo-partial derivative and uncertain residual term in a dynamic linearization data model based on a neural network, Yu et al. (2020a, b) developed a data-driven predictive ILC scheme for unknown nonlinear discrete-time systems subject to nonrepetitive disturbances. In contrast, another data-driven ILC (DDILC) was developed for unknown nonlinear systems by applying the dynamic linearization technique to not only the controlled system but also the nonlinear ideal learning controller (Yu et al. 2022). To enhance the output tracking performance, a high-order DDILC was given in Chi et al. (2018) for nonlinear repetitive control systems with unknown dynamics, by incorporating the historical output tracking errors and control inputs into the ILC updating law.

However, when applied to batch processes with a priori closed-loop control loops, the aforementioned DDILC methods based on updating the control input may not ensure the convergence of the established learning system. An indirect-type adaptive DDILC scheme was developed to improve the output tracking performance by regulating the set-point command along the batch direction (Chi et al. 2023) for ease of application to linear or nonlinear batch processes configured with a inner-loop P-type feedback control structure. Although it had been analyzed that a PI or PID controller has capable to stabilize a nonlinear system with uncertainty (Zhao and Guo 2022; Yu et al. 2020a, b; Xiong and Hou 2022), combining a widely used PI controller with a data-driven set-point learning control (DDSPLC) scheme for batch

T. Liu et al., *Robust Iterative Learning Control of Industrial Batch Systems*, Intelligent Control and Learning Systems 22, https://doi.org/10.1007/978-981-96-9778-6_11

processes with unknown dynamics subject to nonrepetitive uncertainties is an open issue to be addressed.

In this chapter, a robust PI-based adaptive DDSPLC (PI-ADDSPLC) scheme is presented to realize batch performance optimization for unknown nonlinear processes configured with a PI feedback control loop, using the available system I/O data. To facilitate engineering practice, an ADDSPLC is offered to adjust the set-point command for the closed-loop PI control loop to realize batch performance optimization (Hao et al. 2025). By mathematical induction, robust convergence analysis of the output tracking error, together with the boundedness analysis of the set-point command and adaptive learning gain, is conducted. Furthermore, an iterative ESO (IESO) based PI-ADDSPLC is also presented to enhance the output tracking performance against nonrepetitive external disturbances and initial shifts, together with the corresponding convergence and boundedness analysis. The effectiveness and superiority of these data-driven ILC schemes are validated by two illustrative examples.

11.2 Adaptive Data-Driven Set-Point Learning Control (ADDSPLC) Design

Consider an industrial batch process involved with unknown dynamics and external disturbances described by

$$\begin{aligned} y_k(t+1) =& f(y_k(t), y_k(t-1), \ldots, y_k(t-n_y), u_k(t), u_k(t-1), \ldots, u_k(t-n_u)) \\ &+ \omega_k(t), \end{aligned} \tag{11.1}$$

where $k \in \mathbb{Z}_+$ and $t \in \mathbb{Z}_{N-1}$ represent the batch number and time step, respectively; $y_k(t) \in \Re$ and $u_k(t) \in \Re$ denote the process output and control input at time t of batch k, respectively; N is the total length in each batch; $\omega_k(t)$ is the nonrepetitive disturbances; $f(\cdot)$ represents an unknown linear or nonlinear function; n_u and n_y are, respectively, the unknown input and output orders. The initial resetting condition in (11.1) is considered as $y_k(0) = y_0 + \delta_k$ with y_0 being an identical initial condition in each batch run and δ_k being the initial shift. For notational brevity, the unknown function $f(\cdot)$ in (11.1) is simply reformulated by $f(x_1, x_2, \ldots, x_{n_u+n_y+2})$, whose i-th variable is denoted by $x_i \in \Re$, $i = 1, 2, \ldots, n_u + n_y + 2$.

Based on only the available I/O data for a batch process in (11.1) with an inner-loop PI feedback control structure, it aims at designing a robust adaptive set-point learning control scheme to steer the process output to track the desired output trajectory as closely as possible against nonrepetitive external disturbances and initial shifts of the operation conditions along the batch direction, that is, for any $t \in \mathbb{Z}_N/\{0\}$, there stands

$$\sup_{k\in\mathbb{Z}_+} |e_k(t)| \leq \beta_e, \ \limsup_{k\to\infty} |e_k(t)| \leq \beta_{e_{\sup}}, \tag{11.2}$$

where $e_k(t) \triangleq y_d(t) - y_k(t)$ is the output tracking error and $y_d(t)$ is the desired output trajectory satisfying $|y_d(t)| \leq \beta_d < \infty$ with β_d being a finite positive constant, β_e and $\beta_{e_{\sup}}$ with $\beta_e > \beta_{e_{\sup}} \geq 0$ are associated with the upper bounds of initial shift and the batch-wise variation of the nonrepetitive external disturbance. Also, the bounded set-point command $y_k^s(t)$ is needed for practical implementation, that is

$$\sup_{k \in \mathbb{Z}_+} \max_{t \in \mathbb{Z}_{N-1}} \left| y_k^s(t) \right| \leq \beta_s < \infty, \tag{11.3}$$

where $\beta_s > 0$ is a finite positive constant.

For the simplicity of analysis, the following assumptions are made along with a technical lemma.

Assumption 11.1 $u_0(t) = 0$ for any $t \in \mathbb{Z}_{N-1}$ and $y_k(t) = 0$ if $t < 0$ or $k < 0$.

Assumption 11.2 (Meng and Zhang 2022) The function f is continuously differentiable and satisfies

$$\left| \frac{\partial f}{\partial x_i} f(x_1, x_2, \ldots, x_{n_y+n_u+2}) \right| \leq \beta_{\bar{f}}, \ \forall x_i \in \Re, \ i = 1, 2, \ldots, n_y + n_u + 2, \ \forall t \in \mathbb{Z}_{N-1},$$

where $\beta_{\bar{f}} > 0$ is a finite positive constant. Moreover, let the sign of $f/\partial x_{n_y+2}$ be fixed, which, without loss of generality, is assumed to be positive.

Assumption 11.3 (Hou and Jin 2013; Hou and Xiong 2019) An unknown batch process in (11.1) satisfies the generalized *Lipschitz* condition along the time direction

$$|\delta_t y_k(t+1)| \leq \beta_\Gamma \|\delta_t H_k(t)\|$$

for $\|\delta_t H_k(t)\| \neq 0$, where $\delta_t H_k(t) = \left[\delta_t y_k(t) \ \delta_t u_k(t) \right]^{\mathrm{T}}$ and $\beta_\Gamma > 0$ is a finite positive constant.

Assumption 11.4 (Hao et al. 2020; Meng and Zhang 2021) The initial shifts and nonrepetitive external disturbances are bounded, that is

$$|\delta_k| \leq \beta_\delta, \forall k \in \mathbb{Z}_+,$$

$$|\omega_k(t)| \leq \beta_\omega, \forall t \in \mathbb{Z}_{N-1}, \forall k \in \mathbb{Z}_+,$$

where $\beta_\delta > 0$ and $\beta_\omega > 0$ are two finite constants.

Remark 11.1 Assumption 11.1 defines the initial values of the process output and control input for the ease of control design and stability analysis. In contrast, Assumptions 11.2 and 11.3 have been widely adopted in nonlinear system control and data-driven control fields (e.g., Hou and Jin 2013; Chi et al. 2022; Hou et al. 2017; Benosman 2018; Meng and Zhang 2020; Zhao and Guo 2022; Yu et al. 2020a, b;

Xiong and Hou 2022; Hou and Xiong 2019) to describe the process nonlinearity, namely, the variation of process output is limited by the variations of all the control input and past process output. Also, Assumption 11.4 is widely used for robust ILC design of batch processes or repetitive systems subject to nonrepetitive uncertainties in the literature (see, e.g., Meng and Moore 2017; Meng and Zhang 2021), due to the physical constraints of the practical process operation. ◊

Lemma 11.1 (Meng and Zhang 2021) Under Assumption 11.2, the batch process described by (11.1) is equally transformed into the extended dynamic linearization data model (DLDM) below

$$\boldsymbol{y}_i - \boldsymbol{y}_j = \boldsymbol{\Phi}_{i,j}(\boldsymbol{u}_i - \boldsymbol{u}_j) + \boldsymbol{\Upsilon}_{i,j}(\boldsymbol{\omega}_i - \boldsymbol{\omega}_j) + \boldsymbol{\vartheta}_{i,j}(\delta_i - \delta_j), \forall i, j \in \mathbb{Z}_+, \tag{11.4}$$

where

$$\boldsymbol{\Phi}_{i,j} = \begin{bmatrix} \phi_{i,j}^0(0) & 0 & \cdots & 0 \\ \phi_{i,j}^1(0) & \phi_{i,j}^1(1) & \cdots & 0 \\ \vdots & \vdots & \ddots & \vdots \\ \phi_{i,j}^{N-1}(0) & \phi_{i,j}^{N-1}(1) & \cdots & \phi_{i,j}^{N-1}(N-1) \end{bmatrix}, \quad \boldsymbol{\Upsilon}_{i,j} = \begin{bmatrix} 1 & 0 & \cdots & 0 \\ \upsilon_{i,j}^1(0) & 1 & \cdots & 0 \\ \vdots & \vdots & \ddots & \vdots \\ \upsilon_{i,j}^{N-1}(0) & \upsilon_{i,j}^{N-1}(1) & \cdots & 1 \end{bmatrix},$$

$$\boldsymbol{\vartheta}_{i,j} = \left[\vartheta_{i,j}^0 \; \vartheta_{i,j}^1 \; \cdots \; \vartheta_{i,j}^{N-1} \right]^{\mathrm{T}}, \qquad \boldsymbol{y}_i = \left[y_i(1) \; y_i(2) \; \cdots \; y_i(N) \right]^{\mathrm{T}},$$

$$\boldsymbol{u}_i = \left[u_i(0) \; u_i(1) \; \cdots \; u_i(N-1) \right]^{\mathrm{T}}, \qquad \boldsymbol{\omega}_i = \left[\omega_i(0) \; \omega_i(1) \; \cdots \; \omega_i(N-1) \right]^{\mathrm{T}},$$

and all nonzero variables in $\boldsymbol{\Phi}_{i,j}$, $\boldsymbol{\Upsilon}_{i,j}$ and $\boldsymbol{\vartheta}_{i,j}$ are bounded, that is, there is a finite positive bound $\beta_\phi > 0$ satisfying

$$\begin{aligned} \left| \phi_{i,j}^t(\xi) \right| &\le \beta_\phi, \forall \xi \in \mathbb{Z}_t, \forall t \in \mathbb{Z}_{N-1}, \forall i, j \in \mathbb{Z}_+, \\ \left| \upsilon_{i,j}^t(\xi) \right| &\le \beta_\phi, \forall \xi \in \mathbb{Z}_t, \forall t \in \mathbb{Z}_{N-1}, \forall i, j \in \mathbb{Z}_+, \\ \left| \vartheta_{i,j}^t \right| &\le \beta_\phi, \forall t \in \mathbb{Z}_{N-1}, \forall i, j \in \mathbb{Z}_+. \end{aligned} \tag{11.5}$$

The above extended DLDM further generalizes the relationship between the process output and control input to any two batches, compared to the conventional DLDM characterizing such a relationship at two sequential batches. Besides, the parameter matrix $\boldsymbol{\Phi}_{i,j}$ and the uncertainty information consisting of initial shifts and external disturbances can be estimated in the above extended DLDM separately.

Taking $i = k$ and $j = k - 1$ in (11.4) yields

$$\Delta y_k(t+1) = \boldsymbol{\phi}_{k,k-1}^{\mathrm{T}}(t) \Delta \boldsymbol{u}_k(t) + \chi_k(t), \tag{11.6}$$

where $\chi_k(t) = \Delta \boldsymbol{\omega}_k^{\mathrm{T}}(t) \boldsymbol{\upsilon}_{k,k-1}(t) + \vartheta_{k,k-1}^t \Delta \delta_k$ and

$$\boldsymbol{\phi}_{k,k-1}(t) = \left[\phi_{k,k-1}^t(0) \; \phi_{k,k-1}^t(1) \; \cdots \; \phi_{k,k-1}^t(t) \right]^{\mathrm{T}},$$

$$\Delta \boldsymbol{u}_k(t) = \left[\, \Delta u_k(0)\ \Delta u_k(1)\ \cdots\ \Delta u_k(t) \,\right]^{\mathrm{T}},$$
$$\Delta \boldsymbol{\omega}_k(t) = \left[\, \Delta \omega_k(0)\ \Delta \omega_k(1)\ \cdots\ \Delta \omega_k(t) \,\right]^{\mathrm{T}},$$
$$\boldsymbol{\upsilon}_{k,k-1}(t) = \left[\, \upsilon^t_{k,k-1}(0)\ \upsilon^t_{k,k-1}(1)\ \cdots\ \upsilon^t_{k,k-1}(t-1)\ 1 \,\right]^{\mathrm{T}}.$$

Using Assumption 11.4 and Lemma 11.1, one has $\chi_k(t)$ is bounded for any $t \in \mathbb{Z}_{N-1}$ and $k \in \mathbb{Z}_+$, and thus satisfies $\sup_{t\in\mathbb{Z}_{N-1},k\in\mathbb{Z}_+} |\chi_k(t)| \le \beta_\chi(\beta_{\Delta\omega}, \beta_{\Delta\delta})$, where $\beta_\chi(\beta_{\Delta\omega}, \beta_{\Delta\delta}) > 0$ is a finite positive constant relying on $\beta_{\Delta\omega} \triangleq \sup_{t\in\mathbb{Z}_{N-1},k\in\mathbb{Z}_+} |\Delta\omega_k(t)|$ and $\beta_{\Delta\delta} \triangleq \sup_{k\in\mathbb{Z}_+} |\Delta\delta_k|$, respectively. For the simplicity of notations, the symbol "$\bullet$" is used to indicate the relevance to $\beta_{\Delta\omega}$ and $\beta_{\Delta\delta}$ in the later sections.

Figure 11.1 shows the proposed robust ADDSPLC scheme, where the classical PI control loop is within the red dashed-line box, and the rest is the set-point learning control law along with a learning gain estimator and a gradient estimator for adjusting the set-point command of the PI control loop along the batch direction, "Memory" represents a storage for recording output tracking error ($e_k(t)$), set-point command ($y^s_k(t)$), process input ($u_k(t)$), process output ($y_k(t)$) and gradient estimation ($\hat{\boldsymbol{\phi}}_{k,k-1}(t)$) in the current batch, and provide the historical batch information (e_{k-1}, y^s_{k-1}, $\Delta \boldsymbol{u}_{k-1}$, Δy_{k-1}, $\hat{\boldsymbol{\phi}}_{k-1,k-2}$) together with the batch information in the current batch ($\Delta y_k(t-1)$, $\Delta u_k(t-1)$, $\Delta \boldsymbol{y}^s_k(t-1)$, $\hat{\boldsymbol{\phi}}_{k,k-1}(t)$). Notice that the inner-loop PI controller remains unchanged in each batch, and thus is independent of the design of the set-point learning control for batch performance optimization. In the initial batch, only the PI feedback control is implemented, where the set-point command is simply set as the desired output trajectory y_{d}.

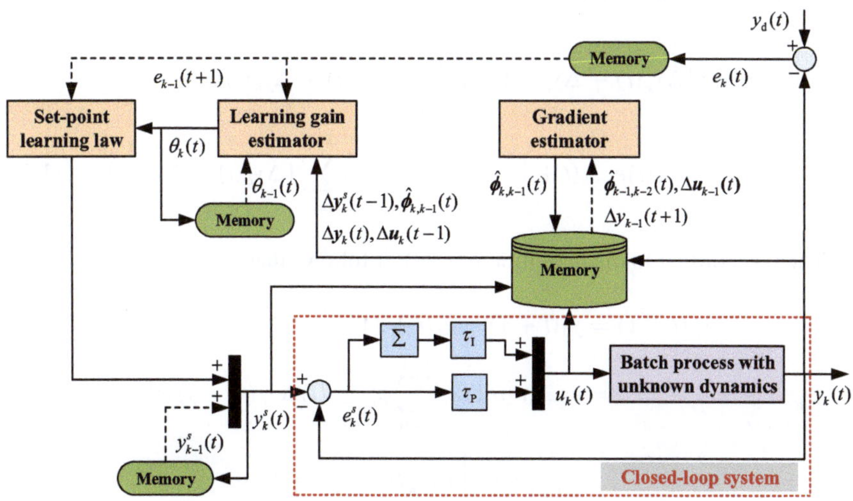

Fig. 11.1 Block diagram of the new PI-ADDSPLC scheme

By using a model-free tuning method presented in Sect. 2.2 of Chap. 2 to determine the PI controller gains, the control input accordingly takes the following form

$$u_k(t) = \tau_P e_k^s(t) + \tau_I \sum_{i=0}^{t} e_k^s(i) = \tau e_k^s(t) + \tau_I \sum_{i=0}^{t-1} e_k^s(i), \tag{11.7}$$

where $\tau = \tau_P + \tau_I$, τ_P and τ_I are the respective PI controller gains, and $e_k^s(t) \triangleq y_k^s(t) - y_k(t)$ represents the set-point tracking error, which is obviously different from the previously defined output tracking error, i.e., $e_k(t)$.

The proposed adaptive learning law for the set-point command takes the following form

$$y_k^s(t) = y_{k-1}^s(t) + \alpha\theta_k(t)e_{k-1}(t+1), \tag{11.8}$$

where α is a user-specified tuning parameter responsible for adjusting the convergence speed of the resulting learning system along the batch direction, and $\theta_k(t)$ denotes the time and batch-varying set-point learning gain.

The following cost function is adopted to estimate the unavailable learning gain $\theta_k(t)$ for control execution

$$J(\theta_k(t)) = |e_k(t+1)|^2 + \lambda|\theta_k(t) - \theta_{k-1}(t)|^2, \tag{11.9}$$

where $\lambda > 0$ is a specified tuning parameter to mitigate the large variation of $\theta_k(t)$ along the batch direction, and therefore adjust the convergence speed of the parameter estimation algorithm.

It follows from (11.7) and (11.8) that

$$\begin{aligned}\Delta u_k(t) &= \tau\big(\Delta y_k^s(t) - \Delta y_k(t)\big) + \tau_I \sum_{i=0}^{t-1} \big(\Delta y_k^s(i) - \Delta y_k(i)\big)\\ &= \tau\alpha\theta_k(t)e_{k-1}(t+1) - \tau\Delta y_k(t) + \tau_I \sum_{i=0}^{t-1} \big(\Delta y_k^s(i) - \Delta y_k(i)\big).\end{aligned} \tag{11.10}$$

Then, according to the definition of $e_k(t)$, it follows that

$$\begin{aligned}e_k(t+1) &= y_d(t+1) - y_k(t+1)\\ &= e_{k-1}(t+1) - \sum_{j=0}^{t-1} \phi_{k,k-1}^t(j)\Delta u_k(j) - \chi_k(t)\\ &\quad - \phi_{k,k-1}^t(t)\Big\{\tau\alpha\theta_k(t)e_{k-1}(t+1) - \tau\Delta y_k(t)\end{aligned}$$

$$+ \tau_{\mathrm{I}} \sum_{i=0}^{t-1} \left(\Delta y_k^s(i) - \Delta y_k(i) \right) \Big\}. \tag{11.11}$$

By differentiating $J(\theta_k(t))$ with respect to $\theta_k(t)$ and letting it be zero, one has

$$\theta_k(t) = \theta_{k-1}(t) + \frac{\rho \nu_k(t)}{\lambda + \nu_k^2(t)} \Big\{ e_{k-1}(t+1) - \tau_{\mathrm{I}} \phi_{k,k-1}^t(t) \sum_{i=0}^{t-1} \left[\Delta y_k^s(i) - \Delta y_k(i) \right]$$
$$- \sum_{i=0}^{t-1} \phi_{k,k-1}^t(i) \Delta u_k(i) + \tau \phi_{k,k-1}^t(t) \Delta y_k(t) - \theta_{k-1}(t) \nu_k(t) - \chi_k(t) \Big\}, \tag{11.12}$$

where $\nu_k(t) = \tau \alpha \phi_{k,k-1}^t(t) e_{k-1}(t+1)$ and $\rho \in (0, 2)$ is a step size to make the parameter estimation algorithm flexible.

Owing to that both $\phi_{k,k-1}^t(i)$, $i = 0, 1, \ldots, t$ and $\chi_k(t)$ are unavailable, the updating of $\theta_k(t)$ cannot be realized. To cope with this obstacle, $\chi_k(t)$ is simply omitted and meanwhile, $\phi_{k,k-1}^t(i)$ is substituted by its estimation obtained using the following parameter estimation algorithm (Chi et al. 2015)

$$\hat{\boldsymbol{\phi}}_{k,k-1}(t) = \hat{\boldsymbol{\phi}}_{k-1,k-2}(t) + \frac{\eta \Delta \boldsymbol{u}_{k-1}(t)}{\mu + \|\Delta \boldsymbol{u}_{k-1}(t)\|^2} \left(\Delta y_{k-1}(t+1) - \hat{\boldsymbol{\phi}}_{k-1,k-2}^{\mathrm{T}}(t) \Delta \boldsymbol{u}_{k-1}(t) \right), \tag{11.13}$$

where $\mu > 0$ and $\eta \in (0, 2)$ are two positive tuning parameters and $\hat{\boldsymbol{\phi}}_{k,k-1}(t) = \left[\hat{\phi}_{k,k-1}^t(0) \cdots \hat{\phi}_{k,k-1}^t(t) \right]^{\mathrm{T}}$ is the estimation of $\boldsymbol{\phi}_{k,k-1}(t)$. Note that the boundedness of $\hat{\boldsymbol{\phi}}_{k,k-1}(t)$ has been established in Chi et al. (2015) and its upper bound is denoted by $\beta_{\hat{\phi}}$, that is, $\sup_{t \in \mathbb{Z}_{N-1}, k \in \mathbb{Z}_+} \left\| \hat{\boldsymbol{\phi}}_{k,k-1}(t) \right\| \leq \beta_{\hat{\phi}}$, to facilitate the subsequent analysis.

For clarity, the proposed set-point learning control algorithm termed PI-ADDSPLC is summarized in the following Algorithm I.

Algorithm I (PI-ADDSPLC)

Input: Tuning parameters ρ, λ, η, μ, and α, user-specified thresholds ε_1 and ε_2, PI controller parameters τ_{P} and τ_{I}, desired output trajectory y_{d}, maximum batch number $k_{\max}$ and batch length *N*, initial values of control input $u_0(t)$, process output $y_0(t)$, gradient vector $\hat{\boldsymbol{\phi}}_0(t)$, and learning gain $\theta_0(t)$.

Output: Process output and ATE performance index.

for $k = 1,2,\ldots,k_{\max}$ **do**

  for $t = 1,2,\ldots,N$ **do**

    Update the gradient vector $\hat{\boldsymbol{\phi}}_{k,k-1}(t)$ by (11.13). If $\left\|\hat{\boldsymbol{\phi}}_{k,k-1}(t)\right\| \le \varepsilon_1$ or $\mathrm{sign}(\hat{\boldsymbol{\phi}}_{k,k-1}(t)) \ne \mathrm{sign}(\hat{\boldsymbol{\phi}}_0(t))$, $\hat{\boldsymbol{\phi}}_{k,k-1}(t)$ is reset to its initial value $\hat{\boldsymbol{\phi}}_0(t)$;

    Update the adaptive learning gain $\theta_k(t)$ by (11.12) with $\phi^t_{k,k-1}(i)$, $i = 0,1, \ ,t$ substituted by its estimation $\hat{\phi}^t_{k,k-1}(i)$ and $\chi_k(t)$ set as zero. If $|\theta_k(t)| \le \varepsilon_2$ or $\mathrm{sign}(\theta_k(t)) \ne \mathrm{sign}(\theta_0(t))$, $\theta_k(t)$ is reset to its initial value $\theta_0(t)$;

    Update the set-point command $y^s_k(t)$ by (11.8) and calculate the set-point tracking error $e^s_k(t)$ by;

    Apply the PI controller in (11.7) to a batch process in (11.1);

  end for

    Calculate the output tracking error $e_k(t)$ and ATE index defined by $\mathrm{ATE}(k) = \frac{1}{N}\sum_{t=1}^{N}|e_k(t)|$;

end for

In Algorithm I, ε_1 and ε_2 are two positive and small enough thresholds for initialization. Note that the resetting algorithms for $\hat{\boldsymbol{\phi}}_{k,k-1}(t)$ and $\theta_k(t)$ are adopted to enhance the parameter estimation performance, especially for estimating time-varying parameters, as mentioned in Yu et al. (2022) and Chi et al. (2023).

Remark 11.4 The proposed PI-ADDSPLC scheme could be extended to a PID-type ADDSPLC design, by the following PID controller

$$u_k(t) = \tau_P e_k^s(t) + \tau_I \sum_{i=0}^{t} e_k^s(i) + \tau_D\big[e_k^s(t) - e_k^s(t-1)\big]$$

as $u_k(t) = (\tau_P + \tau_I + \tau_D)e_k^s(t) + \tau_I \sum_{i=0}^{t-2} e_k^s(i) + (\tau_I - \tau_D)e_k^s(t-1)$. Then, the aforementioned design procedure could be straightforwardly applied, and therefore is omitted. ◊

11.3 Iterative Extended State Observer Based ADDSPLC Scheme

Although robust convergence and boundedness of the established learning system by the above PI-ADDSPLC scheme could be maintained, the adverse effect of nonrepetitive external disturbance and initial shifts cannot be attenuated actively. Another IESO-based PI-ADDSPLC design, as depicted in Fig. 11.2, is therefore presented by embedding an IESO into the above PI-ADDSPLC design for proactively estimating and offsetting $\chi_k(t)$ overlooked in the PI-ADDSPLC scheme, in order to acquire further improved disturbance attenuation performance but at the expense of the implemental complexity. It is noted that the uncertainty term $\chi_k(t)$ could be estimated by IESO along the batch direction, using only the measured process data, as depicted in Fig. 11.2.

For the subsequent analysis, the following assumption is imposed.

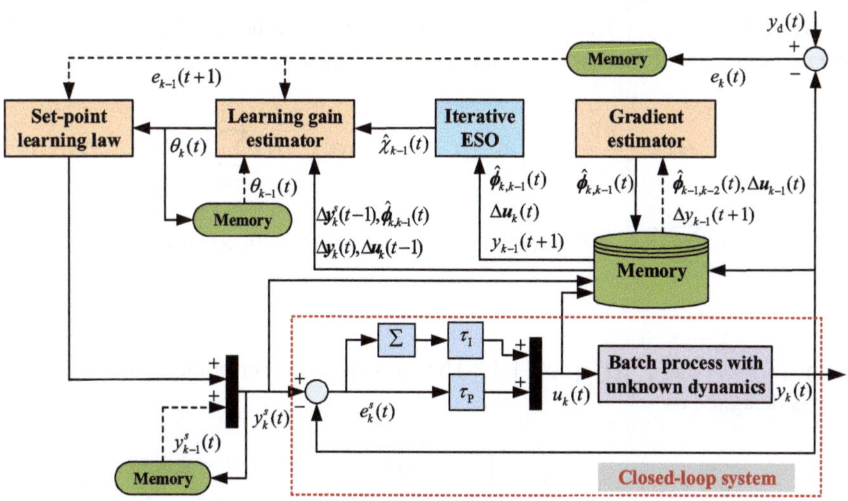

Fig. 11.2 Block diagram of the new IESO-based PI-ADDSPLC scheme

Assumption 11.5 The batch-direction variation of the set-point command $y_k^s(t)$ is bounded, that is, $\left|\Delta y_k^s(t)\right| \leq \delta_{\text{sp}}$ for any $k \in \mathbb{Z}_+$ and $t \in \mathbb{Z}_{N-1}$, where $\delta_{\text{sp}} > 0$ is a finite positive constant.

By augmenting the so-called *total disturbance* $\chi_k(t)$ as an extended system state in (11.6), one has

$$\begin{cases} \mathcal{Y}_k(t+1) = A\mathcal{Y}_{k-1}(t+1) + \mathcal{I}\boldsymbol{\phi}_{k,k-1}^{\text{T}}(t)\Delta\boldsymbol{u}_k(t) + \Omega_k(t), \\ y_{k-1}(t+1) = C\mathcal{Y}_{k-1}(t+1), \end{cases} \tag{11.14}$$

where $C = [\,1\ 0\,]$, $\mathcal{Y}_k(t+1) = [\,y_k(t+1)\ \chi_k(t)\,]^{\text{T}}$ and

$$A = \begin{bmatrix} 1 & 1 \\ 0 & 1 \end{bmatrix}, \mathcal{I} = \begin{bmatrix} 1 \\ 0 \end{bmatrix}, \Omega_k(t) = \begin{bmatrix} \Delta\chi_k(t) \\ \Delta\chi_k(t) \end{bmatrix}.$$

Due to the boundedness of $\chi_k(t)$, $\Omega_k(t)$ is ensured to be bounded for any $t \in \mathbb{Z}_{N-1}$ and $k \in \mathbb{Z}_+$, and satisfies $\sup_{t\in\mathbb{Z}_{N-1},k\in\mathbb{Z}_+}\|\Omega_k(t)\| \leq \beta_\Omega(\bullet)$, where $\beta_\Omega(\bullet)$ is a finite constant relying on $\beta_{\Delta\omega}$ and $\beta_{\Delta\delta}$.

An IESO for the above augmented data model in (11.14) is constructed as

$$\begin{aligned} \hat{\mathcal{Y}}_k(t+1) = {} & A\hat{\mathcal{Y}}_{k-1}(t+1) + \mathcal{I}\hat{\boldsymbol{\phi}}_{k,k-1}^{\text{T}}(t)\Delta\boldsymbol{u}_k(t) \\ & + L\big(y_{k-1}(t+1) - C\hat{\mathcal{Y}}_{k-1}(t+1)\big), \end{aligned} \tag{11.15}$$

where $L = [\,l_1\ l_2\,]^{\text{T}}$ denotes the IESO gain vector to be determined, $\hat{\mathcal{Y}}_{k-1}(t+1) = \left[\hat{y}_{k-1}(t+1)\ \hat{\chi}_{k-1}(t)\right]^{\text{T}}$ and $\hat{\boldsymbol{\phi}}_{k,k-1}(t)$ represents the respective estimations of $\mathcal{Y}_{k-1}(t+1)$ and $\boldsymbol{\phi}_{k,k-1}(t)$.

By replacing $\phi_{k,k-1}^t(i), i = 1, 2, \ldots, t$ and $\chi_k(t)$ using their estimations in (11.12), the updating law of the learning gain $\theta_k(t)$ is rectified as

$$\begin{aligned} \theta_k(t) = \theta_{k-1}(t) + \frac{\rho\hat{\upsilon}_k(t)}{\lambda + \hat{\upsilon}_k^2(t)} \Bigg\{ & e_{k-1}(t+1) - \tau_1\hat{\phi}_{k,k-1}^t(t)\sum_{i=0}^{t-1}\left[\Delta y_k^s(i) - \Delta y_k(i)\right] \\ & - \sum_{i=0}^{t-1}\hat{\phi}_{k,k-1}^t(i)\Delta u_k(i) + \tau\hat{\phi}_{k,k-1}^t(t)\Delta y_k(t) - \theta_{k-1}(t)\hat{\upsilon}_k(t) - \hat{\chi}_k(t)\Bigg\}, \end{aligned} \tag{11.16}$$

with $\hat{\upsilon}_k(t) = \tau\alpha\hat{\phi}_{k,k-1}^t(t)e_{k-1}(t+1)$. The corresponding set-point learning updating law is designed as

$$y_k^s(t) = y_{k-1}^s(t) + \begin{cases} \zeta_k(t), & \text{if}|\zeta_k(t)| \leq \delta_{\text{sp}} \\ \delta_{\text{sp}}\text{sign}(\zeta_k(t)), & \text{otherwise}, \end{cases} \tag{11.17}$$

where $\zeta_k(t) = \alpha\theta_k(t)e_{k-1}(t+1)$ and $\text{sign}(\cdot)$ denotes the sign function.

The above IESO based PI-ADDSPLC design is summarized in Algorithm II.

Algorithm II (IESO based PI-ADDSPLC)

Input: Tuning parameters ρ, λ, η, μ and α, user specified thresholds ε_1 and ε_2, PI controller parameters τ_{P} and τ_{I}, desired output trajectory y_{d}, maximum batch number $k_{\max}$ and batch length *N*, initial values of control input $u_0(t)$, process output $y_0(t)$, gradient vector $\boldsymbol{\phi}_0(t)$, learning gain $\theta_0(t)$ and IESO state $\hat{\mathcal{Y}}_0(t)$.

Output: Process output and ATE index.

1: **for** $k = 1, 2, \ldots, k_{\max}$ **do**

2: **for** $t = 1, 2, \ldots, N$ **do**

3: Update the gradient vector $\hat{\boldsymbol{\phi}}_{k,k-1}(t)$ by (11.13). If $\left\|\hat{\boldsymbol{\phi}}_{k,k-1}(t)\right\| \le \varepsilon_1$ or $\mathrm{sign}(\hat{\boldsymbol{\phi}}_{k,k-1}(t)) \ne \mathrm{sign}(\hat{\boldsymbol{\phi}}_0(t))$, $\hat{\boldsymbol{\phi}}_{k,k-1}(t)$ is reset as its initial value $\hat{\boldsymbol{\phi}}_0(t)$;

4: Update the IESO state $\hat{\mathcal{Y}}_k(t+1)$ by (11.15);

5: Update the adaptive learning gain $\theta_k(t)$ by (11.16). If $|\theta_k(t)| \le \varepsilon_2$ or $\mathrm{sign}(\theta_k(t)) \ne \mathrm{sign}(\theta_0(t))$, $\theta_k(t)$ is reset as its initial value $\theta_0(t)$;

6: Update the set-point command $y_k^s(t)$ by (11.17) and calculate the set-point tracking error $e_k^s(t)$ by **Error! Reference source not found.**;

7: Apply the PI controller in (11.7) to a batch process in (11.1);

8: **end for**

9: Calculate the output tracking error $e_k(t)$ and ATE index;

10: **end for**

11.4 Robust Convergence Analysis

The following theorem characterizes sufficient conditions to guarantee the robust convergence of the output tracking error, together with the boundedness of the adaptive learning gain and set-point command along the batch direction, under the proposed PI-ADDSPLC scheme.

Theorem 11.1 Consider a batch process in (11.1) under Assumptions 11.1, 11.2 and 11.4 where Algorithm I is applied. Then the bounded tracking objective in (11.2) along with the bounded set-point command in (11.3), and the adaptive learning gain in (11.8) are satisfied, if the tuning parameter α is appropriately chosen such that

$$|\tau\alpha| < \frac{2}{\beta_{\phi}\iota}, \tag{11.18}$$

$$\operatorname{sign}(\tau\alpha) = \operatorname{sign}(\theta_0(t)), \tag{11.19}$$

where $\iota < \infty$ denotes a uniform upper bound of the adaptive learning gain $\theta_k(t)$ for any $t \in \mathbb{Z}_{N-1}$ and $k \in \mathbb{Z}_+$.

Proof It follows from (11.11) that

$$e_k(t+1) = \mathcal{A}_k^e(t)e_{k-1}(t+1) + r_{1,k}(t), \tag{11.20}$$

where $\mathcal{A}_k^e(t) \triangleq 1 - \tau\alpha\phi_{k,k-1}^t(t)\theta_k(t)$ and

$$\begin{aligned} r_{1,k}(t) \triangleq & -\sum_{j=0}^{t-1}\phi_{k,k-1}^t(j)\Delta u_k(j) + \tau\phi_{k,k-1}^t(t)\Delta y_k(t) \\ & - \tau_1\phi_{k,k-1}^t(t)\sum_{j=0}^{t-1}\left[\Delta y_k^s(j) - \Delta y_k(j)\right] - \chi_k(t). \end{aligned}$$

Letting $i = k-1$ and $j = 0$ in (11.4) gives

$$\begin{aligned} y_{k-1}(t+1) =& y_0(t+1) + \phi_{k-1,0}^t(t)u_{k-1}(t) \\ & + \sum_{j=0}^{t-1}\phi_{k-1,0}^t(j)u_{k-1}(j) - \sum_{j=0}^{t}\phi_{k-1,0}^t(j)u_0(j) \\ & + \sum_{j=0}^{t-1}\upsilon_{k-1,0}^t(j)\left[\omega_{k-1}(j) - \omega_0(j)\right] \\ & + \left[\omega_{k-1}(t) - \omega_0(t)\right] + \vartheta_{k-1,0}^t(\delta_{k-1} - \delta_0). \end{aligned} \tag{11.21}$$

Substituting (11.21) into the updating law of the set-point command in (11.8) yields

$$y_k^s(t) = \mathcal{A}_k^s(t) y_{k-1}^s(t) + r_{2,k}(t), \tag{11.22}$$

where $\mathcal{A}_k^s(t) \triangleq 1 - \alpha\tau\theta_k(t)\phi_{k-1,0}^t(t)$ and

$$\begin{aligned} r_{2,k}(t) \triangleq & \alpha\theta_k(t)\big\{e_0(t+1) + \tau\phi_{k-1,0}^t(t) y_{k-1}(t) \\ & - \tau_{\mathrm{I}}\phi_{k-1,0}^t(t)\sum_{i=0}^{t-1}\big[y_{k-1}^s(i) - y_{k-1}(i)\big] - \sum_{j=0}^{t-1}\phi_{k-1,0}^t(j)u_{k-1}(j) \\ & + \sum_{j=0}^{t}\phi_{k-1,0}^t(j)u_0(j) - \big[\omega_{k-1}(t) - \omega_0(t)\big] - \vartheta_{k-1,0}^t(\delta_{k-1} - \delta_0) \\ & - \sum_{j=0}^{t-1}\upsilon_{k-1,0}^t(j)\big[\omega_{k-1}(j) - \omega_0(j)\big]. \end{aligned}$$

Also, setting $\chi_k(t) = 0$ in (11.12) and replacing $\phi_{k,k-1}^t(t)$ by its estimation result in

$$\theta_k(t) = \left[1 - \frac{\rho\hat{\upsilon}_k^2(t)}{\lambda + \hat{\upsilon}_k^2(t)}\right]\theta_{k-1}(t) + \frac{\rho\hat{\upsilon}_k(t)}{\lambda + \hat{\upsilon}_k^2(t)} e_{k-1}(t+1) + r_{3,k}(t), \tag{11.23}$$

where

$$\begin{aligned} r_{3,k}(t) \triangleq \frac{\rho\hat{\upsilon}_k(t)}{\lambda + \hat{\upsilon}_k^2(t)}\Bigg\{ & \tau\hat{\phi}_{k,k-1}^t(t)\Delta y_k(t) - \tau_{\mathrm{I}}\hat{\phi}_{k,k-1}^t(t)\sum_{i=0}^{t-1}\big[\Delta y_k^s(i) - \Delta y_k(i)\big] \\ & - \sum_{i=0}^{t-1}\hat{\phi}_{k,k-1}^t(i)\Delta u_k(i)\Bigg\}. \end{aligned}$$

In the following, the bounded convergence of the output tracking error $e_k(t)$ along with the boundedness of the set-point command $y_k^s(t)$ and the adaptive learning gain $\theta_k(t)$ is proved by the mathematical induction for any $t \in \mathbb{Z}_{N-1}$ and $k \in \mathbb{Z}_+$.

Step 1 For $t = 0$, one has

$$\big|r_{3,k}(0)\big| = \left|\frac{\rho\hat{\upsilon}_k(0)}{\lambda + \hat{\upsilon}_k^2(0)}\tau\hat{\phi}_{k,k-1}^t(0)\Delta y_k(0)\right| \le \frac{\rho}{2\sqrt{\lambda}}|\tau|\beta_{\hat{\phi}}\beta_{\Delta\delta} < \infty.$$

It is easy to verify that for any t and k the second term at the right-hand side of (11.23) is always bounded since

$$\left|\frac{\rho\hat{\upsilon}_k(t)}{\lambda+\hat{\upsilon}_k^2(t)}e_{k-1}(t+1)\right| \leq \left|\frac{\rho\tau\alpha\hat{\phi}_{k,k-1}^t(t)e_{k-1}^2(t+1)}{\lambda+\left[\tau\alpha\hat{\phi}_{k,k-1}^t(t)e_{k-1}(t+1)\right]^2}\right|$$
$$\leq \left|\frac{\rho}{\tau\alpha\hat{\phi}_{k,k-1}^t(t)}\right| \leq \frac{\rho}{\tau|\alpha|\varepsilon_1}. \tag{11.24}$$

Owing to $\rho \in (0, 2)$, one has

$$\left|1-\frac{\rho\hat{\upsilon}_k^2(t)}{\lambda+\hat{\upsilon}_k^2(t)}\right| < 1, \forall t \in \mathbb{Z}_{N-1}, k \in \mathbb{Z}_+. \tag{11.25}$$

Moreover, it follows from (11.23) that $\theta_k(0)$ is bounded for any $k \in \mathbb{Z}_+$, and satisfies $\sup_{k\in\mathbb{Z}_+}|\theta_k(0)| \leq \beta_\theta(0,\bullet) < \infty$, where $\beta_\theta(0,\bullet) > 0$ is a finite constant.

It is not difficult to verify that

$$\left|r_{1,k}(0)\right| = \left|\tau\phi_{k,k-1}^t(0)\Delta y_k(0) - \chi_k(0)\right| \leq |\tau|\beta_\phi\beta_{\Delta\delta} + \beta_\chi(\bullet) < \infty.$$

By selecting $\iota = \beta_\theta(0,\bullet)$ along with the conditions in (11.18) and (11.19), one has $\left|\mathcal{A}_k^e(0)\right| < 1$ for any $k \in \mathbb{Z}_+$, implying that $e_k(1)$ is boundedly convergent and satisfies

$$\sup_{k\in\mathbb{Z}_+}|e_k(1)| \leq \beta_e(1,\bullet), \limsup_{k\to\infty}|e_k(1)| \leq \beta_{e_{\sup}}(1,\bullet),$$

where two finite positive constants $\beta_e(1,\bullet)$ and $\beta_{e_{\sup}}(1,\bullet)$ satisfying $\beta_e(1,\bullet) > \beta_{e_{\sup}}(1,\bullet) > 0$ are associated with the upper bound of $r_{1,k}(0)$, thus relying on both $\beta_{\Delta\omega}$ and $\beta_{\Delta\delta}$.

As a result, there follow

$$\left|\Delta y_k^s(0)\right| = |\alpha\theta_k(0)e_{k-1}(1)| \leq |\alpha|\beta_\theta(0,\bullet)\beta_e(0,\bullet) < \infty$$

and

$$|\Delta u_k(0)| = \left|\tau\left(\Delta y_k^s(0) - \Delta y_k(0)\right)\right| \leq |\tau|[|\alpha|\beta_\theta(0,\bullet)\beta_e(0,\bullet) + \beta_{\Delta\delta}] < \infty.$$

By Assumptions 11.1 and 11.4, and Lemma 11.1, together with the boundedness of $\theta_k(0)$ and $e_k(1)$, there follows

$$\begin{aligned}\left|r_{2,k}(0)\right| &= \left|\alpha\theta_k(0)\left\{e_0(1) + \tau\phi_{k-1,0}^t(0)y_{k-1}(0) + \phi_{k-1,0}^t(0)u_0(0)\right.\right.\\ &\quad \left.\left.-\left[\omega_{k-1}(0) - \omega_0(0)\right] - \vartheta_{k-1,0}^t(\delta_{k-1} - \delta_0)\right\}\right|\\ &\leq |\alpha|\beta_\theta(0,\bullet)\left\{\beta_e(1,\bullet) + |\tau|\beta_\phi(|y_0| + \beta_\delta)\right.\\ &\quad \left.+2\left(\beta_\omega + \beta_\phi\beta_\delta\right)\right\} < \infty, \ \forall k \in \mathbb{Z}_+.\end{aligned}$$

Using the conditions in (11.18) and (11.19) again with $\iota = \beta_\theta(0, \bullet)$, one has $\left|\mathcal{A}_k^s(0)\right| < 1$ for any $k \in \mathbb{Z}_+$. Therefore, $y_k^s(0)$ is bounded, that is, $\sup_{k\in\mathbb{Z}_+}\left|y_k^s(0)\right| \le \beta_s(0, \bullet) < \infty$, where $\beta_s(0, \bullet) > 0$ is a finite positive constant.

Step 2 Assume that for any $t \in \mathbb{Z}_{T-1}$ with $T \in \mathbb{Z}_{N-1}$, the following conclusions are satisfied

$$\sup_{k\in\mathbb{Z}_+} |\theta_k(t)| \le \beta_\theta(t, \bullet) < \infty, \tag{11.26}$$

$$\sup_{k\in\mathbb{Z}_+} |e_k(t+1)| \le \beta_e(t+1, \bullet) < \infty, \tag{11.27}$$

$$\lim_{k\to\infty} \sup|e_k(t+1)| \le \beta_{e_{\text{sup}}}(t+1, \bullet) < \infty, \tag{11.28}$$

$$\sup_{k\in\mathbb{Z}_+} \left|y_k^s(t)\right| \le \beta_s(t, \bullet) < \infty, \tag{11.29}$$

where $\beta_\theta(t, \bullet)$, $\beta_s(t, \bullet)$, $\beta_e(t+1, \bullet)$ and $\beta_{e_{\text{sup}}}(t+1, \bullet)$ are finite positive constants for any $t \in \mathbb{Z}_{T-1}$, and satisfy $\beta_e(t+1, \bullet) > \beta_{e_{\text{sup}}}(t+1, \bullet) > 0$ relying on both $\beta_{\Delta\omega}$ and $\beta_{\Delta\delta}$.

When $t = T$, one has

$$\begin{aligned}
\left|r_{3,k}(T)\right| \le& \frac{\rho|\hat{\upsilon}_k(T)|}{2\sqrt{\lambda}|\hat{\upsilon}_k(T)|}\Bigg|\tau_{\text{I}}\hat{\phi}_{k,k-1}^t(T)\sum_{i=0}^{T-1}\left[\Delta y_k^s(t) - \Delta y_k(i)\right] \\
&+\sum_{i=0}^{T-1}\hat{\phi}_{k,k-1}^t(i)\Delta u_k(i) - \tau\hat{\phi}_{k,k-1}^t(T)\Delta y_k(T)\Bigg| \\
\le& \frac{\rho}{2\sqrt{\lambda}}\Bigg\{|\tau_{\text{I}}|\beta_{\hat{\phi}}T\max_{i\in\mathbb{Z}_{T-1}}\left[\left|\Delta y_k^s(i)\right| + |\Delta y_k(i)|\right] \\
&+T\beta_{\hat{\phi}}\max_{i\in\mathbb{Z}_{T-1}}|\Delta u_k(i)| + |\tau|\beta_{\hat{\phi}}|\Delta y_k(T)|\Bigg\}.
\end{aligned}$$

Additionally, it follows from (11.26)–(11.29) that

$$\begin{aligned}
&\max_{i\in Z_{T-1}}\left|\Delta y_k^s(i)\right| \le 2\beta_{s,T-1}^{\max}(\bullet), \\
&\max_{i\in Z_{T-1}}|\Delta y_k(i)| \le 2\beta_{e,T-1}^{\max}(\bullet), \\
&\max_{i\in Z_{T-1}}|\Delta u_k(i)| \le 2|\tau|\max_{i\in Z_{T-1}}\{\beta_s(i, \bullet) + \beta_e(i, \bullet)\} \\
&\qquad + \max_{i\in Z_{T-1}} 2|\tau_{\text{I}}|i\max_{j\in Z_{i-1}}\{\beta_s(j, \bullet) + \beta_e(j, \bullet)\} \\
&\qquad \le 2[|\tau| + |\tau_{\text{I}}|(T-1)]\big(\beta_{s,T-1}^{\max}(\bullet) + \beta_{e,T-1}^{\max}(\bullet)\big),
\end{aligned}$$

where

$$\beta_{s,T-1}^{\max}(\bullet) \triangleq \max_{i \in Z_{T-1}} \beta_s(i, \bullet), \beta_{e,T-1}^{\max}(\bullet) \triangleq \max_{i \in Z_{T-1}} \beta_e(i, \bullet).$$

Therefore,

$$\begin{aligned} \left|r_{3,k}(T)\right| \leq \frac{\rho}{2\sqrt{\lambda}}&\left[2|\tau_{\mathrm{I}}|\beta_{\hat{\phi}}T\left(\beta_{s,T-1}^{\max}(\bullet) + \beta_{e,T-1}^{\max}(\bullet)\right) + 2|\tau|\beta_{\hat{\phi}}\beta_e(T, \bullet)\right. \\ &\left.+2T\beta_{\hat{\phi}}[|\tau| + |\tau_{\mathrm{I}}|(T-1)]\left(\beta_{s,T-1}^{\max}(\bullet) + \beta_{e,T-1}^{\max}(\bullet)\right)\right] < \infty, \end{aligned} \tag{11.30}$$

and hence the boundedness of $r_{3,k}(T)$ for any $k \in \mathbb{Z}_+$ is ensured, and its upper bound is associated with both $\beta_{\Delta\omega}$ and $\beta_{\Delta\delta}$.

According to the inequalities (11.24) and (11.25), it follows from (11.23) and (11.30) that $\sup_{k \in \mathbb{Z}_+} |\theta_k(T)| \leq \beta_\theta(T, \bullet) < \infty$ with $\beta_\theta(T, \bullet)$ being a finite positive constant. Therefore, one has

$$\begin{aligned} \left|r_{1,k}(T)\right| &\leq \beta_\phi \sum_{j=0}^{T-1} |\Delta u_k(j)| + |\tau|\beta_\phi|\Delta y_k(T)| \\ &\quad + |\chi_k(T)| + |\tau_{\mathrm{I}}|\beta_\phi \sum_{j=0}^{T-1} \left[\left|\Delta y_k^s(j)\right| + |\Delta y_k(j)|\right] \\ &\leq 2\beta_\phi T(|\tau| + |\tau_{\mathrm{I}}|(T-1))\left(\beta_{s,T-1}^{\max}(\bullet) + \beta_{e,T-1}^{\max}(\bullet)\right) + 2|\tau|\beta_\phi\beta_e(T, \bullet) \\ &\quad + 2|\tau_{\mathrm{I}}|\beta_\phi T\left(\beta_{s,T-1}^{\max}(\bullet) + \beta_{e,T-1}^{\max}(\bullet)\right) + \beta_\chi(\bullet) < \infty. \end{aligned}$$

Analogously, by the premises given in (11.18) and (11.19) with $\iota = \beta_\theta(T, \bullet)$, one has $\left|\mathcal{A}_k^e(T)\right| < 1$ for any $k \in \mathbb{Z}_+$, ensuring that $e_k(T+1)$ is boundedly convergent for any $k \in \mathbb{Z}_+$, i.e.

$$\sup_{k \in \mathbb{Z}_+} |e_k(T+1)| \leq \beta_e(T+1, \bullet) < \infty,$$

$$\limsup_{k \to \infty} |e_k(T+1)| \leq \beta_{e_{\sup}}(T+1, \bullet) < \beta_e(T+1, \bullet)$$

with $\beta_e(T+1, \bullet)$ and $\beta_{e_{\sup}}(T+1, \bullet)$ being two finite constants relying on both $\beta_{\Delta\omega}$ and $\beta_{\Delta\delta}$.

Therefore, one has

$$\left|\Delta y_k^s(T)\right| = |\alpha\theta_k(T)e_{k-1}(T+1)| \leq |\alpha|\beta_\theta(T, \bullet)\beta_e(T+1, \bullet) < \infty$$

and

$$\begin{aligned} |\Delta u_k(T)| &= \left|\tau\Delta e_k^s(T) + \tau_{\mathrm{I}} \sum_{i=0}^{T-1} \Delta e_k^s(i)\right| \\ &\leq |\tau|[|\alpha|\beta_\theta(T, \bullet)\beta_e(T+1, \bullet) + 2\beta_e(T, \bullet)] \\ &\quad + 2|\tau_{\mathrm{I}}|T\left(\beta_{s,T-1}^{\max}(\bullet) + \beta_{e,T-1}^{\max}(\bullet)\right) < \infty. \end{aligned}$$

By virtue of the boundedness of $u_{k-1}(t)$ for any $t \in \mathbb{Z}_{T-1}$ and $y_{k-1}(t)$ for any $t \in \mathbb{Z}_T$, it follows that

$$|y_{k-1}(t)| \le |y_{\mathrm{d}}(t)| + |e_{k-1}(t)| \le \beta_{\mathrm{d}} + \beta_{e,T}^{\max}(\bullet),$$

$$\begin{aligned} |u_{k-1}(t)| &\le |\tau|\left(\left|y_{k-1}^{s}(t)\right| + |y_{k-1}(t)|\right) + |\tau_{\mathrm{I}}| \sum_{i=0}^{t-1} \left[\left|y_{k-1}^{s}(i)\right| + |y_{k-1}(i)|\right], \\ &\le (|\tau| + (T-1)|\tau_{\mathrm{I}}|)\left(\beta_{s,T-1}^{\max}(\bullet) + \beta_{e,T}^{\max}(\bullet) + \beta_{\mathrm{d}}\right), \end{aligned}$$

where $\beta_{e,T}^{\max}(\bullet) \triangleq \max_{i \in Z_T} \beta_e(i, \bullet)$. Further, one has

$$\begin{aligned} \left|r_{2,k}(T)\right| \le |\alpha|\beta_\theta(T, \bullet)\big[&\beta_e(T+1, \bullet) + |\tau|\beta_\phi\left(\beta_{\mathrm{d}} + \beta_{e,T}^{\max}(\bullet)\right) \\ &+ \beta_\phi|\tau_{\mathrm{I}}|T\left(\beta_{s,T-1}^{\max}(\bullet) + \beta_{e,T}^{\max}(\bullet) + \beta_{\mathrm{d}}\right) \\ &+ \beta_\phi T(|\tau| + (T-1)|\tau_{\mathrm{I}}|)\left(\beta_{s,T-1}^{\max}(\bullet) + \beta_{e,T}^{\max}(\bullet) + \beta_{\mathrm{d}}\right) \\ &+ 2(\beta_\omega + T\beta_\phi\beta_\omega + \beta_\phi\beta_\delta)\big] < \infty, \end{aligned}$$

which along with the conditions in (11.18) and (11.19) with $\iota = \beta_\theta(T, \bullet)$ ensures that $\left|\mathcal{A}_k^s(T)\right| < 1$ for any $k \in \mathbb{Z}_+$. Hence, $y_k^s(T)$ is bounded, and satisfies $\sup_{k \in \mathbb{Z}_+} \left|y_k^s(T)\right| \le \beta_s(T, \bullet)$ with $\beta_s(T, \bullet) > 0$ being a finite positive constant.

Based on mathematical induction, the theorem is proved. Due to that $t \in \mathbb{Z}_{N-1}$ is finite, a uniform $\iota = \max_{i \in \mathbb{Z}_{N-1}} \beta_\theta(i, \bullet)$ can be attained. The proof is complete. □

Remark 11.5 Owing to that $\theta_k(t)$ and $\boldsymbol{\phi}_{k,k-1}^t(t)$ are iteratively estimated, the tuning parameter α plays a vital role in the premises of (11.18) and (11.19) for the set-point learning control. Besides, in terms of the boundedness of the set-point command and the output tracking error established in Theorem 11.1, together with the specific form of the PI controller in (11.7), the control input by the above PI-ADDSPLC scheme is guaranteed to be bounded, as needed by the previous DDILC methods (e.g., Meng and Zhang, 2022; Yu et al. 2020a, b; Yu et al. 2022; Chi et al. 2018). ◊

If both initial shifts and nonrepetitive external disturbances converge ultimately along the batch direction, that is

$$\begin{aligned} &\lim_{k \to \infty} \left[\omega_k(t) - \omega_{k-1}(t)\right] = 0, \quad \forall t \in \mathbb{Z}_{N-1}, \\ &\lim_{k \to \infty} (\delta_k - \delta_{k-1}) = 0, \end{aligned} \tag{11.31}$$

exact output tracking could be achieved in spite of the presence of nonrepetitive external disturbances and initial condition shifts, as claimed by the following corollary.

Corollary 11.1 Consider a batch process in (11.1) under Assumptions 11.1 and 11.2, and the condition in (11.31) where Algorithm I is applied. Then, the output tracking error is asymptotically convergent along the batch direction, and the set-point command in (11.3) and the adaptive learning gain in (11.8) are ensured to be bounded provided that the tuning parameter α is chosen to obey conditions in (11.18) and (11.19).

Proof Based on the condition (11.31), it is concluded from (11.6) that

$$\lim_{k\to\infty} \chi_k(t) = 0.$$

Similar to the proof of Theorem 11.1, the proof of this corollary follows straightforwardly, and thus is omitted. □

Sufficient conditions for guaranteeing the robust convergence of the output tracking error under the proposed Algorithm II is established in the following theorem, when nonrepetitive uncertainties occur.

Theorem 11.2 Consider a batch process in (11.1) under Assumptions 11.1, 11.2, 11.4 and 11.5 where the Algorithm II is applied. If the tuning parameter α is chosen to obey conditions in (11.18) and (11.19), and the IESO gain l_1 and l_2 are taken to satisfy

$$\max\left\{\frac{\left|2 - l_1 + \sqrt{l_1^2 - 4l_2}\right|}{2}, \frac{\left|2 - l_1 - \sqrt{l_1^2 - 4l_2}\right|}{2}\right\} < 1, \tag{11.32}$$

it is concluded that the bounded tracking objective in (11.2), the bounded set-point command in (11.3) and the adaptive learning gain in (11.8), along with the bounded convergence of IESO in (11.15) hold true.

Proof Using the boundedness of $\chi_k(t)$ for any $t \in \mathbb{Z}_{N-1}$ and $k \in \mathbb{Z}_+$, one has $\hat{\boldsymbol{\phi}}_{k,k-1}(t)$ is bounded, whose proof is analogous to that for Theorem 6 in Hao et al. (2021) and therefore is omitted.

Rewrite the evolution of the IESO estimation error as

$$\tilde{\mathcal{Y}}_k(t+1) = (A - LC)\tilde{\mathcal{Y}}_{k-1}(t+1) + \mathcal{I}\tilde{\boldsymbol{\phi}}_{k,k-1}^{\mathrm{T}}(t)\Delta \boldsymbol{u}_k(t) + \Omega_k(t), \tag{11.33}$$

where $\tilde{\boldsymbol{\phi}}_{k,k-1}(t) = \boldsymbol{\phi}_{k,k-1}(t) - \hat{\boldsymbol{\phi}}_{k,k-1}(t)$ and $\tilde{\mathcal{Y}}_k(t) = \mathcal{Y}_k(t) - \hat{\mathcal{Y}}_k(t)$. According to the boundedness of $\boldsymbol{\phi}_{k,k-1}(t)$ and $\hat{\boldsymbol{\phi}}_{k,k-1}(t)$, one has $\tilde{\boldsymbol{\phi}}_{k,k-1}(t)$ is bounded for any $t \in \mathbb{Z}_{N-1}$ and $k \in \mathbb{Z}_+$, and satisfies $\left\|\tilde{\boldsymbol{\phi}}_{k,k-1}(t)\right\| \le \beta_{\tilde{\phi}} < \infty$, where $\beta_{\tilde{\phi}}$ is a finite positive constant.

Applying the PI control law in (11.7) into (11.33) gives

$$\tilde{\mathcal{Y}}_k(t+1) = (A - LC)\tilde{\mathcal{Y}}_{k-1}(t+1) + r_{4,k}(t), \tag{11.34}$$

where

$$r_{4,k}(t) = \mathcal{I}\tau\tilde{\phi}^t_{k,k-1}(t)\big[\Delta y^s_k(t) - \Delta y_k(t)\big] + \mathcal{I}\tau_1\tilde{\phi}^t_{k,k-1}(t)\sum_{i=0}^{t-1}\big[\Delta y^s_k(i) - \Delta y_k(i)\big]$$
$$+\mathcal{I}\sum_{j=0}^{t-1}\tilde{\phi}^t_{k,k-1}(j)\Delta u_k(j) + \Omega_k(t).$$

Denote $h_k(t) = \Delta y^s_k(t)/\zeta_k(t)$. Then one has that $h_k(t) \in (0, 1)$ if $|\zeta_k(t)| > \delta_{\text{sp}}$ and $h_k(t) = 1$ if $|\zeta_k(t)| \le \delta_{\text{sp}}$. Hence, the dynamics of the output tracking error along the batch direction is written as

$$e_k(t+1) = \overline{\mathcal{A}}^e_k(t)e_{k-1}(t+1) + r_{1,k}(t), \tag{11.35}$$

where $\overline{\mathcal{A}}^e_k(t) \triangleq 1 - \tau\alpha h_k(t)\phi^t_{k,k-1}(t)\theta_k(t)$ and $r_{1,k}(t)$ is descried by (11.20).

In addition, it follows that

$$\begin{aligned} y^s_k(t) &= y^s_{k-1}(t) + h_k(t)\zeta_k(t) \\ &= y^s_{k-1}(t) + h_k(t)\alpha\theta_k(t)e_{k-1}(t+1) \\ &= y^s_{k-1}(t) + h_k(t)\alpha\theta_k(t)\big[y_{\text{d}}(t+1) - y_{k-1}(t+1)\big], \end{aligned}$$

which can be combined with (11.21) to get

$$y^s_k(t) = \overline{\mathcal{A}}^s_k(t)y^s_{k-1}(t) + h_k(t)r_{2,k}(t), \tag{11.36}$$

where $\overline{\mathcal{A}}^s_k(t) \triangleq 1 - \alpha\tau h_k(t)\theta_k(t)\phi^t_{k-1,0}(t)$ and $r_{2,k}(t)$ is described by (11.22).

Reformulating the updating law of $\theta_k(t)$ in (11.16) gives

$$\theta_k(t) = \left[1 - \frac{\rho\hat{v}^2_k(t)}{\lambda + \hat{v}^2_k(t)}\right]\theta_{k-1}(t) + \frac{\rho\hat{v}_k(t)}{\lambda + \hat{v}^2_k(t)}e_{k-1}(t+1)$$
$$+ r_{3,k}(t) - \frac{\rho\hat{v}_k(t)}{\lambda + \hat{v}^2_k(t)}\hat{\chi}_k(t), \tag{11.37}$$

where $r_{3,k}(t)$ is described as per (11.23).

Similar to the proof of Theorem 11.1, the main conclusion of this theorem is also proved by using the mathematical induction.

Step 1 When $t = 0$, one has

$$r_{4,k}(0) = \mathcal{I}\tau\tilde{\phi}^t_{k,k-1}(0)\big[\Delta y^s_k(0) - \Delta y_k(0)\big] + \Omega_k(0).$$

Owing to that $\sup_{t\in\mathbb{Z}_{N-1},k\in\mathbb{Z}_+}\left|\Delta y^s_k(0)\right| \le \delta_{\text{sp}}$ along with $\Delta y_k(0) = y_k(0) - y_{k-1}(0) = \delta_k - \delta_{k-1}$, it follows that

$$\left|r_{4,k}(0)\right| \le |\tau|\beta_{\tilde{\phi}}\left(\delta_{\rm sp} + \beta_{\Delta\delta}\right) + \beta_{\Omega}(\bullet) < \infty. \tag{11.38}$$

Consequently, when the premise in (11.32) is satisfied and hence $A - LC$ is Schur stable, $\tilde{\mathcal{Y}}_k(1)$ is boundedly convergent along the batch direction. This further indicates that $\hat{\chi}_k(0)$ is bounded for any $k \in \mathbb{Z}_+$, and satisfies $\sup_{k\in\mathbb{Z}_+}|\hat{\chi}_k(0)| \le \beta_{\hat{\chi}}(0, \bullet) < \infty$, where $\beta_{\hat{\chi}}(0, \bullet)$ is a finite positive constant associated with both $\beta_{\Delta\omega}$ and $\beta_{\Delta\delta}$.

It follows the boundedness of $\hat{\chi}_k(0)$ that

$$\begin{aligned}\left|r_{3,k}(0) - \frac{\rho\hat{\upsilon}_k(0)}{\lambda + \hat{\upsilon}_k^2(0)}\hat{\chi}_k(0)\right| &= \left|\frac{\rho\hat{\upsilon}_k(0)}{\lambda + \hat{\upsilon}_k^2(0)}\left[\tau\hat{\phi}_{k,k-1}^t(0)\Delta y_k(0) - \hat{\chi}_k(0)\right]\right| \\ &\le \frac{\rho}{2\sqrt{\lambda}}\left(\beta_{\hat{\chi}}(0, \bullet) + |\tau|\beta_{\hat{\phi}}\beta_{\Delta\delta}\right) < \infty,\end{aligned}$$

which along with the boundedness of $\frac{\rho\hat{\upsilon}_k(t)}{\lambda+\hat{\upsilon}_k^2(t)}e_{k-1}(t+1)$ proved in Theorem 11.1 and the condition in (11.25), ensures that $\theta_k(0)$ is bounded and satisfies $\sup_{k\in\mathbb{Z}_+}|\theta_k(0)| \le \beta_\theta(0, \bullet) < \infty$, where $\beta_\theta(0, \bullet) > 0$ is a finite constant associated with both $\beta_{\Delta\omega}$ and $\beta_{\Delta\delta}$.

Hence, one has

$$\left|r_{1,k}(0)\right| = \left|\tau\phi_{k,k-1}^t(0)\Delta y_k(0) - \chi_k(0)\right| \le |\tau|\beta_\phi\beta_{\Delta\delta} + \beta_\chi(\bullet) < \infty, \quad \forall k \in \mathbb{Z}_+.$$

Using the conditions in (11.18) and (11.19) with $\iota = \beta_\theta(0, \bullet)$, one has $\left|\overline{\mathcal{A}}_k^e(0)\right| < 1$ for any $k \in \mathbb{Z}_+$, indicating that $e_k(1)$ is convergent and bounded along the batch direction, that is

$$\sup_{k\in\mathbb{Z}_+}|e_k(1)| \le \beta_e(1, \bullet), \ \lim_{k\to\infty}\sup|e_k(1)| \le \beta_{e_{\rm sup}}(1, \bullet),$$

where $\beta_e(1, \bullet)$ and $\beta_{e_{\rm sup}}(1, \bullet)$ are two finite positive constants associated with both $\beta_{\Delta\omega}$ and $\beta_{\Delta\delta}$.

Consequently, there follows

$$\begin{aligned}\left|h_k(0)r_{2,k}(0)\right| &= \left|\alpha h_k(0)\theta_k(0)\left\{e_0(1) + \tau\phi_{k-1,0}^t(0)y_{k-1}(0) + \phi_{k-1,0}^t(0)u_0(0)\right.\right. \\ &\quad \left.\left. - \left[\omega_{k-1}(0) - \omega_0(0)\right] - \vartheta_{k-1,0}^t(\delta_{k-1} - \delta_0)\right\}\right| \\ &\le |\alpha|\beta_\theta(0, \bullet)\left[\beta_e(1, \bullet) + |\tau|\beta_\phi(|y_0| + \beta_\delta) + 2\left(\beta_\omega + \beta_\phi\beta_\delta\right)\right] < \infty.\end{aligned}$$

Using the conditions in (11.18) and (11.19) with $\iota = \beta_\theta(0, \bullet)$ again, one has $\left|\overline{\mathcal{A}}_k^s(0)\right| < 1$ for any $k \in \mathbb{Z}_+$. Hence, $y_k^s(0)$ is bounded and satisfies $\sup_{k\in\mathbb{Z}_+}\left|y_k^s(0)\right| \le \beta_s(0, \bullet) < \infty$, where $\beta_s(0, \bullet) > 0$ is a finite constant associated with both $\beta_{\Delta\omega}$ and $\beta_{\Delta\delta}$.

Step 2 Assume that $\tilde{\mathcal{Y}}_k(t)$, $\theta_k(t)$, $e_k(t+1)$ and $y_k^s(t)$ are boundedly convergent for any $t \in \mathbb{Z}_{T-1}$ with $T \in \mathbb{Z}_{N-1}$ and there are

$$\sup_{k\in\mathbb{Z}_+}\left|\tilde{\mathcal{Y}}_k(t)\right| \le \beta_{\tilde{\mathcal{Y}}}(t,\bullet) < \infty,$$

$$\sup_{k\in\mathbb{Z}_+}|\theta_k(t)| \le \beta_\theta(t,\bullet) < \infty,$$

$$\sup_{k\in\mathbb{Z}_+}|e_k(t+1)| \le \beta_e(t+1,\bullet) < \infty,$$

$$\limsup_{k\to\infty}|e_k(t+1)| \le \beta_{e_{\sup}}(t+1,\bullet) < \beta_e(t+1,\bullet),$$

$$\sup_{k\in\mathbb{Z}_+}\left|y_k^s(t)\right| \le \beta_s(t,\bullet) < \infty,$$

where $\beta_{\tilde{\mathcal{Y}}}(t,\bullet)$, $\beta_\theta(t,\bullet)$, $\beta_e(t+1,\bullet)$, $\beta_{e_{\sup}}(t+1,\bullet)$ and $\beta_s(t,\bullet)$ are finite positive constants associated with both $\beta_{\Delta\omega}$ and $\beta_{\Delta\delta}$.

When $t = T$, one has

$$\begin{aligned}\left|r_{4,k}(T)\right| \le{}& |\tau|\beta_{\tilde{\phi}}\big(\delta_{\text{sp}} + 2\beta_{e,T}^{\max}(\bullet)\big) + \beta_\Omega(\bullet) + 2|\tau_{\text{I}}|\beta_{\tilde{\phi}}T\big(\beta_{s,T-1}^{\max}(\bullet) + \beta_{e,T-1}^{\max}(\bullet)\big)\\ &+ 2\beta_{\tilde{\phi}}T(|\tau| + |\tau_{\text{I}}|T)\big(\beta_{s,T-1}^{\max}(\bullet) + \beta_{e,T-1}^{\max}(\bullet)\big) < \infty,\end{aligned}$$

where $\beta_{e,T-1}^{\max}(\bullet) \triangleq \max_{t\in\mathbb{Z}_{T-1}}\beta_e(t,\bullet)$, $\beta_{s,T-1}^{\max}(\bullet) \triangleq \max_{t\in\mathbb{Z}_{T-1}}\beta_s(t,\bullet)$, and thus $\tilde{\mathcal{Y}}_k(T+1)$ is boundedly convergent along the batch direction.

As a result, $\hat{\chi}_k(T)$ is guaranteed to be bounded and satisfies $\sup_{k\in\mathbb{Z}_+}|\hat{\chi}_k(T)| \le \beta_{\hat{\chi}}(T,\bullet)$ with $\beta_{\hat{\chi}}(T,\bullet)$ being a finite positive constant associated with both $\beta_{\Delta\omega}$ and $\beta_{\Delta\delta}$.

Besides, it follows from (11.37) that

$$\begin{aligned}\left|r_{3,k}(T) - \frac{\rho\hat{\mathfrak{v}}_k(T)}{\lambda + \hat{\mathfrak{v}}_k^2(T)}\hat{\chi}_k(T)\right| \le{}& \frac{\rho}{2\sqrt{\lambda}}\{\beta_{\hat{\chi}}(T,\bullet) + 2|\tau|\beta_{\hat{\phi}}\beta_e(T,\bullet)\\ &+ 2T\beta_{\hat{\phi}}\big(\beta_{s,T-1}^{\max}(\bullet) + \beta_{e,T-1}^{\max}(\bullet)\big)[|\tau| + (1+T)|\tau_{\text{I}}|] < \infty.\end{aligned}$$

By using the boundedness of $\frac{\rho\hat{\mathfrak{v}}_k(t)}{\lambda+\hat{\mathfrak{v}}_k^2(t)}e_{k-1}(t+1)$, $\theta_k(T)$ is ensured to be bounded and satisfies $\sup_{k\in\mathbb{Z}_+}|\theta_k(T)| \le \beta_\theta(T,\bullet) < \infty$, where $\beta_\theta(T,\bullet)$ is a finite positive constant.

Therefore, one has

$$\begin{aligned}\left|r_{1,k}(T)\right| \le{}& \Bigg|\tau\phi_{k,k-1}^t(T)\Delta y_k(T) - \sum_{j=0}^{T-1}\phi_{k,k-1}^t(j)\Delta u_k(j) - \chi_k(T)\\ &- \tau_{\text{I}}\phi_{k,k-1}^t(T)\sum_{i=0}^{T-1}\left[\Delta y_k^s(i) - \Delta y_k(i)\right]\Bigg|\\ \le{}& 2T\beta_\phi(|\tau| + |\tau_{\text{I}}|T)\big(\beta_{s,T-1}^{\max}(\bullet) + \beta_{e,T-1}^{\max}(\bullet)\big) + \beta_\chi(\bullet)\\ &+ 2|\tau|\beta_\phi\beta_e(T,\bullet) + 2|\tau_{\text{I}}|\beta_\phi T\big(\beta_{s,T-1}^{\max}(\bullet) + \beta_{e,T-1}^{\max}(\bullet)\big) < \infty.\end{aligned}$$

Using the conditions in (11.18) and (11.19) with $\iota = \beta_\theta(T, \bullet)$, one has $\left|\overline{\mathcal{A}}_k^e(t)\right| < 1$ for any $k \in \mathbb{Z}_+$, which means that the output tracking error $e_k(T+1)$ is boundedly convergent along the batch direction, that is

$$\begin{aligned}&\sup_{k\in\mathbb{Z}_+} |e_k(T+1)| \le \beta_e(T+1, \bullet) < \infty,\\ &\limsup_{k\to\infty} |e_k(T+1)| \le \beta_{e_{\text{sup}}}(T+1, \bullet) < \beta_e(T+1, \bullet).\end{aligned}$$

Based on the facts that $\left|y_{k-1}^s(i)\right| \le \beta_{s,T-1}^{\max}(\bullet)$ and $|y_{k-1}(i)| \le \beta_\text{d} + \beta_{e,T-1}^{\max}(\bullet)$ for any $i \in \mathbb{Z}_{T-1}$ as well as the boundedness of $e_k(T+1)$, one has

$$\begin{aligned}\left|h_k(T) r_{2,k}(T)\right| \le\, & |\alpha|\beta_\theta(T, \bullet)\big\{\beta_e(T+1, \bullet) + |\tau|\beta_\phi[\beta_\text{d} + \beta_e(T, \bullet)]\\ & + |\tau_\text{I}|\beta_\phi T\big(\beta_{s,T-1}^{\max}(\bullet) + \beta_{e,T-1}^{\max}(\bullet) + \beta_\text{d}\big)\\ & + 2T\beta_\phi\beta_\omega + 2\beta_\omega + 2\beta_\phi\beta_\delta\\ & + \beta_\phi T[|\tau| + (T-1)|\tau_\text{I}|]\big(\beta_{s,T-1}^{\max}(\bullet) + \beta_{e,T-1}^{\max}(\bullet) + \beta_\text{d}\big)\big\} < \infty.\end{aligned}$$

According to the conditions in (11.18) and (11.19) with $\iota = \beta_\theta(T, \bullet)$, one has $\left|\overline{\mathcal{A}}_k^s(T)\right| < 1$ for any $k \in \mathbb{Z}_+$. Hence, $y_k^s(T)$ is boundedly convergent, and satisfies $\sup_{k\in\mathbb{Z}_+}\left|y_k^s(T)\right| \le \beta_s(T, \bullet)$ with $\beta_s(T, \bullet) > 0$ being a finite positive constant relying on both $\beta_{\Delta\omega}$ and $\beta_{\Delta\delta}$.

The conclusion of this theorem holds true by using the mathematical induction. Owing to the finite time duration of each batch, a uniform $\iota = \max_{i\in\mathbb{Z}_{N-1}} \beta_\theta(i, \bullet)$ could be attained. The proof is complete. □

Remark 11.6 Not only the bounded convergence of output tracking error along with the bounded adaptive learning gain and the set-point command but also the bounded convergence of IESO along the batch direction could be guaranteed by the sufficient conditions established in Theorem 11.2. In addition, the sufficient condition established in (11.32) for guaranteeing the bounded convergence of IESO is independent of those in (11.18) and (11.19), thus simplifying the tuning of IESO gains. ◊

11.5 Illustrative Examples

Two illustrative examples borrowed from the existing literature are adopted to validate the effectiveness of the new designs in this chapter. Note that process models in both examples are only used to generate process I/O data for the control design.

Example 11.1 Consider the following nonlinear system subject to unknown dynamics (Hou and Jin 2013; Chi et al. 2023)

$$y_k(t+1) = \begin{cases} \frac{y_k(t)}{1+y_k^2(t)} + u_k^3(t) + \omega_k(t), & t \in [0, 50], \\ \frac{y_k(t)y_k(t-1)y_k(t-2)u_k(t-1)[y_k(t-2)-1]+a(t)u_k(t)}{1+y_k^2(t-1)+y_k^2(t-2)} + \omega_k(t), & t \in (50, 100], \end{cases}$$

where $a(t) = \text{round}(t/50) + 1$, $\omega_k(t) = 0.2\sin(t/70 + k/50) + \delta_{1,k}(t)$ and $y_k(0) = -1 + \delta_k$, $\delta_{1,k}(t)$ and δ_k are nonrepetitive and randomly vary within the interval $[-0.05, 0.05]$.

Due to that the process input and output have no abrupt changes, Assumptions 11.2 and 11.3 are obviously satisfied. Moreover, Assumption 11.4 is always satisfied based on the disturbances and initial shifts mentioned above.

The desired trajectory to be tracked is taken as the same to that in Chi et al. (2023)

$$y_d(t+1) = \begin{cases} 0.5 \times (-1)^{\text{round}(t/10)}, & t \in [0, 30], \\ 0.5\sin(t\pi/10) + 0.3\cos(t\pi/10), & t \in (30, 70], \\ 0.5 \times (-1)^{\text{round}(t/10)}, & t \in (70, 100]. \end{cases}$$

By taking $\lambda_t = 0.15$, $\rho_t = 0.75$, $\eta_t = 1$, $\mu_t = 0.2$ and $\varepsilon_t = 10^{-8}$ in (2.21) and (2.23) of Chap. 2 for model-free PI tuning, the feasible tuning regions for the PI controller gains are determined as $\tau_P \in [0.01, 0.87]$ and $\tau_I \in [0.30, 0.97]$. Applying Algorithm I in Chap. 2 with $\Delta\tau_P = \Delta\tau_I = 0.01$ and $\kappa = 0.3$, the desired PI controller gains are computed as $\tau_P = 0.01$ and $\tau_I = 0.46$. The rest parameter settings of the new PI-ADDSPLC scheme are listed in Table 11.1. For comparison, the iterative dynamic linearization based indirect adaptive ILC (IDL-iAILC) recently developed in Chi et al. (2023) and the optimization-based adaptive ILC (OAILC) (Meng and Zhang 2022) are also conducted, where the parameter settings therein are correspondingly shown in Table 11.1. It should be noted that the iAILC method developed in Chi et al. (2023) could not be used to the nonlinear processes, and thus is not considered in this example.

The tracking results are shown in Fig. 11.3, while the ATE performance indices along with the set-point command evaluated by $\max_{1\le t\le 99}\left|y_k^s(t)\right|$ are shown in Fig. 11.4, in the absence of initial shifts and external disturbances. It can be seen that perfect tracking is achieved after almost 20 batches under the proposed PI-ADDSPLC scheme. However, almost 80 batches are needed to obtain a similar tracking precision under the OAILC design given in Meng and Zhang (2022), and the IDL-iAILC design in Chi et al. (2023) results in a relatively larger steady-state tracking error. Moreover, the time- and batch-evolutions of the adaptive learning gain $\theta_k(t)$ and the derivative estimation $\hat{\phi}_{k,k-1}(t)$ depicted in Fig. 11.5 well validate their boundedness, which could be utilized to numerically determine a feasible region, i.e., (0, 1.75) for the tuning parameter α. Hence, it is reasonable to choose $\alpha = 0.7$ for execution.

When the nonrepetitive disturbances $\omega_k(t)$ and initial shifts $y_k(0)$ occur, the proposed IESO-based PI-ADDSPLC scheme is also assessed, for which the gains of IESO are chosen as $l_1 = 0.4$ and $l_2 = 0.03$ to meet the condition in (11.32), a constraint for $\Delta y_k^s(t)$ is taken as $\delta_{sp} = 10$, and the rest parameters are chosen as the same as those in the above nominal case. The corresponding ATE indices

Table 11.1 Parameter settings of different methods for Example 11.1

Parameters	OAILC	IDL-iAILC	Proposed
Feedback controller	–	$K = 0.15.$	$\tau_P = 0.01,$ $\tau_I = 0.46.$
Learning controller	$\mu_1 = 1,$ $\mu_2 = 0.001,$ $\lambda = 1,$ $\gamma_1 = 0.8,$ $\gamma_2 = 0.16,$ $\gamma_3 = 0.04,$ $\varepsilon_3 = 0.01.$	$\eta = 0.1,$ $\mu = 0.6,$ $\alpha = 3.6,$ $\eta_{nl} = 1,$ $\mu_{nl} = 1,$ $\varepsilon_2 = 1e - 7,$ $\varepsilon_3 = 1e - 7.$	$\eta = 0.1,$ $\mu = 0.6,$ $\alpha = 0.7,$ $\rho = 1,$ $\lambda = 1,$ $\varepsilon_1 = 1e - 7,$ $\varepsilon_2 = 1e - 7.$
Initial conditions	$\hat{\theta}_0(t) = 1.$	$\hat{\phi}_0(t) = 1,$ $\theta_0(t) = 1.$	$\hat{\phi}_0(t) = 1,$ $\theta_0(t) = 1.$

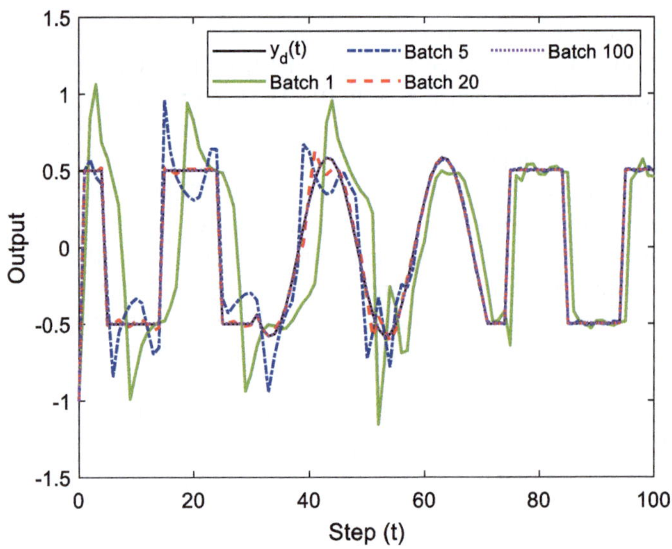

Fig. 11.3 Tracking results by the new design applied to Example 11.1 without external disturbances and initial shifts

along with the set-point command evaluated by $\max_{1 \le t \le 99} |y_k^s(t)|$ are plotted in Fig. 11.6. It is seen that the tracking performance by both new schemes are obviously better than those by the OAILC method in Meng and Zhang (2022) and IDL-iAILC method in Chi et al. (2023) in aspects of the tracking speed and steady-state tracking error. Besides, the tracking performance by the proposed IESO-based PI-ADDSPLC scheme is further improved compared to that by the PI-ADDSPLC scheme.

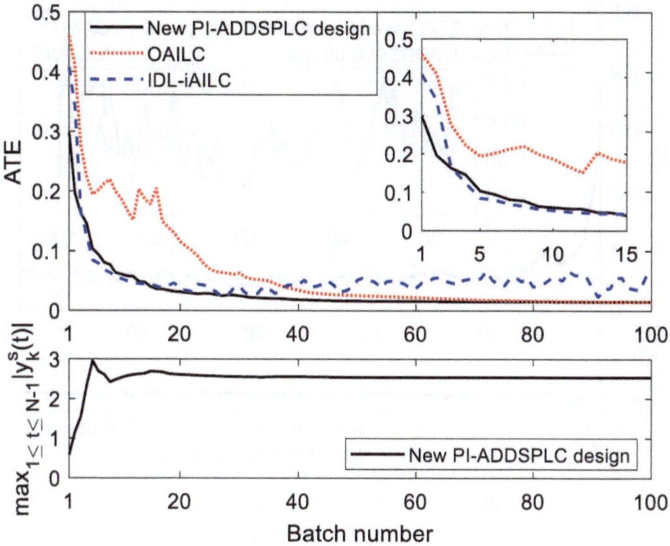

Fig. 11.4 Plot of ATE indices by different methods and the set-point command assessed by $\max_{1\leq t\leq N-1}|y_k^s(t)|$ for the new design applied to Example 11.1 without external disturbance and initial shifts

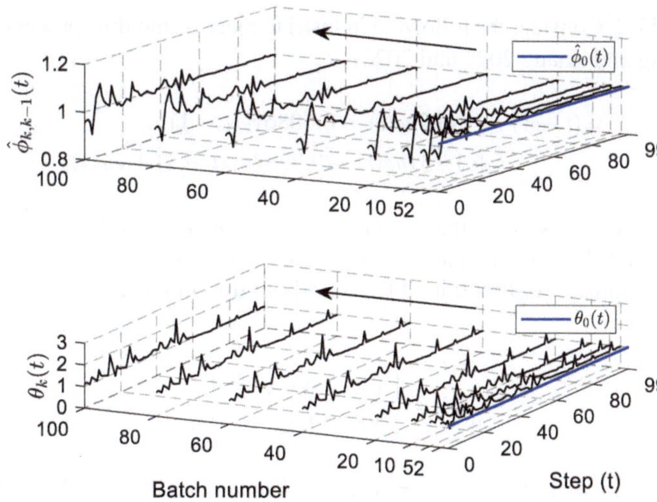

Fig. 11.5 Evolutions of the derivative estimation $\hat{\boldsymbol{\phi}}_{k,k-1}(t)$ and adaptive learning gain $\theta_k(t)$ by the new design applied to Example 11.1 without external disturbances and initial shifts

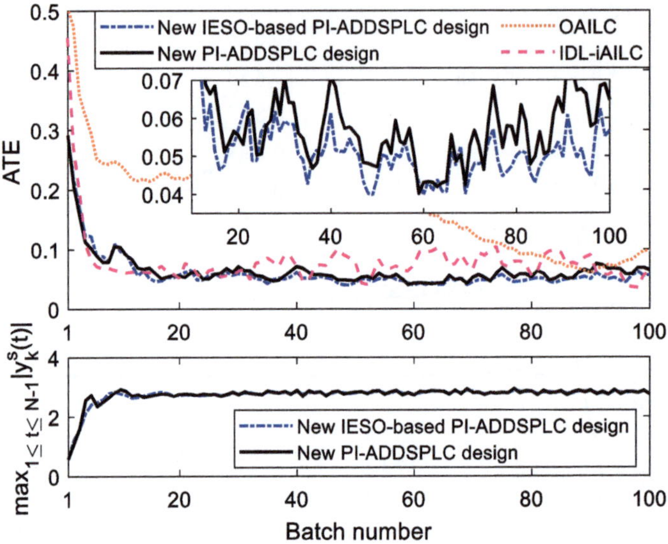

Fig. 11.6 ATE indices under different methods and the set-point command assessed by $\max_{1\le t\le N-1}|y_k^s(t)|$ for the new designs applied to Example 11.1 with nonrepetitive uncertainties

Example 11.2 Consider the following industrial injection molding process (Liu et al. 2014; Meng and Zhang 2021 and 2022)

$$\begin{aligned} y_k(t+1) &= 1.607y_k(t) - 0.6086y_k(t-1) \\ &\quad + 1.239u_k(t) - 0.9282u_k(t-1) + \omega_k(t), \end{aligned}$$

where the initial process output is set as $y_k(0) = \delta_k$ and $\omega_k(t) = 10\sin(t/10 + k/30) + \delta_{1,k}(t)$, δ_k and $\delta_{1,k}(t)$ are uncertain parameters which vary randomly within $[-1, 1]$. Obviously, Assumptions 11.2–11.4 hold true in this example. The desired output profile is

$$y_{\mathrm{d}}(t) = \begin{cases} 150, & 0 \le t \le 50, \\ 300, & 51 \le t \le 100. \end{cases}$$

By taking $\lambda_t = 1$, $\rho_t = 1$, $\eta_t = 1$, $\mu_t = 1$ and $\varepsilon_t = 10^{-8}$ in (2.21) and (2.23) of Chap. 2, the feasible tuning regions of the PI controller gains are, respectively, computed as $\tau_{\mathrm{P}} \in [0.14, 0.5]$ and $\tau_{\mathrm{I}} \in [0.48, 0.5]$. Owing to the desired output profile features a piecewise function structure, the PI controller gains have two groups of steady-state values, namely, $\tau_{\mathrm{P}} = 0.44$, $\tau_{\mathrm{I}} = 0.5$ and $\tau_{\mathrm{P}} = 0.34$, $\tau_{\mathrm{I}} = 0.48$. By comparing the resulting ATE performance indices by these two groups of parameters, $\tau_{\mathrm{P}} = 0.44$ and $\tau_{\mathrm{I}} = 0.5$ are adopted for design of the set-point learning control.

The parameters used in the proposed PI-ADDSPLC design are shown in Table 11.2 for clarity. Similarly, the indirect adaptive ILC (iAILC) and IDL-iAILC methods in

Chi et al. (2023) along with OAILC method in Meng and Zhang (2022) are also carried out for comparison, where the corresponding parameters therein are also shown in Table 11.2 in terms of the guidelines given therein.

When there are no initial shifts and external disturbances, namely, $y_k(0) = 0$ and $\omega_k(t) = 0$, Fig. 11.7 shows the tracking results, and Fig. 11.8 shows the ATE performance indices under different schemes along with the set-point command evaluated by $\max_{1\leq t\leq 99}\left|y_k^s(t)\right|$. It is seen that the proposed design achieves perfect tracking after almost 10 batches, and the proposed scheme delivers faster tracking speed than that by iAILC and IDL-iAILC methods (Chi et al. 2023) and OAILC (Meng and Zhang 2022). Besides, the set-point command is ensured to be bounded, as depicted in the bottom plot of Fig. 11.8.

Compared to the existing OAILC method in Meng and Zhang (2022), and iAILC and IDL-iAILC methods in Chi et al. (2023), the output tracking error of the initial batch under the proposed scheme is obviously reduced. Meanwhile, the time- and batch-evolutions of the learning gain $\theta_k(t)$ and the derivative estimation $\hat{\boldsymbol{\phi}}_{k,k-1}(t)$ shown in Fig. 11.9 well illustrate their boundedness. According to the simulation bounds of $\hat{\boldsymbol{\phi}}_{k,k-1}(t)$ and $\theta_k(t)$, a feasible region for the parameter α can be approximately determined as (0, 0.16), validating the appropriateness of taking $\alpha = 0.1$ in the new design.

When nonrepetitive external disturbances and initial shifts occur, the proposed IESO-based PI-ADDSPLC scheme is also performed for comparison, for which the gains of IESO are chosen as $l_1 = 1$ and $l_2 = 0.8$ to meet the condition in (11.32) and a constraint for $\Delta y_k^s(t)$ is selected as $\delta_{\text{sp}} = 50$. Correspondingly, The ATE performance indices along with the set-point command evaluated by $\max_{1\leq t\leq 99}\left|y_k^s(t)\right|$ are plotted in Fig. 11.10. It is seen that both new designs maintain the robust convergence and boundedness of the resulting ILC system well. Also, the output tracking performance

Table 11.2 Parameter settings of different methods for Example 11.2

Parameters	OAILC	iAILC	IDL-iAILC	Proposed
FC	–	$K = 1.$	$K = 1.$	$\tau_{\text{P}} = 0.44,$ $\tau_{\text{I}} = 0.5.$
LC	$\mu_1 = 1,$ $\mu_2 = 0.001,$ $\lambda = 1,$ $\gamma_1 = 0.95,$ $\gamma_2 = 0.05,$ $\varepsilon_3 = 1e-3.$	$\eta_{nl} = 0.01,$ $\mu_l = 2,$ $\alpha = 0.1,$ $\varepsilon_1 = 1e-7.$	$\eta = 0.01,$ $\mu = 2,$ $\alpha = 0.1,$ $\eta_{nl} = 0.1,$ $\mu_{nl} = 1,$ $\varepsilon_2 = 1e-7,$ $\varepsilon_3 = 1e-7.$	$\eta = 0.01,$ $\mu = 2,$ $\alpha = 0.1,$ $\rho = 0.1,$ $\lambda = 1,$ $\varepsilon_1 = 1e-7,$ $\varepsilon_2 = 1e-7.$
IC	$\hat{\theta}_0(t) = 1.$	$\theta_0(t) = 1.$	$\hat{\phi}_0(t) = 1,$ $\theta_0(t) = 1.$	$\hat{\phi}_0(t) = 1,$ $\theta_0(t) = 1.$

[a] FC, LC, and IC represent feedback controller, learning controller, and initial conditions, respectively

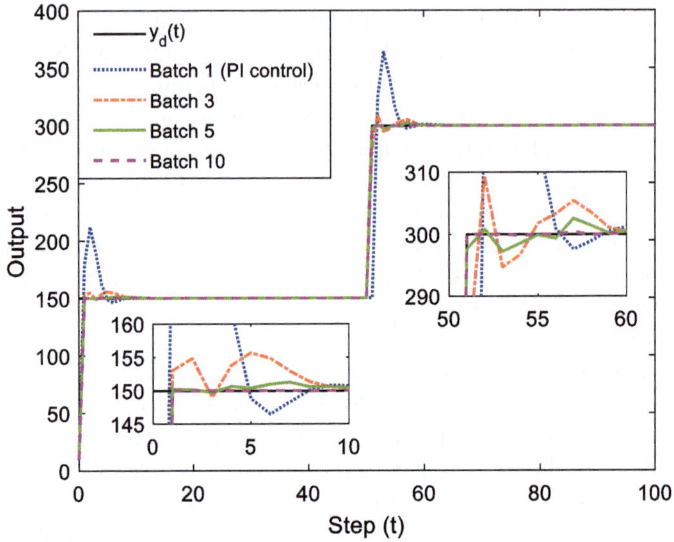

Fig. 11.7 Tracking results by the new design applied to Example 11.2 without external disturbances and initial shifts

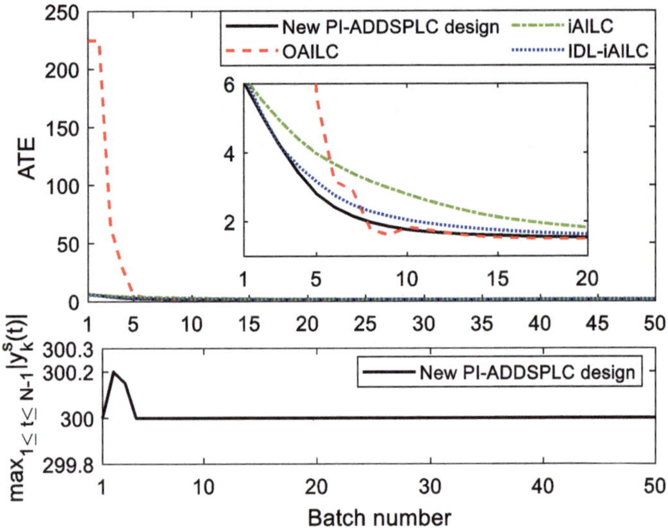

Fig. 11.8 ATE indices under different methods and the set-point command assessed by $\max_{1 \le t \le N-1} |y_k^s(t)|$ for the new design applied to Example 11.2 without external disturbance and initial shifts

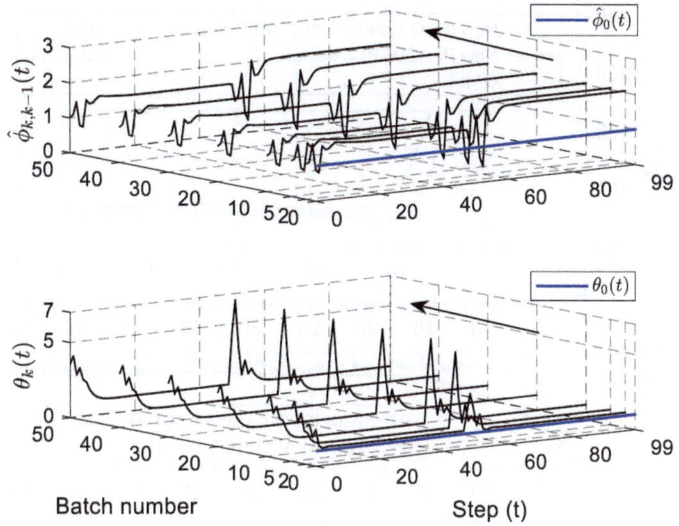

Fig. 11.9 Evolutions of the derivative estimation $\hat{\boldsymbol{\phi}}_{k,k-1}(t)$ and adaptive learning gain $\theta_k(t)$ by the new design applied to Example 11.2 without external disturbances and initial shifts

is obviously enhanced in comparison with the data-driven ILC methods in Meng and Zhang (2022) and Chi et al. (2023). It should be noted that owing to the disturbance compensation, the proposed IESO-based PI-ADDSPLC scheme outperforms the PI-ADDSPLC one. Besides, the boundedness of the set-point command is guaranteed, as per Theorem 11.2.

11.6 Summary

For unknown linear or nonlinear batch processes with nonrepetitive external disturbances and initial shifts, this chapter has presented a robust PI-ADDSPLC method for batch optimization (Hao et al. 2025), by using a widely adopted PI feedback control loop and only the process I/O data. Compared with the existing DDILC methods based on a P-type feedback control structure, significantly improved tracking performance is realized by the proposed scheme from the initial batch. Using mathematical induction, robust convergence analysis of the output tracking error, as well as the boundedness analysis of the adaptive learning gain and the set-point command, has been conducted. In addition, to further improve the output tracking performance against nonrepetitive uncertainties, an IESO-based PI-ADDSPLC has also been presented by embedding an IESO into the new PI-ADDSPLC. A nonlinear numerical example, along with an application to the injection molding process, has been adopted to validate the effectiveness and advantages of both proposed ADDSPLC schemes.

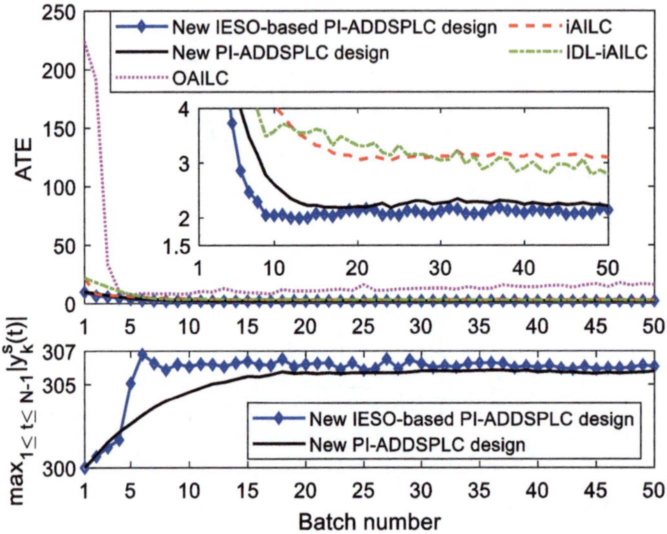

Fig. 11.10 ATE indices under different methods and the set-point command assessed by $\max_{1\le t\le N-1}|y_k^s(t)|$ for the new designs applied to Example 11.2 with nonrepetitive uncertainties

Note that both the proposed ADDSPLC schemes are based on the classical PI feedback control loop for designing a set-point learning control system. The extension to other advanced closed-loop feedback control structures deserves further investigation, especially for nonlinear batch processes with unknown dynamics subject to input and/or output delays.

References

Benosman M (2018) Model-based vs data-driven adaptive control: an overview. Int J Adap Control Signal Process 32(1):1–24

Chi R, Hou Z, Huang B, Jin S (2015) A unified data-driven design framework of optimality-based generalized iterative learning control. Comput Chem Eng 77:10–23

Chi R, Hou Z, Jin S, Huang B (2018) Computationally efficient data-driven high order optimal iterative learning control. IEEE Trans Neural Netw Learn Syst 29(12):5971–5980

Chi R, Li H, Shen D, Hou Z, Huang B (2023) Enhanced P-type control: indirect adaptive learning from set-point updates. IEEE Trans Autom Control 68(3):1600–1613

Chi R, Hui Y, Hou Z (2022) Data-driven iterative learning control for discrete-time systems, Springer Nature

Hao S, Liu T, Rogers E (2020) Extended state observer based indirect-type ILC for single-input single-output batch processes with time- and batch-varying uncertainties. Automatica 112(108673):1–7

Hao S, Liu T, Paszke W (2021) PIO based data-driven iterative learning control for nonlinear batch processes with nonrepetitive disturbances subject to input constraints. IFAC PapersOnLine 54(3):25–30

Hao S, Liu T, Rogers E, Wang Y, Paszke W (2025) PI-based set-point learning control for batch processes with unknown dynamics and nonrepetitive uncertainties. IEEE Trans Autom Control 70(6):3649–3664

Hou Z, Jin S (2013) Model free adaptive control: theory and application. CPC Press, New York, NY, USA

Hou Z, Xiong S (2019) On model-free adaptive control and its stability analysis. IEEE Trans Autom Control 64(11):4555–4569

Hou Z, Chi R, Gao H (2017) An overview of dynamic-linearization-based data-driven control and applications. IEEE Trans Ind Electron 64(5):4076–4090

Liu T, Wang X, Chen J (2014) Robust PID based indirect-type iterative learning control for batch processes with time-varying uncertainties. J Process Control 24(12):95–106

Meng D, Moore KL (2017) Robust iterative learning control for nonrepetitive uncertain systems. IEEE Trans Autom Control 62(2):907–913

Meng D, Zhang J (2021) Robust optimization-based iterative learning control for nonlinear systems with nonrepetitive uncertainties. IEEE/CAA J Automatica Sinica 8(5):1001–1014

Meng D, Zhang J (2022) Design and analysis of data-driven learning control: an optimization-based approach. IEEE Trans Neural Netw Learn Syst 33(10):5527–5541

Xiong S, Hou Z (2022) Stabilizing regions of PID controller for a class of unknown nonlinear non-affine discrete-time systems. Int J Robust Nonlinear Control 32(18):9421–9437

Yu H, Guan Z, Chen T, Yamamoto T (2020a) Design of data-driven PID controllers with adaptive updating rules. Automatica 121(109185):1–10

Yu Q, Hou Z, Bu X, Yu Q (2020b) RBFNN-based data-driven predictive iterative learning control for nonaffine nonlinear systems. IEEE Trans Neural Netw Learn Syst 31(4):1170–1182

Yu X., Hou Z., Polycarpou M. M. (2022). Controller-dynamic-linearization-based data-driven ILC for nonlinear discrete-time systems with RBFNN. IEEE Trans Syst Man Cybern Syst 52(8):4981–4992

Zhao C, Guo L (2022) Towards a theoretical foundation of PID control for uncertain nonlinear systems. Automatica 142(110360):1–12

Index

T. Liu et al., *Robust Iterative Learning Control of Industrial Batch Systems*, Intelligent Control and Learning Systems 22, https://doi.org/10.1007/978-981-96-9778-6

MIX
Papier aus verantwortungsvollen Quellen
Paper from responsible sources
FSC® C105338

If you have any concerns about our products,
you can contact us on
ProductSafety@springernature.com

In case Publisher is established outside the EU,
the EU authorized representative is:
Springer Nature Customer Service Center GmbH
Europaplatz 3, 69115 Heidelberg, Germany

Printed by Libri Plureos GmbH
in Hamburg, Germany